TExES 143 Mathematics – Physics 8-12

Teacher Certification Exam

By: Sharon Wynne, M.S
Southern Connecticut State University

"And, while there's no reason yet to panic, I think it's only prudent that we make preparations to panic."

XAMonline, INC.
Boston

XAMonline, Inc.
21 Orient Ave.
Melrose, MA 02176
Toll Free 1-800-509-4128
Email: info@xamonline.com
Web www.xamonline.com
Fax: 1-781-662-9268

Library of Congress Cataloging-in-Publication Data

Wynne, Sharon A.
Mathematics – Physics 8-12 143: Teacher Certification / Sharon A. Wynne. -2nd ed.
ISBN 978-1-58197-939-8
1. Mathematics – Physics 8-12 143. 2. Study Guides. 3. TExES
4. Teachers' Certification & Licensure. 5. Careers

Printed in the United States of America **œ-1**

TExES: Mathematics – Physics 8-12 143
ISBN: 978-1-58197-939-8

Table of Contents

Great Study and Testing Tips!

What to study in order to prepare for the subject assessments is the focus of this study guide but equally important is *how* you study.

You can increase your chances of truly mastering the information by taking some simple, but effective steps.

Study Tips:

1. Some foods aid the learning process. Foods such as milk, nuts, seeds, rice, and oats help your study efforts by releasing natural memory enhancers called CCKs (*cholecystokinin*) composed of *tryptophan*, *choline*, and *phenylalanine*. All of these chemicals enhance the neurotransmitters associated with memory. Before studying, try a light, protein-rich meal of eggs, turkey, and fish. All of these foods release the memory enhancing chemicals. The better the connections, the more you comprehend.

Likewise, before you take a test, stick to a light snack of energy boosting and relaxing foods. A glass of milk, a piece of fruit, or some peanuts all release various memory-boosting chemicals and help you to relax and focus on the subject at hand.

2. Learn to take great notes. A by-product of our modern culture is that we have grown accustomed to getting our information in short doses (i.e. TV news sound bites or USA Today style newspaper articles.)

Consequently, we've subconsciously trained ourselves to assimilate information better in neat little packages. If your notes are scrawled all over the paper, it fragments the flow of the information. Strive for clarity. Newspapers use a standard format to achieve clarity. Your notes can be much clearer through use of proper formatting. A very effective format is called the *"Cornell Method."*

Take a sheet of loose-leaf lined notebook paper and draw a line all the way down the paper about 1-2" from the left-hand edge.

Draw another line across the width of the paper about 1-2" up from the bottom. Repeat this process on the reverse side of the page.

Look at the highly effective result. You have ample room for notes, a left hand margin for special emphasis items or inserting supplementary data from the textbook, a large area at the bottom for a brief summary, and a little rectangular space for just about anything you want.

3. **Get the concept then the details.** Too often we focus on the details and don't gather an understanding of the concept. However, if you simply memorize only dates, places, or names, you may well miss the whole point of the subject. A key way to understand things is to put them in your own words. If you are working from a textbook, automatically summarize each paragraph in your mind. If you are outlining text, don't simply copy the author's words.

Rephrase them in your own words. You remember your own thoughts and words much better than someone else's, and subconsciously tend to associate the important details to the core concepts.

4. **Ask Why?** Pull apart written material paragraph by paragraph and don't forget the captions under the illustrations.

Example: If the heading is "Stream Erosion", flip it around to read "Why do streams erode?" Then answer the questions.

If you train your mind to think in a series of questions and answers, not only will you learn more, but it also helps to lessen the test anxiety because you are used to answering questions.

5. **Read for reinforcement and future needs.** Even if you only have 10 minutes, put your notes or a book in your hand. Your mind is similar to a computer; you have to input data in order to have it processed. *By reading, you are creating the neural connections for future retrieval.* The more times you read something, the more you reinforce the learning of ideas.

Even if you don't fully understand something on the first pass, *your mind stores much of the material for later recall.*

6. **Relax to learn so go into exile.** Our bodies respond to an inner clock called biorhythms. Burning the midnight oil works well for some people, but not everyone.

If possible, set aside a particular place to study that is free of distractions. Shut off the television, cell phone, pager and exile your friends and family during your study period.

If you really are bothered by silence, try background music. Light classical music at a low volume has been shown to aid in concentration over other types.

Music that evokes pleasant emotions without lyrics are highly suggested. Try just about anything by Mozart. It relaxes you.

7. **Use arrows not highlighters.** At best, it's difficult to read a page full of yellow, pink, blue, and green streaks.

Try staring at a neon sign for a while and you'll soon see my point, the horde of colors obscure the message.

8. **Budget your study time.** Although you shouldn't ignore any of the material, ***allocate your available study time in the same ratio that topics may appear on the test.***

Testing Tips:

1. **Get smart, play dumb. Don't read anything into the question.** Don't make an assumption that the test writer is looking for something else than what is asked. Stick to the question as written and don't read extra things into it.

2. **Read the question and all the choices *twice* before answering the question.** You may miss something by not carefully reading, and then re-reading both the question and the answers.

If you really don't have a clue as to the right answer, leave it blank on the first time through. Go on to the other questions, as they may provide a clue as to how to answer the skipped questions.

If later on, you still can't answer the skipped ones . . . ***Guess.*** The only penalty for guessing is that you *might* get it wrong. Only one thing is certain; if you don't put anything down, you will get it wrong!

3. **Turn the question into a statement.** Look at the way the questions are worded. The syntax of the question usually provides a clue. Does it seem more familiar as a statement rather than as a question? Does it sound strange?

By turning a question into a statement, you may be able to spot if an answer sounds right, and it may also trigger memories of material you have read.

4. **Look for hidden clues.** It's actually very difficult to compose multiple-foil (choice) questions without giving away part of the answer in the options presented.

In most multiple-choice questions you can often readily eliminate one or two of the potential answers. This leaves you with only two real possibilities and automatically your odds go to Fifty-Fifty for very little work.

5. **Trust your instincts.** For every fact that you have read, you subconsciously retain something of that knowledge. On questions that you aren't really certain about, go with your basic instincts. **Your first impression on how to answer a question is usually correct.**

6. **Mark your answers directly on the test booklet.** Don't bother trying to fill in the optical scan sheet on the first pass through the test.

Just be very careful not to miss-mark your answers when you eventually transcribe them to the scan sheet.

7. **Watch the clock!** You have a set amount of time to answer the questions. Don't get bogged down trying to answer a single question at the expense of 10 questions you can more readily answer.

DOMAIN I. NUMBER CONCEPTS

Competency 1.0 The teacher understands the real number system and its structure, operations, algorithms, and representations.

To convert a fraction to a decimal, simply divide the numerator (top) by the denominator (bottom). Use long division if necessary.

If a decimal has a fixed number of digits, the decimal is said to be terminating. To write such a decimal as a fraction, first determine what place value the farthest right digit is in, for example: tenths, hundredths, thousandths, ten thousandths, hundred thousands, etc. Then drop the decimal and place the string of digits over the number given by the place value.

If a decimal continues forever by repeating a string of digits, the decimal is said to be repeating. To write a repeating decimal as a fraction, follow these steps.

a. Let $x =$ the repeating decimal
(ex. $x = .716716716...$)
b. Multiply x by the multiple of ten that will move the decimal just to the right of the repeating block of digits.
(ex. $1000x = 716.716716...$)
c. Subtract the first equation from the second.
(ex. $1000x - x = 716.716.716... - .716716...$)
d. Simplify and solve this equation. The repeating block of digits will subtract out.
(ex. $999x = 716$ so $x = {}^{716}\!/_{999}$)
e. The solution will be the fraction for the repeating decimal.

A. Natural numbers--the counting numbers, $1, 2, 3, ...$

B. Whole numbers--the counting numbers along with zero, $0, 1, 2 ...$

C. Integers--the counting numbers, their opposites, and zero, $..., {}^{-}1, 0, 1, ...$

D. Rationals--all of the fractions that can be formed from the whole numbers. Zero cannot be the denominator. In decimal form, these numbers will either be terminating or repeating decimals. Simplify square roots to determine if the number can be written as a fraction.

E. **Irrationals**--real numbers that cannot be written as a fraction. The decimal forms of these numbers are neither terminating nor repeating. Examples: $\pi, e, \sqrt{2}$, etc.

F. **Real numbers**--the set of numbers obtained by combining the rationals and irrationals. Complex numbers, i.e. numbers that involve i or $\sqrt{^-1}$, are not real numbers.

The **real number properties** are best explained in terms of a small set of numbers. For each property, a given set will be provided.

Axioms of Addition

Closure—For all real numbers *a* and *b*, *a* + *b* is a unique real number.

Associative—For all real numbers *a, b,* and *c, (a + b) + c = a + (b + c).*

Additive Identity—There exists a unique real number 0 (zero) such that *a* + 0 = 0 + *a* = *a* for every real number *a.*

Additive Inverses—For each real number *a*, there exists a real number –*a* (the opposite of *a*) such that *a + (-a) = (-a) + a* = 0.

Commutative—For all real numbers *a* and *b, a + b = b +a.*

Axioms of Multiplication

Closure—For all real numbers *a* and *b, ab* is a unique real number.

Associative—For all real numbers *a, b,* and *c, (ab)c = a(bc).*

Multiplicative Identity—There exists a unique nonzero real number 1 (one) such that $1 \bullet a = a \bullet 1 = a$.

Multiplicative Inverses—For each nonzero real number, there exists a real number 1/*a* (the reciprocal of a) such that *a*(1/*a*) = (1/*a*)*a* = 1.

Commutative—For all real numbers *a* and *b*, *ab* = *ba.*

The Distributive Axiom of Multiplication over Addition

For all real numbers *a, b,* and *c, a*(*b* + *c*) = *ab* + *ac.*

Mathematical operations include addition, subtraction, multiplication and division.

Addition can be indicated by the expressions: sum, greater than, and, more than, increased by, added to.

Subtraction can be expressed by: difference, fewer than, minus, less than, decreased by.

Multiplication is shown by: product, times, multiplied by, twice.

Division is used for: quotient, divided by, ratio.

Examples:		
	7 added to a number	$n + 7$
	a number decreased by 8	$n - 8$
	12 times a number divided by 7	$12n \div 7$
	28 less than a number	$n - 28$
	the ratio of a number to 55	$\frac{n}{55}$
	4 times the sum of a number and 21	$4(n + 21)$

To find the **amount of sales tax** on an item, change the percent of sales tax into an equivalent decimal number. Then multiply the decimal number times the price of the object to find the sales tax. The total cost of an item will be the price of the item plus the sales tax.

Example: A guitar costs $120 plus 7% sales tax. How much are the sales tax and the total bill?

7% = .07 as a decimal (.07)(120) = $8.40 sales tax
$120 + $8.40 = $128.40 ← total price

Example: A suit costs $450 plus 6½% sales tax. How much are the sales tax and the total bill?

6½% = .065 as a decimal
(.065)(450) = $29.25 sales tax
$450 + $29.25 = $479.25 ← total price

The Order of Operations are to be followed when evaluating algebraic expressions. Follow these steps in order:

1. Simplify inside grouping characters such as parentheses, brackets, square root, fraction bar, etc.

2. Multiply out expressions with exponents.

3. Do multiplication or division, from left to right.

4. Do addition or subtraction, from left to right.

Samples of simplifying expressions with exponents:

$(^-2)^3 = -8$ $\quad$ $^-2^3 = {}^-8$

$(^-2)^4 = 16$ $\quad$ $^-2^4 = 16$ $\quad$ Note change of sign.

$(2/3)^3 = 8/27$

$5^0 = 1$

$4^{-1} = 1/4$

Competency 2.0 **The teacher understands the complex number system and its structure, operations, algorithms, and representations.**

Complex numbers are of the form $a+b\,\mathbf{i}$, where a and b are real numbers and $i=\sqrt{-1}$. When **i** appears in an answer, it is acceptable unless it is in a denominator. When $\mathbf{i}^2$ appears in a problem, it is always replaced by **–1**. Remember, $\mathbf{i^2 = -1}$.

To add or subtract complex numbers, add or subtract the real parts . Then add or subtract the imaginary parts and keep the **i** (just like combining like terms).

Examples: Add $(2+3i)+(^-7-4i)$.

$2+{}^-7={}^-5 \qquad 3i+{}^-4i={}^-i$ so,

$(2+3i)+(^-7-4i)={}^-5-i$

Subtract $(8-5i)-(^-3+7i)$

$8-5i+3-7i=11-12i$

To multiply 2 complex numbers, F.O.I.L. (First, Outer, Inner, Last) the 2 numbers together. Replace $\mathbf{i}^2$ with **–1** and finish combining like terms. Answers should have the form $a+b\,\mathbf{i}$.

Example: Multiply $(8+3i)(6-2i)$ F.O.I.L. this.

$48-16i+18i-6i^2$ Let $i^2={}^-1$.

$48-16i+18i-6(^-1)$

$48-16i+18i+6$

$54+2i$ This is the answer.

Example: Multiply $(5+8i)^2$ ←Write this out twice.

$(5+8i)(5+8i)$ F.O.I.L. this

$25+40i+40i+64i^2$ Let $i^2={}^-1$.

$25+40i+40i+64(^-1)$

$25+40i+40i-64$

${}^-39+80i$ This is the answer.

When dividing 2 complex numbers, you must eliminate the complex number in the denominator. If the complex number in the denominator is of the form b **i**, multiply both the numerator and denominator by **i**. Remember to replace **i²** with **–1** and then continue simplifying the fraction.

Example:

$$\frac{2+3i}{5i} \qquad \text{Multiply this by } \frac{i}{i}$$

$$\frac{2+3i}{5i}\times\frac{i}{i}=\frac{(2+3i)\,i}{5i\cdot i}=\frac{2i+3i^2}{5i^2}=\frac{2i+3(^-1)}{^-5}=\frac{^-3+2i}{^-5}=\frac{3-2i}{5}$$

If the complex number in the denominator is of the form $a+b$ **i**, multiply both the numerator and denominator by **the conjugate of the denominator**. **The conjugate of the denominator** is the same 2 terms with the opposite sign between the 2 terms (the real term does not change signs). The conjugate of $2-3i$ is $2+3i$. The conjugate of –6+11i is –6 – 11i. Multiply together the factors on the top and bottom of the fraction. Remember to replace **i²** with **–1**, combine like terms, and then continue simplifying the fraction.

Example:

$$\frac{4+7i}{6-5i} \qquad \text{Multiply by } \frac{6+5i}{6+5i}\text{, the conjugate.}$$

$$\frac{(4+7i)}{(6-5i)}\times\frac{(6+5i)}{(6+5i)}=\frac{24+20i+42i+35i^2}{36+30i-30i-25i^2}=\frac{24+62i+35(^-1)}{36-25(^-1)}=\frac{^-11+62i}{61}$$

Example:

$$\frac{24}{^-3-5i} \qquad \text{Multiply by } \frac{^-3+5i}{^-3+5i}\text{, the conjugate.}$$

$$\frac{24}{^-3-5i}\times\frac{^-3+5i}{^-3+5i}=\frac{^-72+120i}{9-25i^2}=\frac{^-72+120i}{9+25}=\frac{^-72+120i}{34}=\frac{^-36+60i}{17}$$

Divided everything by 2.

One way to graph points is in the rectangular coordinate system. In this system, the point (a,b) describes the point whose distance along the x-axis is "a" and whose distance along the y-axis is "b." The other method used to locate points is the **polar plane coordinate system**. This system consists of a fixed point called the pole or origin (labeled O) and a ray with O as the initial point called the polar axis. The ordered pair of a point P in the polar coordinate system is (r,θ), where $|r|$ is the distance from the pole and θ is the angle measure from the polar axis to the ray formed by the pole and point P. The coordinates of the pole are $(0,\theta)$, where θ is arbitrary. Angle θ can be measured in either degrees or in radians.

Sample problem:

1. Graph the point P with polar coordinates $(^-2, ^-45 \text{ degrees})$.

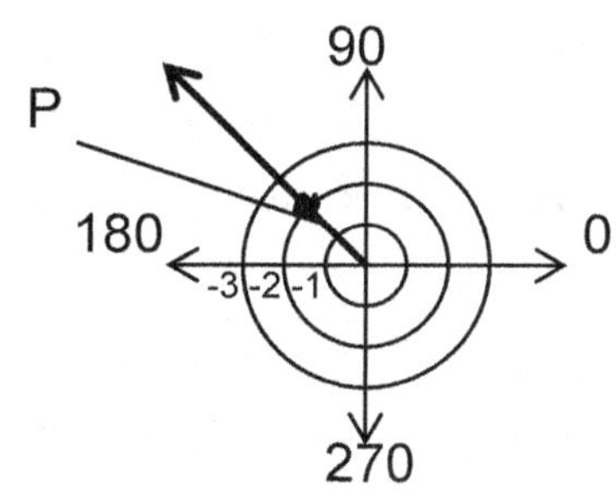

Draw $\theta = {}^-45$ degrees in standard position. Since r is negative, locate the point $|{}^-2|$ units from the pole on the ray opposite the terminal side of the angle. Note that P can be represented by $(^-2, ^-45 \text{ degrees} + 180 \text{ degrees}) = (2, 135 \text{ degrees})$ or by $(^-2, ^-45 \text{ degrees} - 180 \text{ degrees}) = (2, ^-225 \text{ degrees})$.

2. Graph the point $P = \left(3, \frac{\pi}{4}\right)$ and show another graph that also represents the same point P.

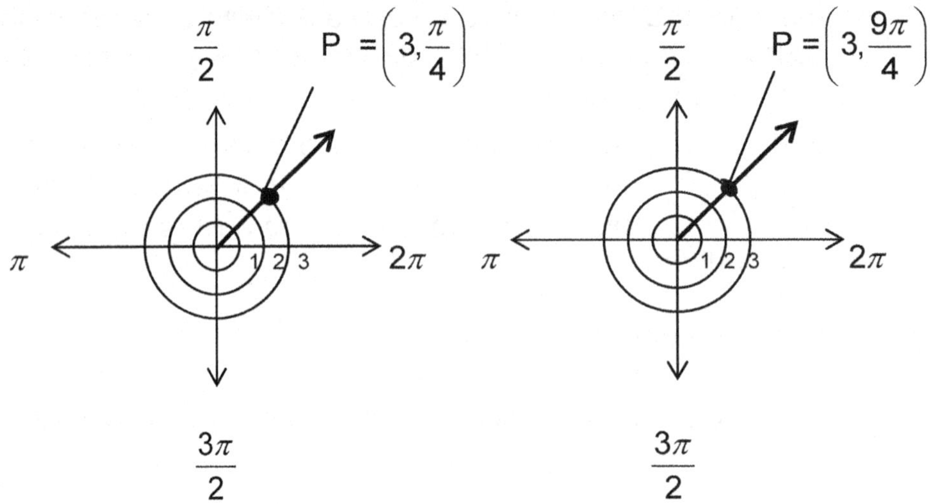

In the second graph, the angle 2π is added to $\frac{\pi}{4}$ to give the point $\left(3, \frac{9\pi}{4}\right)$.

It is possible that *r* is negative. Now instead of measuring $|r|$ units along the terminal side of the angle, we would locate the point $|{}^{-}3|$ units from the pole on the ray opposite the terminal side. This would give the points $\left({}^{-}3, \frac{5\pi}{4}\right)$ and $\left({}^{-}3, \frac{{}^{-}3\pi}{4}\right)$.

Complex numbers are numbers of the form *a* + *bi,* where *a* and *b* are real numbers and *i* is the imaginary unit $\sqrt{-1}$. Complex numbers attach meaning to and allow calculations with the square root of negative numbers. When graphing complex numbers, we plot the real number *a* on the x-axis (labeled R for real) and *b* on the y-axis (labeled I for imaginary). The modulus is the length of the **vector** from the origin to the position of the complex number on the graph. The argument is the angle the vector makes with the horizontal axis R.

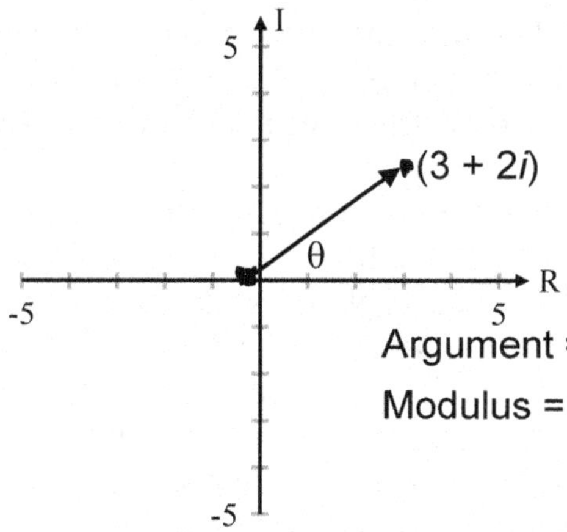

Argument = θ

Modulus = $\sqrt{(3^2)+(2^2)} = \sqrt{13}$

When graphing complex numbers in vector form, it is often desirable to present the numbers as **ordered pairs**. Thus, in the above example, the ordered pair of the plotted point is (3,2) when graphed in the complex plane.

An alternative representation of a complex number is the **polar form**. We can represent any complex number *z* with the following equation:

$z = r(\cos\theta + i\sin\theta)$, where *r* is the modulus and θ is the argument.

Thus, from the example above, the polar form of 3 + 2*i* is

$$\sqrt{13}(\cos(\tan^{-1}\frac{2}{3}) + i\sin(\tan^{-1}\frac{2}{3})).$$

The final alternative representation of a complex number is the **exponential form**. We can represent any complex number *z* with the following equation:

$$z = re^{i\vartheta}$$

Thus, from the example above, the exponential form of 3 + 2*i* is

$\sqrt{13}e^{0.588i}$ (with the argument, θ, given in radians).

Writing complex numbers in polar form simplifies multiplication and division applications. Conversely, Cartesian notation (*a* + *bi*) makes addition and subtraction applications easier to manage. In addition, representing complex numbers as vectors enables addition and subtraction in graphical form. Finally, writing complex numbers in exponential form is often more convenient than polar form because it eliminates cumbersome trigonometric relations.

Competency 3.0 The teacher understands number theory concepts and principles and uses numbers to model and solve problems in a variety of situations.

Prime numbers are numbers that can only be factored into 1 and the number itself. When factoring into prime factors, all the factors must be numbers that cannot be factored again (without using 1). Initially numbers can be factored into any 2 factors. Check each resulting factor to see if it can be factored again. Continue factoring until all remaining factors are prime. This is the list of prime factors. Regardless of what way the original number was factored, the final list of prime factors will always be the same.

Example: Factor 30 into prime factors.

Factor 30 into any 2 factors.

$5 \cdot 6$	Now factor the 6.
$5 \cdot 2 \cdot 3$	These are all prime factors.

Factor 30 into any 2 factors.

$3 \cdot 10$	Now factor the 10.
$3 \cdot 2 \cdot 5$	These are the same prime factors even though the original factors were different.

Example: Factor 240 into prime factors.

Factor 240 into any 2 factors.

$24 \cdot 10$	Now factor both 24 and 10.
$4 \cdot 6 \cdot 2 \cdot 5$	Now factor both 4 and 6.
$2 \cdot 2 \cdot 2 \cdot 3 \cdot 2 \cdot 5$	These are prime factors.

This can also be written as $2^4 \cdot 3 \cdot 5$.

Number theory concerns questions about numbers. Usually, meaning whole numbers or rational numbers (fractions).

Examples of number theory include:

Elementary number theory involves divisibility among integers
Modular arithmetic
Euclid's Algorithm: Given two numbers not prime to one another, find their greatest common denominator (GCD).
Euclid proved that the formula 2n−1(2n − 1) gives an even perfect number whenever 2n − 1 is prime.
Elementary properties of primes (the unique factorization theorem, the infinitude of primes),
Congruence classes (the sets Z/nZ as commutative rings)
Fermat's Last Theorem $x^n + y^n = z^n$ where n is bigger than 2.

Mersenne prime number are prime numbers one less than a prime power of 2. For example: $2^3 - 1 = 7$, a prime number. All Mersenne primes are perfect numbers.

The Fundamental Theorem of Arithmetic states that every composite (non-prime) number can be written as a product of primes in one, and only one way.

Divisibility Tests

a. A number is divisible by 2 if that number is an even number (which means it ends in 0,2,4,6 or 8).

1,354 ends in 4, so it is divisible by 2. 240,685 ends in a 5, so it is not divisible by 2.

b. A number is divisible by 3 if the sum of its digits is evenly divisible by 3.

The sum of the digits of 964 is 9+6+4 = 19. Since 19 is not divisible by 3, neither is 964. The digits of 86,514 is 8+6+5+1+4 = 24. Since 24 is divisible by 3, 86,514 is also divisible by 3.

c. A number is divisible by 4 if the number in its last 2 digits is evenly divisible by 4.

The number 113,336 ends with the number 36 in the last 2 columns. Since 36 is divisible by 4, then 113,336 is also divisible by 4.

The number 135,627 ends with the number 27 in the last 2 columns. Since 27 is not evenly divisible by 4, then 135,627 is also not divisible by 4.

d. A number is divisible by 5 if the number ends in either a 5 or a 0. 225 ends with a 5 so it is divisible by 5. The number 470 is also divisible by 5 because its last digit is a 0. 2,358 is not divisible by 5 because its last digit is an 8, not a 5 or a 0.

e. A number is divisible by 6 if the number is even and the sum of its digits is evenly divisible by 3.

4,950 is an even number and its digits add to 18. (4+9+5+0 = 18) Since the number is even and the sum of its digits is 18 (which is divisible by 3), then 4950 is divisible by 6. 326 is an even number, but its digits add up to 11. Since 11 is not divisible by 3, then 326 is not divisible by 6. 698,135 is not an even number, so it cannot possibly be divided evenly by 6.

f. A number is divisible by 8 if the number in its last 3 digits is evenly divisible by 8.

The number 113,336 ends with the 3-digit number 336 in the last 3 places. Since 336 is divisible by 8, then 113,336 is also divisible by 8.
The number 465,627 ends with the number 627 in the last 3 places. Since 627 is not evenly divisible by 8, then 465,627 is also not divisible by 8.

g. A number is divisible by 9 if the sum of its digits is evenly divisible by 9.

The sum of the digits of 874 is 8+7+4 = 19. Since 19 is not divisible by 9, neither is 874. The digits of 116,514 is 1+1+6+5+1+4 = 18. Since 18 is divisible by 9, 116,514 is also divisible by 9.

h. A number is divisible by 10 if the number ends in the digit 0.

305 ends with a 5 so it is not divisible by 10. The number 2,030,270 is divisible by 10 because its last digit is a 0. 42,978 is not divisible by 10 because its last digit is an 8, not a 0.

i. Why these rules work.

All even numbers are divisible by 2 by definition. A 2-digit number (with T as the tens digit and U as the ones digit) has as its sum of the digits, T + U. Suppose this sum of T + U is divisible by 3. Then it equals 3 times some constant, K. So, T + U = 3K. Solving this for U, U = 3K - T. The original 2 digit number would be represented by 10T + U. Substituting 3K - T in place of U, this 2-digit number becomes 10T + U = 10T + (3K - T) = 9T + 3K. This 2-digit number is clearly divisible by 3, since each term is divisible by 3. Therefore, if the sum of the digits of a number is divisible by 3, then the number itself is also divisible by 3. Since 4 divides evenly into 100, 200, or 300, 4 will divide evenly into any amount of hundreds. The only part of a number that determines if 4 will divide into it evenly is the number in the last 2 places. Numbers divisible by 5 end in 5 or 0. This is clear if you look at the answers to the multiplication table for 5. Answers to the multiplication table for 6 are all even numbers. Since 6 factors into 2 times 3, the divisibility rules for 2 and 3 must both work. Any number of thousands is divisible by 8. Only the last 3 places of the number determine whether or not it is divisible by 8. A 2 digit number (with T as the tens digit and U as the ones digit) has as its sum of the digits, T + U. Suppose this sum of T + U is divisible by 9. Then it equals 9 times some constant, K. So, T + U = 9K. Solving this for U, U = 9K - T. The original 2-digit number would be represented by 10T + U. Substituting 9K - T in place of U, this 2-digit number becomes 10T + U = 10T + (9K - T) = 9T + 9K. This 2-digit number is clearly divisible by 9, since each term is divisible by 9. Therefore, if the sum of the digits of a number is divisible by 9, then the number itself is also divisible by 9. Numbers divisible by 10 must be multiples of 10 which all end in a zero.

The **unit rate for purchasing** an item is its price divided by the number of pounds/ ounces, etc. in the item. The item with the lower unit rate is the lower price.

Example: Find the item with the best unit price:

$1.79 for 10 ounces
$1.89 for 12 ounces
$5.49 for 32 ounces

$$\frac{1.79}{10} = .179 \text{ per ounce} \quad \frac{1.89}{12} = .1575 \text{ per ounce} \quad \frac{5.49}{32} = .172 \text{ per ounce}$$

$1.89 for 12 ounces is the best price.

A second way to find the better buy is to make a proportion with the price over the number of ounces, etc. Cross multiply the proportion, writing the products above the numerator that is used. The better price will have the smaller product.

Example: Find the better buy:

$8.19 for 40 pounds or $4.89 for 22 pounds

Find the unit price.

$$\frac{40}{8.19}=\frac{1}{x} \qquad \frac{22}{4.89}=\frac{1}{x}$$

$$40x=8.19 \qquad 22x=4.89$$

$$x=.20475 \qquad x=.222\overline{27}$$

Since $.20475 < .222\overline{27}$, $8.19 is less and is a better buy.

Vectors and matrices are mathematical representations that share many properties with number systems, but differ in several key ways. For example, matrices, vectors, and real numbers share the properties and limitations of the existence of inverses. Conversely, certain operations of vectors and matrices differ from operations of real numbers .

The inverses of matrices, vectors, and real numbers are similar in nature and share several limitations. The inverse of any number *a* is the number *b* such that *ab* = 1. All real numbers have an inverse with the exception of zero (any number multiplied by zero is zero, not one). Similarly, all square matrices (i.e. matrices with an equal number of rows and columns) with a determinant that is not equal to zero and all vectors with a magnitude not equal to zero have inverses defined by the equation *ab* = 1, where *a* and *b* are matrices or vectors. One additional limitation on the inverses of matrices is that only square matrices have true inverses.

The properties of vector, matrix, and real number operations differ in several key ways. Vector and matrix addition are, like real number addition, commutative. An operation is commutative if reversing the order of the elements does not affect the outcome of the operation (i.e. *a* + *b* = *b* + *a*).

Conversely, matrix multiplication and vector cross products are non-commutative, while real number multiplication and vector dot products are commutative. Thus, the order of the elements in matrix and vector cross product operations is of great importance.

The difference between **permutations and combinations** is that in permutations all possible ways of writing an arrangement of objects are given while in a combination a given arrangement of objects is listed only once.

Given the set {1, 2, 3, 4}, list the arrangements of two numbers that can be written as a combination and as a permutation.

Combination	Permutation
12, 13, 14, 23, 24, 34	12, 21, 13, 31, 14, 41, 23, 32, 24, 42, 34, 43,
six ways	twelve ways

Using the formulas given below the same results can be found.

$${}_nP_r = \frac{n!}{(n-r)!}$$

The notation ${}_nP_r$ is read "the number of permutations of n objects taken r at a time."

$${}_4P_2 = \frac{4!}{(4-2)!}$$

Substitute known values.

$${}_4P_2 = 12$$

Solve.

$${}_nC_r = \frac{n!}{(n-r)!r!}$$

The number of combinations when r objects are selected from n objects.

$${}_4C_2 = \frac{4!}{(4-2)!2!}$$

Substitute known values.

$${}_4C_2 = 6$$

Solve.

Estimation and approximation may be used to check the reasonableness of answers.

Example: Estimate the answer.

$$\frac{58 \times 810}{1989}$$

58 becomes 60, 810 becomes 800 and 1989 becomes 2000.

$$\frac{60 \times 800}{2000} = 24$$

Word problems: An estimate may sometimes be all that is needed to solve a problem.

Example: Janet goes into a store to purchase a CD on sale for \$13.95. While shopping, she sees two pairs of shoes, prices \$19.95 and \$14.50. She only has \$50. Can she purchase everything?

Solve by rounding:

\$19.95→\$20.00
\$14.50→\$15.00
\$13.95→\$14.00
\$49.00 Yes, she can purchase the CD and the shoes.

DOMAIN II. PATTERNS AND ALGEBRA

Competency 4.0 The teacher uses patterns to model and solve problems and formulate conjectures.

Example:

Kepler discovered a relationship between the average distance of a planet from the sun and the time it takes the planet to orbit the sun.

The following table shows the data for the six planets closest to the sun:

	Mercury	Venus	Earth	Mars	Jupiter	Saturn
Average distance, x	0.387	0.723	1	1.523	5.203	9.541
x^3	0.058	.378	1	3.533	140.852	868.524
Time, y	0.241	0.615	1	1.881	11.861	29.457
y^2	0.058	0.378	1	3.538	140.683	867.715

Looking at the data in the table, we see that $x^3 \simeq y^2$. We can conjecture the following function for Kepler's relationship: $y = \sqrt{x^3}$.

The **iterative process** involves repeated use of the same steps. A **recursive function** is an example of the iterative process. A recursive function is a function that requires the computation of all previous terms in order to find a subsequent term. Perhaps the most famous recursive function is the **Fibonacci sequence**. This is the sequence of numbers 1,1,2,3,5,8,13,21,34 ... for which the next term is found by adding the previous two terms.

Example: Find the recursive formula for the sequence 1, 3, 9, 27, 81...

We see that any term other than the first term is obtained by multiplying the preceding term by 3. Then, we may express the formula in symbolic notation as

$$a_n = 3a_{n-1},\ a_1 = 1,$$

where a represents a term, the subscript n denotes the place of the term in the sequence and the subscript $n-1$ represents the preceding term.

When given a set of numbers where the common difference between the terms is constant, use the following formula:

$a_n = a_1 + (n-1)d$ where $a_1 =$ the first term
$n =$ the nth term (general term)
$d =$ the common difference

Sample problem:

1. Find the 8th term of the arithmetic sequence 5, 8, 11, 14, ...

$a_n = a_1 + (n-1)d$
$a_1 = 5$ Identify 1st term.
$d = 3$ Find d.
$a_8 = 5 + (8-1)3$ Substitute.
$a_8 = 26$

2. Given two terms of an arithmetic sequence find a and d.

$a_4 = 21 \qquad a_6 = 32$
$a_n = a_1 + (n-1)d$
$21 = a_1 + (4-1)d$
$32 = a_1 + (6-1)d$

$21 = a_1 + 3d$ Solve the system of equations.
$32 = a_1 + 5d$

$21 = a_1 + 3d$
$-32 = {}^-a_1 - 5d$ Multiply by $^-1$ and add the equations.

${}^-11 = {}^-2d$
$5.5 = d$

$21 = a_1 + 3(5.5)$ Substitute $d = 5.5$ into one of the equations.
$21 = a_1 + 16.5$
$a_1 = 4.5$

The sequence begins with 4.5 and has a common difference of 5.5 between numbers.

When using geometric sequences consecutive numbers are compared to find the common ratio.

$$r = \frac{a_{n+1}}{a_n}$$

$r =$ the common ratio
$a_n =$ the nth term

The ratio is then used in the geometric sequence formula:

$$a_n = a_1 r^{n-1}$$

Sample problems:

1. Find the 8th term of the geometric sequence 2, 8, 32, 128 ...

$r = \frac{a_{n+1}}{a_n}$ Use the common ratio formula to find ratio r.

$r = \frac{8}{2} = 4$ Substitute $a_n = 2$ $a_{n+1} = 8$

$r = 4$

$a_n = a_1 \times r^{n-1}$ Use $r = 4$ to solve for the 8th term.
$a_8 = 2 \times 4^{8-1}$
$a_8 = 32768$

The sums of terms in a progression is simply found by determining if it is an arithmetic or geometric sequence and then using the appropriate formula.

Sum of first n terms of an arithmetic sequence.

$$S_n = \frac{n}{2}(a_1 + a_n)$$

or

$$S_n = \frac{n}{2}\left[2a_1 + (n-1)d\right]$$

Sum of first n terms of a geometric sequence.

$$S_n = \frac{a_1\left(r^n - 1\right)}{r - 1}, r \neq 1$$

Sample Problems:

1. $\sum_{i=1}^{10}(2i+2)$

This means find the sum of the terms beginning with the first term and ending with the 10th term of the sequence $a = 2i + 2$.

$$a_1 = 2(1) + 2 = 4$$
$$a_{10} = 2(10) + 2 = 22$$

$$S_n = \frac{n}{2}(a_1 + a_n)$$
$$S_n = \frac{10}{2}(4 + 22)$$
$$S_n = 130$$

2. Find the sum of the first 6 terms in an arithmetic sequence if the first term is 2 and the common difference, d is-3.

$n = 6 \qquad a_1 = 2 \qquad d = {}^{-}3$

$$S_n = \frac{n}{2}\left[2a_1 + (n-1)d\right]$$

$S_6 = \frac{6}{2}\left[2 \times 2 + (6-1)\,{}^{-}3\right]$ Substitute known values.

$S_6 = 3\left[4 + \left({}^{-}15\right)\right]$ Solve.

$$S_6 = 3(-11) = -33$$

3. Find $\sum_{i=1}^{5} 4 \times 2^{i}$ This means the sum of the first 5 terms

where $a_i = a \times b^i$ and $r = b$.

$a_1 = 4 \times 2^1 = 8$ Identify a_1, r, n

$r = 2 \qquad n = 5$

$S_n = \dfrac{a_1(r^n - 1)}{r - 1}$ Substitute a, r, n

$S_5 = \dfrac{8(2^5 - 1)}{2 - 1}$ Solve.

$S_5 = \dfrac{8(31)}{1} = 248$

Practice problems:

1. Find the sum of the first five terms of the sequence if $a = 7$ and $d = 4$.

2. $\sum_{i=1}^{7}(2i - 4)$

3. $\sum_{i=1}^{6} {}^{-}3\left(\frac{2}{5}\right)^{i}$

A **recurrence relation** is an equation that defines a sequence recursively; in other words, each term of the sequence is defined as a function of the preceding terms.

A real-life application would be using a recurrence relation to determine how much your savings would be in an account at the end of a certain period of time. For example:

You deposit $5,000 in your savings account. Your bank pays 5% interest compounded annually. How much will your account be worth at the end of 10 years?

Let *V* represent the amount of money in the account and V_n represent the amount of money after *n* years.

The amount in the account after *n* years equals the amount in the account after *n* – 1 years plus the interest for the *n*th year. This can be expressed as the recurrence relation V_0 where your initial deposit is represented by $V_0 = 5,000$.

$$
\begin{aligned}
&V_0 = V_0\\
&V_1 = 1.05V_0\\
&V_2 = 1.05V_1 = (1.05)^2 V_0\\
&V_3 = 1.05V_2 = (1.05)^3 V_0\\
&\ldots\ldots\\
&V_n = (1.05)V_{n-1} = (1.05)^n V_0
\end{aligned}
$$

Inserting the values into the equation, you get $V_{10} = (1.05)^{10}(5,000) = 8,144$.

You determine that after investing $5,000 in an account earning 5% interest, compounded annually for 10 years, you would have $8,144.

Competency 5.0 **The teacher understands attributes of functions, relations, and their graphs.**

A **function** can be defined as a set of ordered pairs in which each element of the domain is paired with one and only one element of the range. The symbol $\text{f}(x)$ is read "f of x." A letter other than "f" can be used to represent a function. The letter "g" is commonly used as in $\text{g}(x)$.

Sample problems:

1. Given $\text{f}(x) = 4x^2 - 2x + 3$, find $\text{f}(^-3)$.

(This question is asking for the range value that corresponds to the domain value of $^-3$).

$\text{f}(x) = 4x^2 - 2x + 3$ $\text{f}(^-3) = 4(^-3)^2 - 2(^-3) + 3$	1. Replace x with $^-3$.
$\text{f}(^-3) = 45$	2. Solve.

2. Find f(3) and f(10), given $\text{f}(x) = 7$.

$\text{f}(x) = 7$ $(3) = 7$	1. There are no x values to substitute for. This is your answer.
$\text{f}(x) = 7$ $\text{f}10) = 7$	2. Same as above.

Notice that both answers are equal to the constant given.

A **relation** is any set of ordered pairs.

The **domain** of the relation is the set of all first co-ordinates of the ordered pairs. (These are the x coordinates.)

The **range** of the relation is the set of all second co-ordinates of the ordered pairs. (These are the y coordinates.)

Practice Problems:

1. If $A = \{(x,y) \mid y = x^2 - 6\}$, find the domain and range.

2. Give the domain and range of set B if:

   $B = \{(1,{}^{-}2),(4,{}^{-}2),(7,{}^{-}2),(6,{}^{-}2)\}$

3. Determine the domain of this function:

   $$f(x) = \frac{5x+7}{x^2-4}$$

4. Determine the domain and range of these graphs.

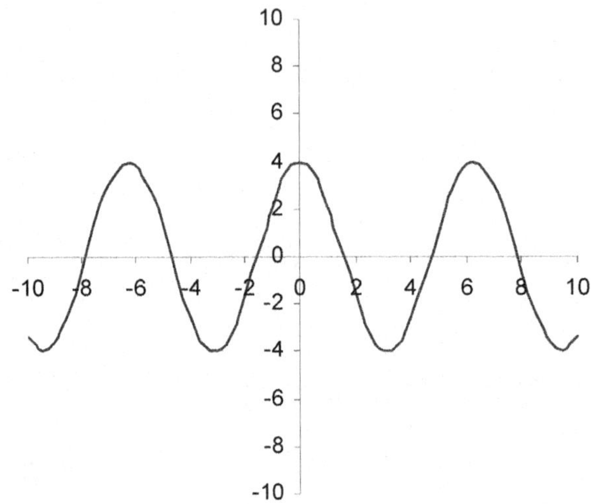

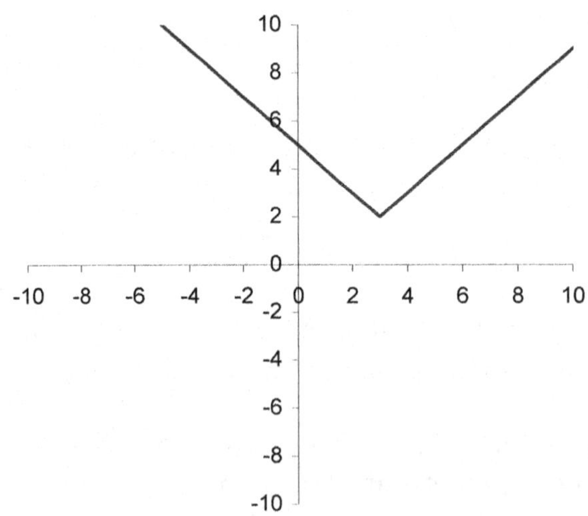

5. If $E = \{(x,y) \mid y = 5\}$, find the domain and range.

6. Determine the ordered pairs in the relation shown in this mapping.

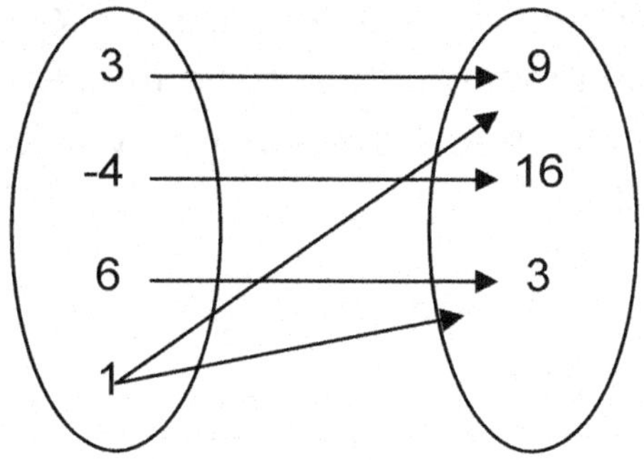

Definition: A function f is even if $f(^-x) = f(x)$ and odd if $f(^-x) = {^-f(x)}$ for all x in the domain of f.

Sample problems:

Determine if the given function is even, odd, or neither even nor odd.

1. $f(x) = x^4 - 2x^2 + 7$
   $f(^-x) = (^-x)^4 - 2(^-x)^2 + 7$
   $f(^-x) = x^4 - 2x^2 + 7$
   f(x) is an even function.

   1. Find $f(^-x)$.
   2. Replace x with ^-x .
   3. Since $f(^-x) = f(x)$, f(x) is an even function.

2. $f(x) = 3x^3 + 2x$
   $f(^-x) = 3(^-x)^3 + 2(^-x)$
   $f(^-x) = {^-3x^3} - 2x$
   f(x) is not an even function.

   $^-f(x) = {^-3x^3} - 2x$
   f(x) is an odd function.

   1. Find $f(^-x)$.
   2. Replace x with ^-x.
   3. Since $f(x)$ is not equal to $f(^-x)$, $^-f(x) = {^-(3x^3 + 2x)}$
   4. Try $^-f(x)$.
   5. Since $f(^-x) = {^-f(x)}$, f(x) is an odd function.

3. $g(x) = 2x^2 - x + 4$
   $g(^-x) = 2(^-x)^2 - (^-x) + 4$
   $g(^-x) = 2x^2 + x + 4$
   g(x) is not an even function.

   $^-g(x) = {^-2x^2} + x - 4$
   $g(x)$ is neither even nor odd.

   1. First find $g(^-x)$.
   2. Replace x with ^-x.
   3. Since $g(x)$ does not equal $g(^-x)$, $^-g(x) = {^-(2x^2 - x + 4)}$
   4. Try $^-g(x)$.
   5. Since $^-g(x)$ does not equal $g(^-x)$, $g(x)$ is not an odd function.

How to write the equation of the inverse of a function.

1. To find the inverse of an equation using x and y, replace each letter with the other letter. Then solve the new equation for y, when possible. Given an equation like $y = 3x - 4$, replace each letter with the other:

   $x = 3y - 4$. Now solve this for y:
   $x + 4 = 3y$
   $1/3\,x + 4/3 = y$ This is the inverse.

   Sometimes the function is named by a letter:
   $f(x) = 5x + 10$

   Temporarily replace $f(x)$ with y.
   $y = 5x + 10$

   Now replace each letter with the other: $x = 5y + 10$
   Solve for the new y: $x - 10 = 5y$
   $1/5\,x - 2 = y$

The inverse of $f(x)$ is denoted as $f^{-1}(x)$, so the answer is $f^{-1}(x) = 1/5\,X - 2$.

If $f(x)$ is a function and the value of 3 is in the domain, the corresponding element **in the range** would be f(3). It is found by evaluating the function for $x = 3$. The same holds true for adding, subtracting, and multiplying in function form.

The symbol f^{-1} is read "the inverse of f". The $^{-}1$ is not an exponent. The inverse of a function can be found by reversing the order of coordinates in each ordered pair that satisfies the function. Finding the inverse functions means switching the place of x and y and then solving for y.

Sample problem:

1. Find $p(a+1)+3\{p(4a)\}$ if $p(x)=2x^2+x+1$.

Find $p(a+1)$.

$p(a+1)=2(a+1)^2+(a+1)+1$	Substitute $(a+1)$ for x.
$p(a+1)=2a^2+5a+4$	Solve.

Find $3\{p(4a)\}$.

$3\{p(4a)\}=3[2(4a)^2+(4a)+1]$	Substitute $(4a)$ for x, multiply by 3.
$3\{p(4a)\}=96a^2+12a+3$	Solve.

$p(a+1)+3\{p(4a)\}=2a^2+5a+4+96a^2+12a+3$	Combine like terms.
$p(a+1)+3\{p(4a)\}=98a^2+17a+7$	

Composition is a process that creates a new function by substituting an entire function into another function. The composition of two functions f(x) and g(x) is denoted by $(f \circ g)(x)$ or f(g(x)). The domain of the composed function, f(g(x)), is the set of all values of x in the domain of g that produce a value for g(x) that is in the domain of f. In other words, f(g(x)) is defined whenever both g(x) and f(g(x)) are defined.

Example 1:

If f(x) = x + 1 and $g(x) = x^3$, find the composition functions $f \circ g$ and $g \circ f$ and state their domains.

Solution:

$(f \circ g)(x) = f(g(x)) = f(x^3) = x^3 + 1$
$(g \circ f)(x) = g(f(x)) = g(x + 1) = (x + 1)^3$

The domain of both composite functions is the set of all real numbers.

Note that f(g(x)) and g(f(x)) are not the same. In general, unlike multiplication and addition, composition is not reversible. Thus, the order of composition is important.

Example 2:

If f(x) = sqrt(x) and g(x) = x +2, find the composition functions $f \circ g$ and $g \circ f$ and state their domains.

Solution:

$(f \circ g)(x)$ = f(g(x)) = f(x + 2) = sqrt(x + 2)
$(g \circ f)(x)$ = g(f(x)) = g(sqrt(x)) = sqrt(x) +2

The domain of f(g(x)) is $x \geq -2$ because x + 2 must be non-negative in order to take the square root.

The domain of g(f(x)) is $x \geq 0$ because x must be non-negative in order to take the square root.

Note that defining the domain of composite functions is important when square roots are involved.

Different types of function transformations affect the graph and characteristics of a function in predictable ways. The basic types of transformation are horizontal and vertical shift, horizontal and vertical scaling, and reflection. As an example of the types of transformations, we will consider transformations of the functions $f(x) = x^2$.

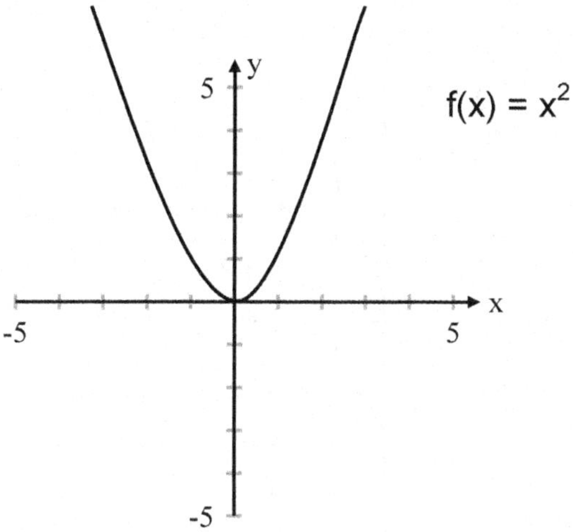

Horizontal shifts take the form $g(x) = f(x \pm c)$. For example, we obtain the graph of the function $g(x) = (x + 2)^2$ by shifting the graph of $f(x) = x^2$ two units to the left. The graph of the function $h(x) = (x - 2)^2$ is the graph of $f(x) = x^2$ shifted two units to the right.

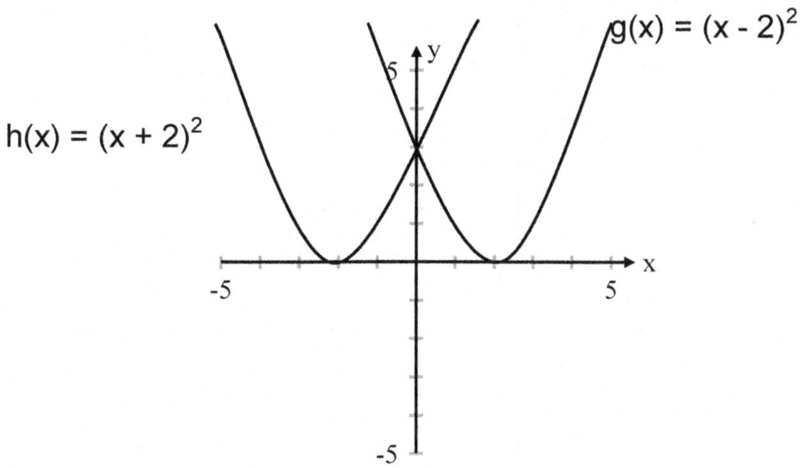

Vertical shifts take the form $g(x) = f(x) \pm c$. For example, we obtain the graph of the function $g(x) = (x^2) - 2$ by shifting the graph of $f(x) = x^2$ two units down. The graph of the function $h(x) = (x^2) + 2$ is the graph of $f(x) = x^2$ shifted two units up.

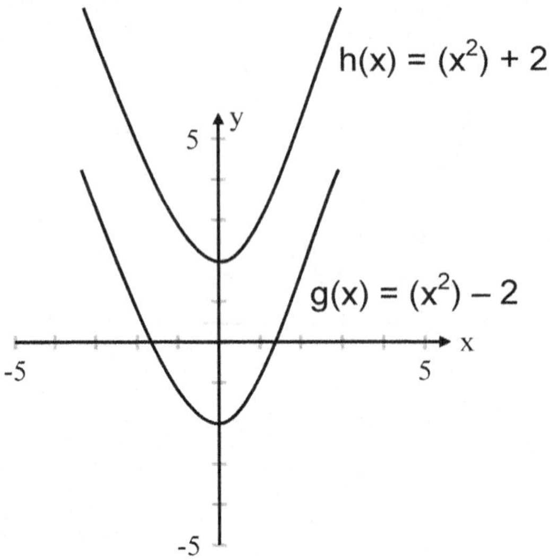

Horizontal scaling takes the form $g(x) = f(cx)$. For example, we obtain the graph of the function $g(x) = (2x)^2$ by compressing the graph of $f(x) = x^2$ in the x-direction by a factor of two. If $c > 1$ the graph is compressed in the x-direction, while if $1 > c > 0$ the graph is stretched in the x-direction.

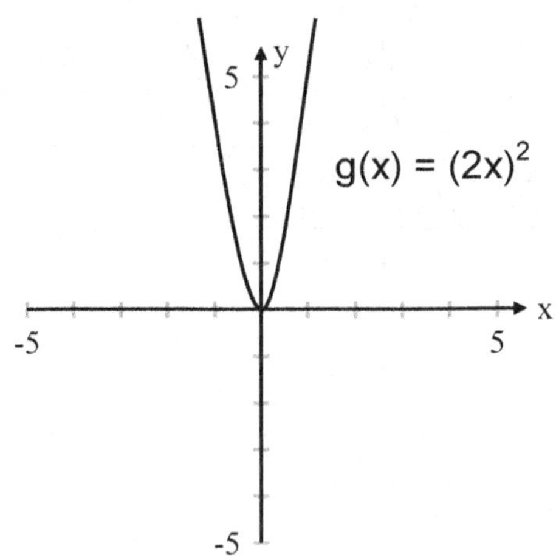

In mathematics, identities are necessary truths. In other words, an identity relates two functions that are equal for all variable values. We can often identify identities by constructing a table of values for each function under consideration and plotting the graph from the table of values. Functions that are identities have identical graphs. The following are three examples of identity functions with corresponding tables of values and graphs.

Example 1:

Consider the functions $y = x^2 - 1$ and $y = (x - 1)(x + 1)$. Determine if the functions represent an identity by graphing.

Using the definition of identity, if the functions represent an identity, $x^2 - 1 = (x - 1)(x + 1)$ for all values of x.

x	$x^2 - 1$	$(x - 1)(x + 1)$
0	-1	-1
1	0	0
-1	0	0
2	3	3
-2	3	3

Note that for all values x, $x^2 - 1 = (x - 1)(x + 1)$. The graph of $y = x^2 - 1$ and $y = (x - 1)(x + 1)$ are identical.

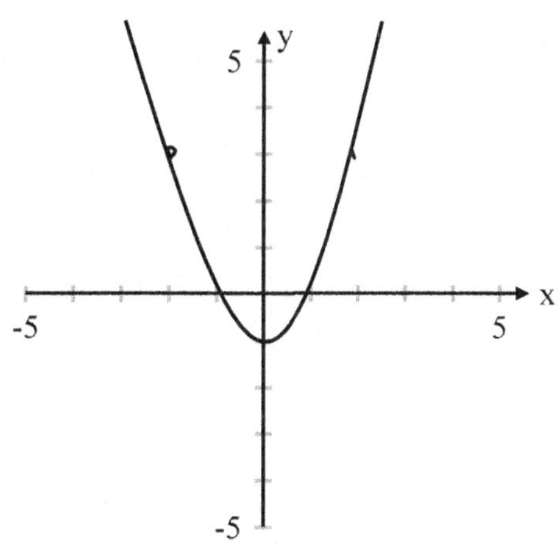

Example 2:

Consider the functions y = log x^3 and y = 3 log x. Determine if the functions represent an identity by graphing.

Determine if log x^3 = 3 log x for all values of x.

x	log x^3	3 log x
1	0	0
10	3	3
100	5	5

The log x^3 = 3 log x for all values of x. The functions form an identity and the graph of y = log x^3 and y = 3 log x is identical.

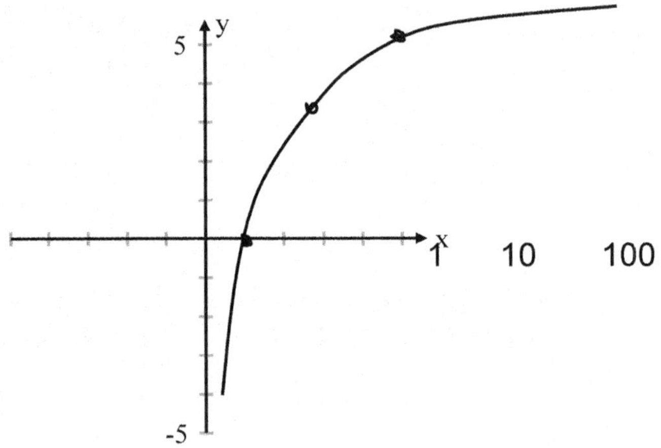

Example 3:

Consider the functions y = sin $(x + \frac{\pi}{2})$ and y = cos x. Determine if the functions represent an identity by graphing.

Determine if sin $(x + \frac{\pi}{2})$ = cos x for all values of x.

x	sin $(x + \frac{\pi}{2})$	cos x
0	1	1
$\frac{\pi}{4}$	0.707	0.707
$\frac{\pi}{2}$	0	0
$\frac{3\pi}{4}$	-0.707	-0.707
π	-1	-1

The functions y = sin $(x + \frac{\pi}{2})$ and y = cos x form an identity because they are equal for all values of x. Shifting the graph of y = sin x to the right by $\frac{\pi}{2}$ units gives the graph of y = cos x.

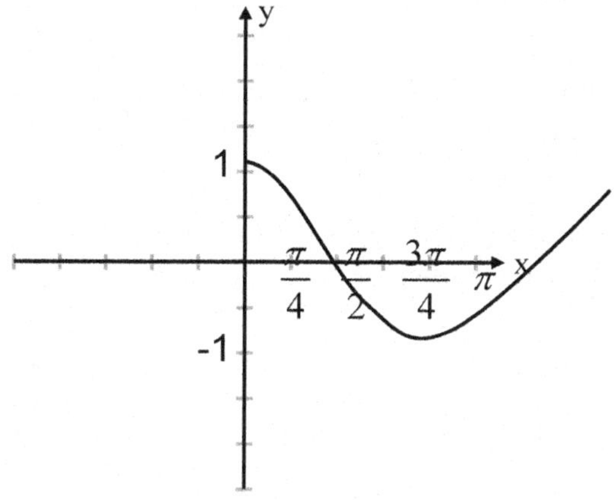

Competency 6.0 The teacher understands linear and quadratic functions, analyzes their algebraic and graphical properties, and uses them to model and solve problems

- A first degree equation has an equation of the form $a\mathbf{x} + b\mathbf{y} = c$. To find the slope of a line, solve the equation for y. This gets the equation into **slope intercept form**, $\mathbf{y} = m\mathbf{x} + b$. The value m is the line's slope.

- To find the y intercept, substitute 0 for x and solve for y. This is the y intercept. The y intercept is also the value of b in $\mathbf{y} = m\mathbf{x} + b$.

- To find the x intercept, substitute 0 for y and solve for x. This is the x intercept.

- If the equation solves to ***x* = any number**, then the graph is a **vertical line**. It only has an x intercept. Its slope is **undefined**.

- If the equation solves to ***y* = any number**, then the graph is a **horizontal line**. It only has a y intercept. Its slope is 0 (zero).

1. Find the slope and intercepts of $3x + 2y = 14$.

$$3x + 2y = 14$$
$$2y = {}^{-}3x + 14$$
$$y = {}^{-}3/2\ x + 7$$

The slope of the line is ${}^{-}3/2$, the value of m.
The y intercept of the line is 7.

The intercepts can also be found by substituting 0 in place of the other variable in the equation.

To find the y intercept:
let $x = 0$; $3(0) + 2y = 14$
$0 + 2y = 14$
$2y = 14$
$y = 7$
(0,7) is the y intercept.

To find the x intercept:
let $y = 0$; $3x + 2(0) = 14$
$3x + 0 = 14$
$3x = 14$
$x = 14/3$
(14/3,0) is the x intercept.

- Given two points on a line, the first thing to do is to find the slope of the line. If 2 points on the graph are (x_1, y_1) and (x_2, y_2), then the slope is found using the formula:

$$\text{slope} = \frac{y_2 - y_1}{x_2 - x_1}$$

The slope will now be denoted by the letter **m**. To write the equation of a line, choose either point. Substitute them into the formula:

$$Y - y_a = \text{m}(X - x_a)$$

Remember (x_a, y_a) can be (x_1, y_1) or (x_2, y_2) If **m**, the value of the slope, is distributed through the parentheses, the equation can be rewritten into other forms of the equation of a line.

Find the equation of a line through $(9, {}^-6)$ and $({}^-1, 2)$.

$$\text{slope} = \frac{y_2 - y_1}{x_2 - x_1} = \frac{2 - {}^-6}{{}^-1 - 9} = \frac{8}{{}^-10} = -\frac{4}{5}$$

$$Y - y_a = \text{m}(X - x_a) \rightarrow Y - 2 = {}^-4/5(X - {}^-1) \rightarrow$$
$$Y - 2 = {}^-4/5(X + 1) \rightarrow Y - 2 = {}^-4/5\,X - 4/5 \rightarrow$$
$$Y = {}^-4/5\ X + 6/5$$ This is the slope-intercept form.

Multiplying by 5 to eliminate fractions, it is:

$$5Y = {}^-4X + 6 \rightarrow 4X + 5Y = 6$$ Standard form.

Write the equation of a line through these two points:

1. $(5,8)$ and $({}^-3,2)$
2. $(11,10)$ and $(11,{}^-3)$
3. $({}^-4,6)$ and $(6,12)$
4. $(7,5)$ and $({}^-3,5)$

When given the following system of equations:

$$ax + by = e$$
$$cx + dy = f$$

the matrix equation is written in the form:

$$\begin{pmatrix} a & b \\ c & d \end{pmatrix}\begin{pmatrix} x \\ y \end{pmatrix} = \begin{pmatrix} e \\ f \end{pmatrix}$$

The solution is found using the inverse of the matrix of coefficients. Inverse of matrices can be written as follows:

$$A^{-1} = \frac{1}{\text{determinant of } A}\begin{pmatrix} d & {}^{-}b \\ {}^{-}c & a \end{pmatrix}$$

Sample Problem:

1. Write the matrix equation of the system.

$$3x - 4y = 2$$
$$2x + y = 5$$

$$\begin{pmatrix} 3 & {}^{-}4 \\ 2 & 1 \end{pmatrix}\begin{pmatrix} x \\ y \end{pmatrix} = \begin{pmatrix} 2 \\ 5 \end{pmatrix}$$ Definition of matrix equation.

$$\begin{pmatrix} x \\ y \end{pmatrix} = \frac{1}{11}\begin{pmatrix} 1 & 4 \\ {}^{-}2 & 3 \end{pmatrix}\begin{pmatrix} 2 \\ 5 \end{pmatrix}$$ Multiply by the inverse of the coefficient matrix.

$$\begin{pmatrix} x \\ y \end{pmatrix} = \frac{1}{11}\begin{pmatrix} 22 \\ 11 \end{pmatrix}$$ Matrix multiplication.

$$\begin{pmatrix} x \\ y \end{pmatrix} = \begin{pmatrix} 2 \\ 1 \end{pmatrix}$$ Scalar multiplication.

The solution is (2,1).

Practice problems:

1. $$x + 2y = 5$$
$$3x + 5y = 14$$

2. $${}^{-}3x + 4y - z = 3$$
$$x + 2y - 3z = 9$$
$$y - 5z = {}^{-}1$$

A **quadratic equation** is written in the form $ax^2 + bx + c = 0$. To solve a quadratic equation by factoring, at least one of the factors must equal zero.

<u>Example</u>:
Solve the equation.

$x^2 + 10x - 24 = 0$	
$(x+12)(x-2) = 0$	Factor.
$x + 12 = 0$ or $x - 2 = 0$	Set each factor equal to 0.
$x = {}^{-}12 \qquad x = 2$	Solve.

Check:
$x^2 + 10x - 24 = 0$

$({}^{-}12)^2 + 10({}^{-}12) - 24 = 0$	$(2)^2 + 10(2) - 24 = 0$
$144 - 120 - 24 = 0$	$4 + 20 - 24 = 0$
$0 = 0$	$0 = 0$

A quadratic equation that cannot be solved by factoring can be solved by **completing the square**.

<u>Example</u>:

Solve the equation.

$x^2 - 6x + 8 = 0$	
$x^2 - 6x = {}^{-}8$	Move the constant to the right side.
$x^2 - 6x + 9 = {}^{-}8 + 9$	Add the square of half the coefficient of x to both sides.
$(x-3)^2 = 1$	Write the left side as a perfect square.
$x - 3 = \pm\sqrt{1}$	Take the square root of both sides.
$x - 3 = 1 \qquad x - 3 = {}^{-}1$	Solve.
$x = 4 \qquad x = 2$	

Check:

$x^2 - 6x + 8 = 0$

$4^2 - 6(4) + 8 = 0$	$2^2 - 6(2) + 8 = 0$
$16 - 24 + 8 = 0$	$4 - 12 + 8 = 0$
$0 = 0$	$0 = 0$

The general technique for **graphing quadratics** is the same as for graphing linear equations. Graphing quadratic equations, however, results in a parabola instead of a straight line.

Example:

Graph $y = 3x^2 + x - 2$.

x	$y = 3x^2 + x - 2$
$^-2$	8
$^-1$	0
0	$^-2$
1	2
2	12

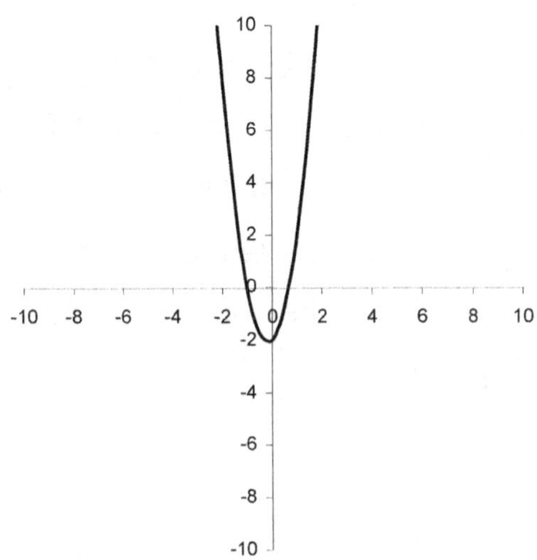

To solve a quadratic equation using the quadratic formula, be sure that your equation is in the form $ax^2 + bx + c = 0$. Substitute these values into the formula:

$$x = \frac{-b \pm \sqrt{b^2 - 4ac}}{2a}$$

Example:

Solve the equation.

$$3x^2 = 7 + 2x \rightarrow 3x^2 - 2x - 7 = 0$$

$$a = 3 \quad b = {}^-2 \quad c = {}^-7$$

$$x = \frac{-({}^-2) \pm \sqrt{({}^-2)^2 - 4(3)({}^-7)}}{2(3)}$$

$$x = \frac{2 \pm \sqrt{4 + 84}}{6}$$

$$x = \frac{2 \pm \sqrt{88}}{6}$$

$$x = \frac{2 \pm 2\sqrt{22}}{6}$$

$$x = \frac{1 \pm \sqrt{22}}{3}$$

Follow these steps to write a quadratic equation from its roots:

1. Add the roots together. The answer is their **sum**. Multiply the roots together. The answer is their **product**.
2. A quadratic equation can be written using the sum and product like this:

$$\mathbf{x^2 + (\text{opposite of the sum})x + \text{product} = 0}$$

3. If there are any fractions in the equation, multiply every term by the common denominator to eliminate the fractions. This is the quadratic equation.
4. If a quadratic equation has only 1 root, use it twice and follow the first 3 steps above.

Example:

Find a quadratic equation with roots of 4 and $^-9$.

Solutions:

The sum of 4 and $^-9$ is $^-5$. The product of 4 and $^-9$ is $^-36$.
The equation would be:

$$\mathbf{x^2 + (opposite\ of\ the\ sum)}x + \mathbf{product = 0}$$
$$\mathbf{x^2 + 5x - 36 = 0}$$

Find a quadratic equation with roots of $5+2i$ and $5-2i$.

Solutions:

The sum of $5+2i$ and $5-2i$ is 10. The product of $5+2i$ and $5-2i$ is $25-4i^2 = 25+4 = 29$.
The equation would be:

$$\mathbf{x^2 + (opposite\ of\ the\ sum)}x + \mathbf{product = 0}$$
$$\mathbf{x^2 - 10x + 29 = 0}$$

Find a quadratic equation with roots of $2/3$ and $^-3/4$.

Solutions:

The sum of $2/3$ and $^-3/4$ is $^-1/12$. The product of $2/3$ and $^-3/4$ is $^-1/2$.
The equation would be :

$$\mathbf{x^2 + (opposite\ of\ the\ sum)}x + \mathbf{product = 0}$$
$$\mathbf{x^2 + 1/12\,x - 1/2 = 0}$$

Common denominator = 12, so multiply by 12.

$$\mathbf{12(x^2 + 1/12\,x - 1/2 = 0}$$
$$\mathbf{12x^2 + 1x - 6 = 0}$$
$$\mathbf{12x^2 + x - 6 = 0}$$

Try these:

1. Find a quadratic equation with a root of 5.
2. Find a quadratic equation with roots of 8/5 and $^-6/5$.
3. Find a quadratic equation with roots of 12 and $^-3$.

Some word problems can be solved by setting up a quadratic equation or inequality. Examples of this type could be problems that deal with finding a maximum area. Examples follow:

Example:

A family wants to enclose 3 sides of a rectangular garden with 200 feet of fence. In order to have a garden with an area of **at least** 4800 square feet, find the dimensions of the garden. Assume that the fourth side of the garden is already bordered by a wall or a fence.

Solution:
Let $x =$ distance from the wall

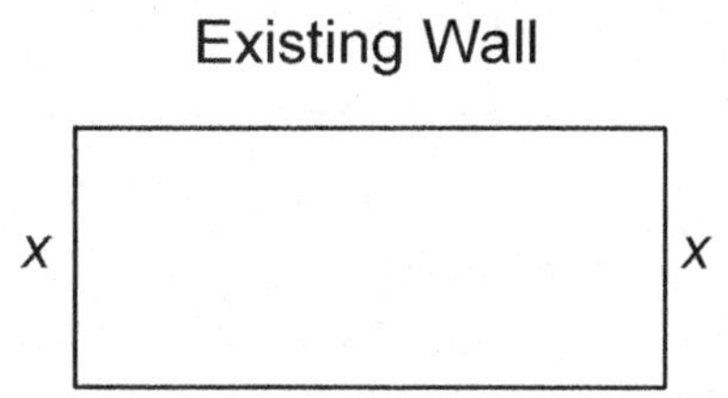

Then 2x feet of fence is used for these 2 sides. The remaining side of the garden would use the rest of the 200 feet of fence, that is, $200-2x$ feet of fence. Therefore the width of the garden is x feet and the length is $200-2x$ ft.

The area, $200x-2x^2$, needs to be greater than or equal to 4800 sq. ft. So, this problem uses the inequality $4800 \le 200x-2x^2$. This becomes $2x^2-200x+4800 \le 0$. Solving this, we get:

$$200x-2x^2 \ge 4800$$
$$-2x^2+200x-4800 \ge 0$$
$$2\left(-x^2+100x-2400\right) \ge 0$$
$$-x^2+100x-2400 \ge 0$$
$$(-x+60)(x-40) \ge 0$$
$$-x+60 \ge 0$$
$$-x \ge -60$$
$$x \le 60$$
$$x-40 \ge 0$$
$$x \ge 40$$

So the area will be at least 4800 square feet if the width of the garden is from 40 up to 60 feet.

Quadratic equations can be used to model different real life situations. The graphs of these quadratics can be used to determine information about this real life situation.

Example:
The height of a projectile fired upward at a velocity of v meters per second from an original height of h meters is $y = h + vx - 4.9x^2$. If a rocket is fired from an original height of 250 meters with an original velocity of 4800 meters per second, find the approximate time the rocket would drop to sea level (a height of 0).
Solution:
The equation for this problem is: $y = 250 + 4800x - 4.9x^2$. If the height at sea level is zero, then $y = 0$ so $0 = 250 + 4800x - 4.9x^2$. Solving this for x could be done by using the quadratic formula. In addition, the approximate time in x seconds until the rocket would be at sea level could be estimated by looking at the graph. When the y value of the graph goes from positive to negative then there is a root (also called the solution or x intercept) in that interval.

$$x = \frac{{}^{-}4800 \pm \sqrt{4800^2 - 4({}^{-}4.9)(250)}}{2({}^{-}4.9)} \approx 980 \text{ or } {}^{-}0.05 \text{ seconds}$$

Since the time has to be positive, it will be about 980 seconds until the rocket is at sea level.

To graph an inequality, graph the quadratic as if it was an equation; however, if the inequality has just a $>$ or $<$ sign, then make the curve itself dotted. Shade above the curve for $>$ or $\geq$. Shade below the curve for $<$ or $\leq$.

Examples:

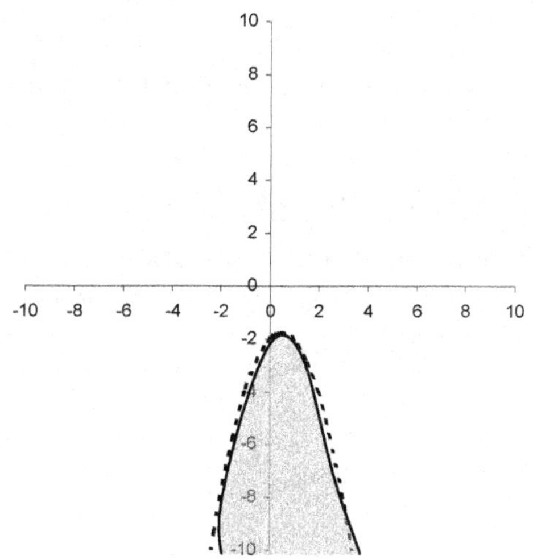
10
8
6
4
2
0
-2
-4
-6
-8
-10
-10
-8
-6
-4
-2
0
2
4
6
8
10

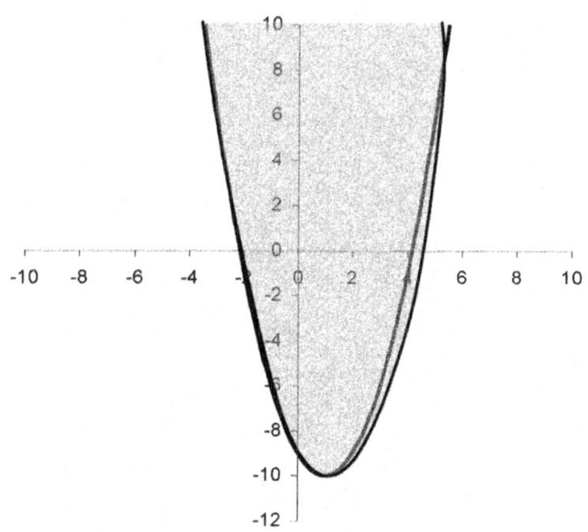
10
8
6
4
2
0
-2
-4
-6
-8
-10
-12
-10
-8
-6
-4
0
2
6
8
10

Competency 7.0 **The teacher understands polynomial, rational, radical, absolute value, and piecewise functions, analyzes their algebraic and graphical properties, and uses them to model and solve problems.**

-The **absolute value function** for a 1st degree equation is of the form:

$y = m(x-h)+k$. Its graph is in the shape of a $\vee$. The point (h,k) is the location of the maximum/minimum point on the graph. "± m" are the slopes of the 2 sides of the $\vee$. The graph opens up if m is positive and down if m is negative.

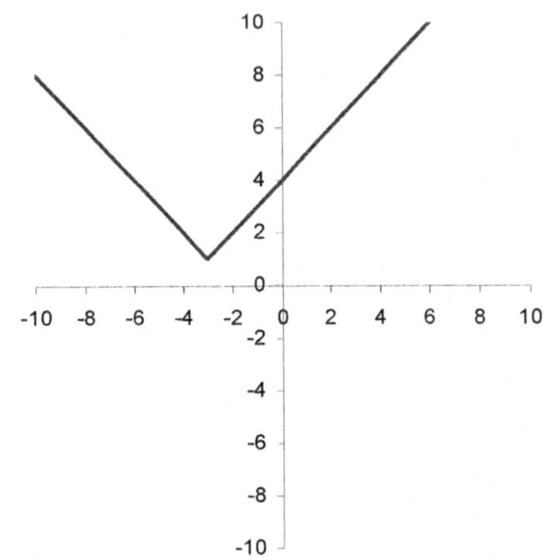

$y = |x+3|+1$

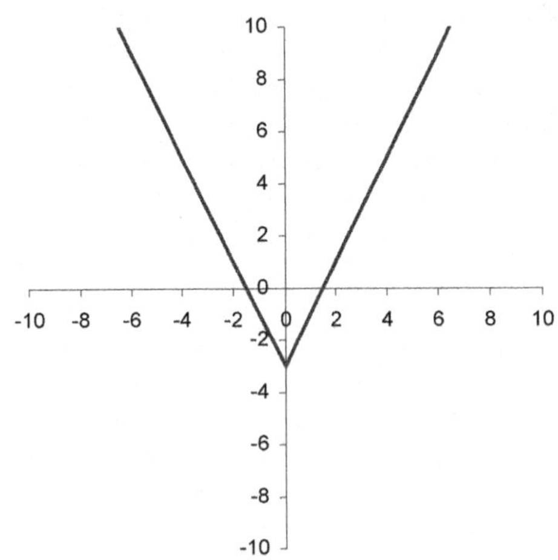

$y = 2|x| - 3$

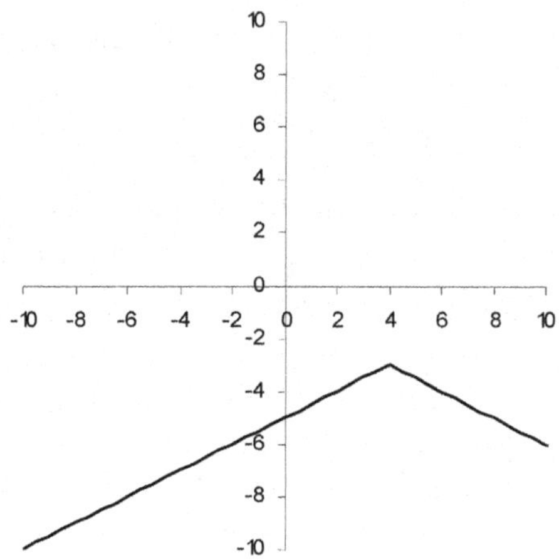

$$y = {}^{-}1/2|x-4|-3$$

-Note that on the first graph above, the graph opens up since m is positive 1. It has $({}^{-}3,1)$ as its minimum point. The slopes of the 2 upward rays are ± 1.

-The second graph also opens up since m is positive. $(0,{}^{-}3)$ is its minimum point. The slopes of the 2 upward rays are ± 2.

-The third graph is a downward $\wedge$ because m is ${}^{-}1/2$. The maximum point on the graph is at $(4,{}^{-}3)$. The slopes of the 2 downward rays are $\pm\ 1/2$.

-The **identity function** is the linear equation $y = x$. Its graph is a line going through the origin (0,0) and through the first and third quadrants at a $45°$ degree angle.

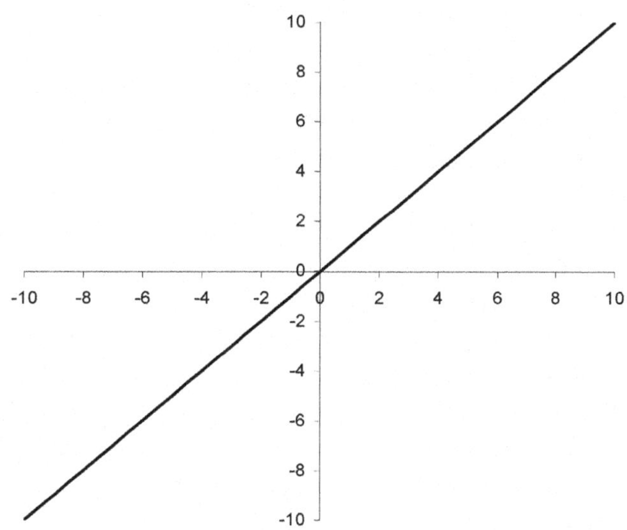

-The **greatest integer function** or **step function** has the equation: $f(x) = j[rx - h] + k$ or $y = j[rx - h] + k$. (h,k) is the location of the left endpoint of one step. j is the vertical jump from step to step. r is the reciprocal of the length of each step. If (x, y) is a point of the function, then when x is an integer, its y value is the same integer. If (x, y) is a point of the function, then when x is not an integer, its y value is the first integer less than x. Points on $y = [x]$ would include:

(3,3), ($^-2$,$^-2$), (0,0), (1.5,1), (2.83,2), ($^-3.2$,$^-4$), ($^-.4$,$^-1$).

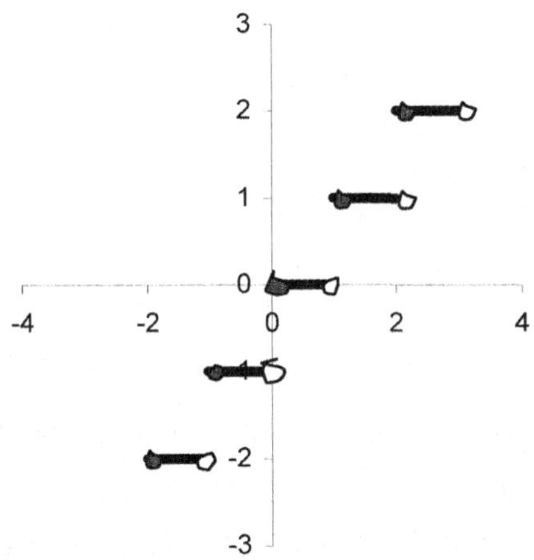

$y = [x]$

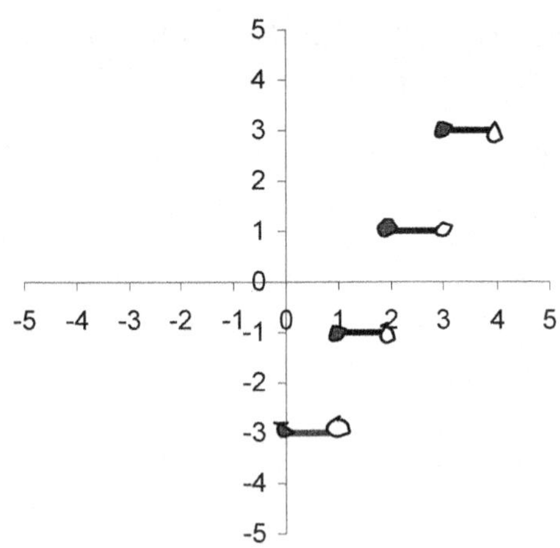

$y = 2[x] - 3$

-Note that in the graph of the first equation, the steps are going up as they move to the right.

Each step is one space wide (inverse of r) with a solid dot on the left and a hollow dot on the right where the jump to the next step occurs. Each step is one square higher $(j = 1)$ than the previous step. One step of the graph starts at $(0,0) \leftarrow$ values of (h,k).
-In the second graph, the graph goes up to the right. One step starts at the point $(0,{}^{-}3) \leftarrow$ values of (h,k). Each step is one square wide $(r = 1)$ and each step is 2 squares higher than the previous step $(j = 2)$.

Practice: Graph the following equations:

1. $f(x) = x$
2. $y = {}^{-}|x - 3| + 5$
3. $y = 3[x]$
4. $y = 2/5|x - 5| - 2$

Functions defined by two or more formulas are **piecewise functions**. The formula used to evaluate piecewise functions varies depending on the value of x. The graphs of piecewise functions consist of two or more pieces, or intervals, and are often discontinuous.

Example 1:

$$f(x) = \begin{cases} x + 1 & \text{if } x > 2 \\ x - 2 & \text{if } x \leq 2 \end{cases}$$

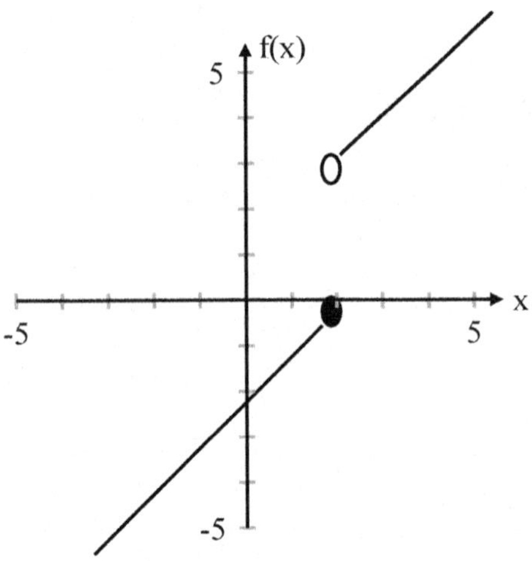

Example 2:

$$f(x) = \begin{cases} x & \text{if } x \geq 1 \\ x^2 & \text{if } x < 1 \end{cases}$$

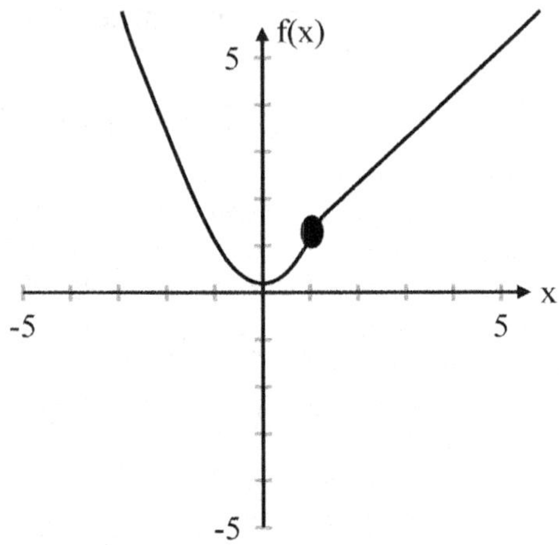

When graphing or interpreting the graph of piecewise functions it is important to note the points at the beginning and end of each interval because the graph must clearly indicate what happens at the end of each interval. Note that in the graph of Example 1, point (2, 3) is not part of the graph and is represented by an empty circle. On the other hand, point (2, 0) is part of the graph and is represented as a solid circle. Note also that the graph of Example 2 is continuous despite representing a piecewise function.

Practice: Graph the following piecewise equations.

1. $f(x) = x^2$ if $x > 0$
   $= x + 4$ if $x \leq 0$

2. $f(x) = x^2 - 1$ if $x > 2$
   $= x^2 + 2$ if $x \leq 2$

A rational function is given in the form $f(x) = p(x)/q(x)$. In the equation, $p(x)$ and $q(x)$ both represent polynomial functions where $q(x)$ does not equal zero. The branches of rational functions approach asymptotes. Setting the denominator equal to zero and solving will give the value(s) of the vertical asymptotes(s) since the function will be undefined at this point. If the value of $f(x)$ approaches b as the $|x|$ increases, the equation $y = b$ is a horizontal asymptote. To find the horizontal asymptote it is necessary to make a table of values for x that are to the right and left of the vertical asymptotes. The pattern for the horizontal asymptotes will become apparent as the $|x|$ increases.

If there are more than one vertical asymptotes, remember to choose numbers to the right and left of each one in order to find the horizontal asymptotes and have sufficient points to graph the function.

Sample problem:

1. Graph $f(x) = \dfrac{3x+1}{x-2}$.

$x - 2 = 0$
$x = 2$

1. Set denominator $= 0$ to find the vertical asymptote.

x	$f(x)$
3	10
10	3.875
100	3.07
1000	3.007
1	$^-4$
$^-10$	2.417
$^-100$	2.93
$^-1000$	2.99

2. Make table choosing numbers to the right and left of the vertical asymptote.

3. The pattern shows that as the $|x|$ increases $f(x)$ approaches the value 3, therefore a horizontal asymptote exists at $y = 3$

Sketch the graph.

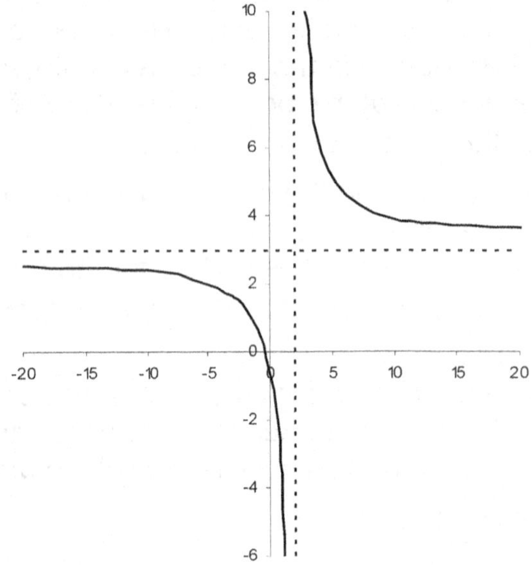

-If two things vary directly, as one gets larger, the other also gets larger. If one gets smaller, then the other gets smaller too. If x and y vary directly, there should be a constant, c, such that $y = cx$. Something can also vary directly with the square of something else, $y = cx^2$.

-If two things vary inversely, as one gets larger, the other one gets smaller. If x and y vary inversely, there should be a constant, c, such that $xy = c$ or $y = c/x$. Something can also vary inversely with the square of something else, $y = c/x^2$.

Example: If $30 is paid for 5 hours work, how much would be paid for 19 hours work?

This is direct variation and $30 = 5c, so the constant is 6 ($6/hour). So $y = 6(19)$ or y = $114.

This could also be done as a proportion:

$$\frac{\$30}{5} = \frac{y}{19}$$

$$5y = 570$$

$$y = 114$$

Example: On a 546 mile trip from Miami to Charlotte, one car drove 65 mph while another car drove 70 mph. How does this affect the driving time for the trip?

This is an inverse variation, since increasing your speed should decrease your driving time. Using the equation: rate $\times$ time = distance, rt = d.

$65t = 546$	and	$70t = 546$
$t = 8.4$	and	$t = 7.8$

slower speed, more time   faster speed, less time

Example: A 14" pizza from Azzip Pizza costs \$8.00. How much would a 20" pizza cost if its price was based on the same price per square inch?

Here the price is directly proportional to the square of the radius. Using a proportion:

$$\frac{\$8.00}{7^2\pi} = \frac{x}{10^2\pi}$$

$$\frac{8}{153.86} = \frac{x}{314}$$

$$16.33 = x$$

\$16.33 would be the price of the large pizza.

There are 2 easy ways to find the **values of a function**. First to find the value of a function when $x = 3$, substitute 3 in place of every letter x. Then simplify the expression following the order of operations. For example, if $f(x) = x^3 - 6x + 4$, then to find f(3), substitute 3 for x.

The equation becomes $f(3) = 3^3 - 6(3) + 4 = 27 - 18 + 4 = 13$.

So (3, 13) is a point on the graph of f(x).

A second way to find the value of a function is to use synthetic division. To find the value of a function when $x = 3$, divide 3 into the coefficients of the function. (Remember that coefficients of missing terms, like x^2 , must be included). The remainder is the value of the function.

If $f(x) = x^3 - 6x + 4$, then to find f(3) using synthetic division:
Note the 0 for the missing x^2 term.

$$\begin{array}{r|rrrr} 3 & 1 & 0 & ^-6 & 4 \\ & & 3 & 9 & 9 \\ \hline & 1 & 3 & 3 & 13 \end{array}$$

← this is the value of the function.

Therefore, (3, 13) is a point on the graph of $f(x) = x^3 - 6x + 4$.

Example: Find values of the function at integer values from x = -3 to x = 3 if $f(x) = x^3 - 6x + 4$.

If $x = {}^-3$:

$$\begin{aligned} f({}^-3) &= ({}^-3)^3 - 6({}^-3) + 4 \\ &= ({}^-27) - 6({}^-3) + 4 \\ &= {}^-27 + 18 + 4 = {}^-5 \end{aligned}$$

synthetic division:

$$\begin{array}{r|rrrr} ^-3 & 1 & 0 & ^-6 & 4 \\ & & ^-3 & 9 & ^-9 \\ \hline & 1 & ^-3 & 3 & ^-5 \end{array}$$

← this is the value of the function if $x = {}^-3$.

Therefore, $({}^-3, {}^-5)$ is a point on the graph.

If $x = {}^-2$:

$$\begin{aligned} f({}^-2) &= ({}^-2)^3 - 6({}^-2) + 4 \\ &= ({}^-8) - 6({}^-2) + 4 \\ &= {}^-8 + 12 + 4 = 8 \end{aligned}$$

← this is the value of the function if $x = {}^-2$.

Therefore, ($^-2$, 8) is a point on the graph.

If $x = {}^-1$:

$$\begin{aligned} f({}^-1) &= ({}^-1)^3 - 6({}^-1) + 4 \\ &= ({}^-1) - 6({}^-1) + 4 \\ &= {}^-1 + 6 + 4 = 9 \end{aligned}$$

synthetic division:

$$\begin{array}{r|rrrr} ^{-}1 & 1 & 0 & ^{-}6 & 4 \\ & & ^{-}1 & 1 & 5 \\ \hline & 1 & ^{-}1 & ^{-}5 & 9 \end{array}$$

$1 \quad ^{-}1 \quad ^{-}5 \quad 9 \leftarrow$ this is the value if the function if $x = {}^{-}1$.
Therefore, ($^{-}1$, 9) is a point on the graph.

If $x = 0$:

$$\begin{aligned} f(0) &= (0)^3 - 6(0) + 4 \\ &= 0 - 6(0) + 4 \\ &= 0 - 0 + 4 = 4 \leftarrow \text{this is the value of the function if } x = 0. \end{aligned}$$

Therefore, (0, 4) is a point on the graph.

If $x = 1$:

$$\begin{aligned} f(1) &= (1)^3 - 6(1) + 4 \\ &= (1) - 6(1) + 4 \\ &= 1 - 6 + 4 = {}^{-}1 \end{aligned}$$

synthetic division:

$$\begin{array}{r|rrrr} 1 & 1 & 0 & ^{-}6 & 4 \\ & & 1 & 1 & ^{-}5 \\ \hline & 1 & 1 & ^{-}5 & ^{-}1 \end{array}$$

$1 \quad 1 \quad ^{-}5 \quad ^{-}1 \leftarrow$ this is the value if the function of $x = 1$.
Therefore, (1, $^{-}1$) is a point on the graph.

If $x = 2$:

$$\begin{aligned} f(2) &= (2)^3 - 6(2) + 4 \\ &= 8 - 6(2) + 4 \\ &= 8 - 12 + 4 = 0 \end{aligned}$$

synthetic division:

$$\begin{array}{r|rrrr} 2 & 1 & 0 & ^{-}6 & 4 \\ & & 2 & 4 & ^{-}4 \\ \hline & 1 & 2 & ^{-}2 & 0 \end{array}$$

$1 \quad 2 \quad ^{-}2 \quad 0 \leftarrow$ this is the value of the function if $x = 2$.
Therefore, (2, 0) is a point on the graph.

If $x = 3$:

$$f(3) = (3)^3 - 6(3) + 4$$
$$= 27 - 6(3) + 4$$
$$= 27 - 18 + 4 = 13$$

synthetic division:

$$\begin{array}{r|rrrr} 3 & 1 & 0 & ^{-}6 & 4 \\ & & 3 & 9 & 9 \\ \hline & 1 & 3 & 3 & 13 \end{array}$$

1 3 3 13 ← this is the value of the function if $x = 3$.
Therefore, (3, 13) is a point on the graph.

The following points are points on the graph:

X	Y
$^{-}3$	$^{-}5$
$^{-}2$	8
$^{-}1$	9
0	4
1	$^{-}1$
2	0
3	13

Note the change in sign of the y value between $x = {}^{-}3$ and $x = {}^{-}2$. This indicates there is a zero between $x = {}^{-}3$ and $x = {}^{-}2$. Since there is another change in sign of the y value between $x = 0$ and $x = {}^{-}1$, there is a second root there. When $x = 2$, $y = 0$ so $x = 2$ is an exact root of this polynomial.

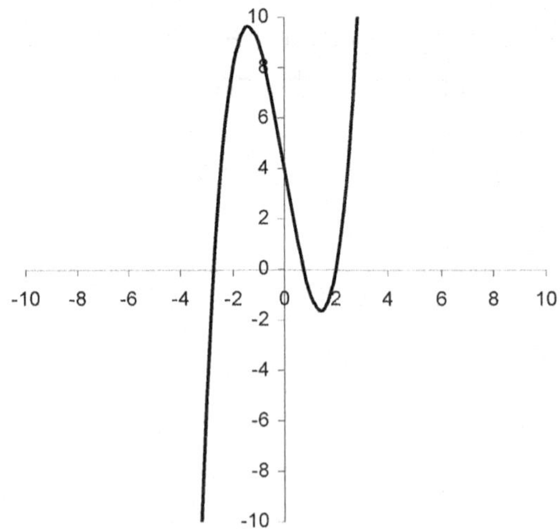

Example: Find values of the function at $x = {}^{-}5, 2,$ and 17 if $f(x) = 2x^5 - 4x^3 + 3x^2 - 9x + 10$.

If $x = {}^{-}5$:

$$\begin{aligned} f({}^{-}5) &= 2({}^{-}5)^5 - 4({}^{-}5)^3 + 3({}^{-}5)^2 - 9({}^{-}5) + 10 \\ &= 2({}^{-}3125) - 4({}^{-}125) + 3(25) - 9({}^{-}5) + 10 \\ &= {}^{-}6250 + 500 + 75 + 45 + 10 = {}^{-}5620 \end{aligned}$$

synthetic division:

$^{-}5$	2	0	$^{-}4$	3	$^{-}9$	10
		$^{-}10$	50	$^{-}230$	1135	$^{-}5630$
	2	$^{-}10$	46	$^{-}227$	$^{-}1126$	$^{-}5620$ ← this is the value of the function if $x = {}^{-}5$.

Therefore, $({}^{-}5, {}^{-}5620)$ is a point on the graph.

If $x = 2$:

$$\begin{aligned} f(2) &= 2(2)^5 - 4(2)^3 + 3(2)^2 - 9(2) + 10 \\ &= 2(32) - 4(8) + 3(4) - 9(2) + 10 \\ &= 64 - 32 + 12 - 18 + 10 = 36 \end{aligned}$$

synthetic division:

2	2	0	$^{-}4$	3	$^{-}9$	10
		4	8	8	22	26
	2	4	4	11	13	36 ← this is the value of the function if $x = 2$.

Therefore, (2, 36) is a point on the graph.

If $x = 17$:

$$\begin{aligned} f(17) &= 2(17)^5 - 4(17)^3 + 3(17)^2 - 9(17) + 10 \\ &= 2(1419857) - 4(4913) + 3(289) - 9(17) + 10 \\ &= 2839714 - 19652 + 867 - 153 + 10 = 2820786 \end{aligned}$$

synthetic division:

$$
\begin{array}{r|rrrrrr}
 & 2 & 0 & ^-4 & 3 & ^-9 & 10 \\
17 & & 34 & 578 & 9758 & 165937 & 2820776 \\
\hline
 & 2 & 34 & 574 & 9761 & 165928 & 2820786
\end{array}
$$

$\leftarrow$ this is the value of the function if $x = 17$.

Therefore, (17, 2820786) is a point on the graph.

To solve an **equation or inequality**, follow these steps:

STEP 1. If there are parentheses, use the distributive property to eliminate them.

STEP 2. If there are fractions, determine their LCD (least common denominator). Multiply every term of the equation by the LCD. This will cancel out all of the fractions while solving the equation or inequality.

STEP 3. If there are decimals, find the largest decimal. Multiply each term by a power of 10(10, 100, 1000,etc.) with the same number of zeros as the length of the decimal. This will eliminate all decimals while solving the equation or inequality.

STEP 4. Combine like terms on each side of the equation or inequality.

STEP 5. If there are variables on both sides of the equation, add or subtract one of those variable terms to move it to the other side. Combine like terms.

STEP 6. If there are constants on both sides, add or subtract one of those constants to move it to the other side. Combine like terms.

STEP 7. If there is a coefficient in front of the variable, divide both sides by this number. This is the answer to an equation. However, remember:

Dividing or multiplying an inequality by a negative number will reverse the direction of the inequality sign.

STEP 8. The solution of a linear equation solves to one single number. The solution of an inequality is always stated including the inequality sign.

Example: Solve: $3(2x+5)-4x=5(x+9)$

$6x+15-4x=5x+45$ ref. step 1

$2x+15=5x+45$ ref. step 4

$^{-}3x+15=45$ ref. step 5

$^{-}3x=30$ ref. step 6

$x={}^{-}10$ ref. step 7

Example: Solve: $1/2(5x+34)=1/4(3x-5)$

$5/2x+17=3/4x-5/4$ ref.step 1

LCD of 5/2, 3/4, and 5/4 is 4.

Multiply by the LCD of 4.

$4(5/2x+17)=(3/4x-5/4)4$ ref.step 2

$10x+68=3x-5$

$7x+68={}^{-}5$ ref.step 5

$7x={}^{-}73$ ref.step 6

$x={}^{-}73/7$ or $^{-}10\ 3/7$ ref.step 7

Check:

$$\frac{1}{2}\left[5\frac{-13}{7}+34\right]=\frac{1}{4}\left[3\left(\frac{-13}{7}\right)-\frac{5}{4}\right]$$

$$\frac{1}{2}\left[\frac{-13(5)}{7}+34\right]=\frac{1}{4}\left[3\left(\frac{-13}{7}\right)-\frac{5}{4}\right]$$

$$\frac{-13(5)}{7}+17=\frac{3(-13)}{28}-\frac{5}{4}$$

$$\frac{-13(5)+17(14)}{14}=\frac{3(-13)}{28}-\frac{5}{4}$$

$$\left[-13(5)+17(14)\right]2=\frac{3(-13)-35}{28}$$

$$\frac{-130+476}{28}=\frac{-219-35}{28}$$

$$\frac{-254}{28}=\frac{-254}{28}$$

Example: Solve: $6x+21<8x+31$

$^{-}2x+21<31$ ref. step 5

$^{-}2x<10$ ref. step 6

$x>{}^{-}5$ ref. step 7

Note that the inequality sign has changed.

To solve an absolute value equation, follow these steps:

1. Get the absolute value expression alone on one side of the equation.

2. Split the absolute value equation into 2 separate equations without absolute value bars. Write the expression inside the absolute value bars (without the bars) equal to the expression on the other side of the equation. Now write the expression inside the absolute value bars equal to the opposite of the expression on the other side of the equation.

3. Now solve each of these equations.

4. **Check each answer by substituting them into the original equation** (with the absolute value symbol). There will be answers that do not check in the original equation. These answers are discarded as they are **extraneous solutions**. If all answers are discarded as incorrect, then the answer to the equation is $\varnothing$, which means the empty set or the null set. (0, 1, or 2 solutions could be correct.)

To solve an absolute value inequality, follow these steps:

1. Get the absolute value expression alone on one side of the inequality. Remember: **Dividing or multiplying by a negative number will reverse the direction of the inequality sign.**

2. Remember what the inequality sign is at this point.

3. Split the absolute value inequality into 2 separate inequalities without absolute value bars. First rewrite the inequality without the absolute bars and solve it. Next write the expression inside the absolute value bar followed by the opposite inequality sign and then by the opposite of the expression on the other side of the inequality. Now solve it.

4. If the sign in the inequality on step 2 is $<$ or $\leq$, the answer is those 2 inequalities connected by the word **and**. The solution set consists of the points between the 2 numbers on the number line. If the sign in the inequality on step 2 is $>$ or $\geq$, the answer is those 2 inequalities connected by the word **or**. The solution set consists of the points outside the 2 numbers on the number line.

If an expression inside an absolute value bar is compared to a negative number, the answer can also be either all real numbers or the empty set ($\varnothing$). For instance,

$$|x+3| < {}^{-}6$$

would have the empty set as the answer, since an absolute value is always positive and will never be less than $^{-}6$. However,

$$|x+3| > {}^{-}6$$

would have all real numbers as the answer, since an absolute value is always positive or at least zero, and will never be less than -6. In similar fashion,

$$|x+3| = {}^{-}6$$

would never check because an absolute value will never give a negative value.

Example: Solve and check:

$$|2x-5|+1=12$$

$$|2x-5|=11 \quad \text{Get absolute value alone.}$$

Rewrite as 2 equations and solve separately.

same equation without absolute value		same equation without absolute value but right side is opposite
$2x-5=11$		$2x-5={}^{-}11$
$2x=16$	and	$2x={}^{-}6$
$x=8$		$x={}^{-}3$

Checks:

$\lvert 2x-5\rvert+1=12$	$\lvert 2x-5\rvert+1=12$
$\lvert 2(8)-5\rvert+1=12$	$\lvert 2({}^{-}3)-5\rvert+1=12$
$\lvert 11\rvert+1=12$	$\lvert {}^{-}11\rvert+1=12$
$12=12$	$12=12$

This time both 8 and $^{-}3$ check.

Example: Solve and check:

$$2|x-7|-13 \geq 11$$

$$2|x-7| \geq 24 \quad \text{Get absolute value alone.}$$

$$|x-7| \geq 12$$

Rewrite as 2 inequalities and solve separately.

same inequality without absolute value		same inequality without absolute value but right side and inequality sign are both the opposite
$x-7 \geq 12$	or	$x-7 \leq {}^{-}12$
$x \geq 19$	or	$x \leq {}^{-}5$

Equations and inequalities can be used to solve various types of word problems. Examples follow.

Example: The YMCA wants to sell raffle tickets to raise at least \$32,000. If they must pay \$7,250 in expenses and prizes out of the money collected from the tickets, how many tickets worth \$25 each must they sell?

Solution: Since they want to raise **at least \$32,000**, that means they would be happy to get 32,000 **or more**. This requires an inequality.

Let $x =$ number of tickets sold
Then $25x =$ total money collected for x tickets

Total money minus expenses is greater than \$32,000.

$$25x-7250 \geq 32000$$

$$25x \geq 39250$$

$$x \geq 1570$$

If they sell **1,570 tickets or more**, they will raise AT LEAST \$32,000.

Example: The Simpsons went out for dinner. All 4 of them ordered the aardvark steak dinner. Bert paid for the 4 meals and included a tip of \$12 for a total of \$84.60. How much was an aardvark steak dinner?

Let $x =$ the price of one aardvark dinner.
So $4x =$ the price of 4 aardvark dinners.

$$4x+12=84.60$$

$$4x=72.60$$

$$x=\$18.15 \text{ for each dinner.}$$

Word problems can sometimes be solved by using a system of two equations in 2 unknowns. This system can then be solved using substitution, the addition-subtraction method, or graphing.

Example: Mrs. Winters bought 4 dresses and 6 pairs of shoes for \$340. Mrs. Summers went to the same store and bought 3 dresses and 8 pairs of shoes for \$360.
If all the dresses were the same price and all the shoes were the same price, find the price charged for a dress and for a pair of shoes.

Let x = price of a dress
Let y = price of a pair of shoes

Then Mrs. Winters' equation would be: $4x+6y=340$
Mrs. Summers' equation would be: $3x+8y=360$

To solve by addition-subtraction:

Multiply the first equation by 4: $4(4x+6y=340)$

Multiply the other equation by $^-3$: $^-3(3x+8y=360)$

By doing this, the equations can be added to each other to eliminate one variable and solve for the other variable.

$$\begin{array}{r} 16x+24y=1360 \\ \underline{-9x-24y={}^-1080} \\ 7x=280 \end{array}$$

$x=40 \leftarrow$ the price of a dress was \$40

solving for y, $y=30$ $\leftarrow$ the price of a pair of shoes,\$30

Example: Aardvark Taxi charges \$4 initially plus \$1 for every mile traveled. Baboon Taxi charges \$6 initially plus \$.75 for every mile traveled. Determine when it is cheaper to ride with Aardvark Taxi or to ride with Baboon Taxi.

Aardvark Taxi's equation: $y=1x+4$
Baboon Taxi's equation : $y=.75x+6$

Using substitution: $.75x+6=x+4$
Multiplying by 4: $3x+24=4x+16$
Solving for x : $8=x$

This tells you that at 8 miles the total charge for the 2 companies is the same. If you compare the charge for 1 mile, Aardvark charges $5 and Baboon charges $6.75. Clearly Aardvark is cheaper for distances up to 8 miles, but Baboon Taxi is cheaper for distances greater than 8 miles.

- Some problems can be solved using equations with rational expressions. First write the equation. To solve it, multiply each term by the LCD of all fractions.
This will cancel out all of the denominators and give an equivalent algebraic equation that can be solved.

1. The denominator of a fraction is two less than three times the numerator. If 3 is added to both the numerator and denominator, the new fraction equals $1/2$.

original fraction: $\frac{x}{3x-2}$ revised fraction: $\frac{x+3}{3x+1}$

$$\frac{x+3}{3x+1}=\frac{1}{2} \qquad 2x+6=3x+1$$

$$x=5$$

original fraction: $\frac{5}{13}$

2. Elly Mae can feed the animals in 15 minutes. Jethro can feed them in 10 minutes. How long will it take them if they work together?

Solution: If Elly Mae can feed the animals in 15 minutes, then she could feed $1/15$ of them in 1 minute, $2/15$ of them in 2 minutes, $x/15$ of them in x minutes. In the same fashion Jethro could feed $x/10$ of them in x minutes. Together they complete 1 job. The equation is:

$$\frac{x}{15}+\frac{x}{10}=1$$

Multiply each term by the LCD of 30:

$$2x+3x=30$$
$$x=6 \text{ minutes}$$

3. A salesman drove 480 miles from Pittsburgh to Hartford. The next day he returned the same distance to Pittsburgh in half an hour less time than his original trip took, because he increased his average speed by 4 mph. Find his original speed.

Since distance = rate x time then time = $\dfrac{\text{distance}}{\text{rate}}$

original time $-1/2$ hour = shorter return time

$$\frac{480}{x} - \frac{1}{2} = \frac{480}{x+4}$$

Multiplying by the LCD of $2x(x+4)$, the equation becomes:

$$480\left[2(x+4)\right] - 1\left[x(x+4)\right] = 480(2x)$$
$$960x + 3840 - x^2 - 4x = 960x$$
$$x^2 + 4x - 3840 = 0$$
$$(x+64)(x-60) = 0$$
$$x = 60$$

60 mph is the original speed
64 mph is the faster return speed

Try these:

1. Working together, Larry, Moe, and Curly can paint an elephant in 3 minutes. Working alone, it would take Larry 10 minutes or Moe 6 minutes to paint the elephant. How long would it take Curly to paint the elephant if he worked alone?

2. The denominator of a fraction is 5 more than twice the numerator. If the numerator is doubled, and the denominator is increased by 5, the new fraction is equal to $1/2$. Find the original number.

3. A trip from Augusta, Maine to Galveston, Texas is 2108 miles. If a car drove 6 mph faster than a truck and got to Galveston 3 hours before the truck, find the speeds of the car and truck.

Competency 8.0 The teacher understands exponential and logarithmic functions, analyzes their algebraic and graphical properties, and uses them to model and solve problems.

Logarithmic and exponential functions have distinctive characteristics and properties that aid in the identification of unknown graphs and the derivation of symbolic equations from known graphs.

Logarithmic Functions

Logarithmic functions of base *a* are of the basic form

$$f(x) = \log_a x, \text{ where } a > 0 \text{ and not equal to } 1.$$

The domain of the function, *f*, is (0, + inf.) and the range is (- inf., + inf.). The x-intercept of the logarithmic function is (1,0) because any number raised to the power of 0 is equal to one. The graph of the function, *f*, has a vertical asymptote at x = 0. As the value of f(x) approaches negative infinity, x becomes closer and closer to 0.

Example:

Graph the function $f(x) = \log_2 (x + 1)$.

The domain of the function is all values of x such that x + 1 > 0. Thus, the domain of f(x) is x > -1.

The range of f(x) is (-inf., +inf.).

The vertical asymptote of f(x) is the value of x that satisfies the equation x + 1 = 0. Thus, the vertical asymptote is x = -1. Note that we can find the vertical asymptote of a logarithmic function by setting the product of the logarithm (containing the variable) equal to 0.

The x-intercept of f(x) is the value of x that satisfies the equation + 1 = 1 because $2^0 = 1$. Thus, the x-intercept of f(x) is (0,0). Note that we can find the x-intercept of a logarithmic function by setting the product of the logarithm equal to 1.

Finally, we find two additional values of f(x), one between the vertical asymptote and the x-intercept and the other to the right of the x-intercept. For example, f(-0.5) = -1 and f(3) = 2.

$f(x) = \log_2 (x + 1)$

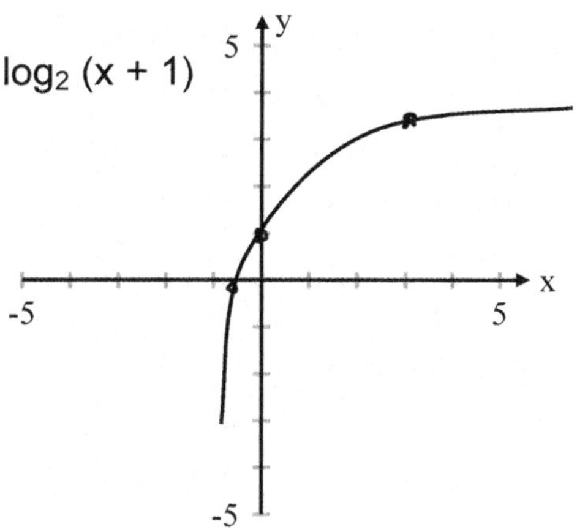

Exponential Functions

The inverse of a logarithmic function is an exponential function. Exponential functions of base *a* take the basic form

$f(x) = a^x$, where a > 0 and not equal to 1.

The domain of the function, *f*, is (-inf., +inf.). The range is the set of all positive real numbers. If a < 1, *f* is a decreasing function and if a > 1 *f* is an increasing function. The y-intercept of f(x) is (0,1) because any base raised to the power of 0 equals 1. Finally, f(x) has a horizontal asymptote at y = 0.

Example:

Graph the function $f(x) = 2^x - 4$.

The domain of the function is the set of all real numbers and the range is y > -4. Because the base is greater than 1, the function is increasing. The y-intercept of f(x) is (0,-3). The x-intercept of f(x) is (2,0). The horizontal asymptote of f(x) is y = -4.

Finally, to construct the graph of f(x) we find two additional values for the function. For example, f(-2) = -3.75 and f(3) = 4.

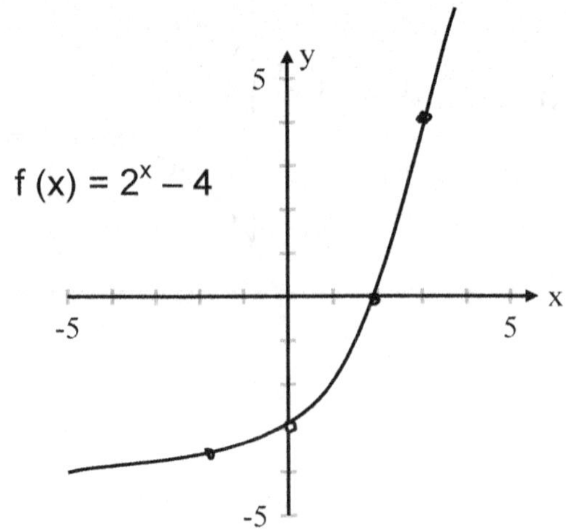

Note that the horizontal asymptote of any exponential function of the form $g(x) = a^x + b$ is $y = b$. Note also that the graph of such exponential functions is the graph of $h(x) = a^x$ shifted b units up or down. Finally, the graph of exponential functions of the form $g(x) = a^{(x + b)}$ is the graph of $h(x) = a^x$ shifted b units left or right.

When changing common logarithms to exponential form,

$$y = \log_b x \quad \text{if and only if } x = b^y$$

Natural logarithms can be changed to exponential form by using,

$$\log_e x = \ln x \quad \text{or} \quad \ln x = y \quad \text{can be written as } e^y = x$$

Practice Problems:

Express in exponential form.

1. $\log_3 81 = 4$

$x = 81 \quad b = 3 \quad y = 4$	Identify values.
$81 = 3^4$	Rewrite in exponential form.

Solve by writing in exponential form.

2. $\log_x 125 = 3$

$x^3 = 125$	Write in exponential form.
$x^3 = 5^3$	Write 125 in exponential form.
$x = 5$	Bases must be equal if exponents are equal.

Use a scientific calculator to solve.

3. Find $\ln 72$.

$\ln 72 = 4.2767$	Use the $\ln x$ key to find natural logs.

4. Find $\ln x = 4.2767$

Find $\ln x = 4.2767$	Write in exponential form.
$e^{4.2767} = x$	Use the key (or 2nd $\ln x$) to find x.
$x = 72.002439$	The small difference is due to rounding.

To solve logarithms or exponential functions it is necessary to use several properties.

Multiplication Property $\log_b mn = \log_b m + \log_b n$

Quotient Property $\log_b \frac{m}{n} = \log_b m - \log_b n$

Powers Property $\log_b n^r = r \log_b n$

Equality Property $\log_b n = \log_b m$ if and only if $n = m$.

Change of Base Formula $\log_b n = \frac{\log n}{\log b}$

$\log_b b^x = x$ and $b^{\log_b x} = x$

Sample problem.

Solve for x.

1. $\log_6(x-5) + \log_6 x = 2$

$\log_6 x(x-5) = 2$	Use product property.
$\log_6 x^2 - 5x = 2$	Distribute.
$x^2 - 5x = 6^2$	Write in exponential form.
$x^2 - 5x - 36 = 0$	Solve quadratic equation.
$(x+4)(x-9) = 0$	
$x = {}^-4 \quad x = 9$	

***Be sure to check results. Remember x must be greater than zero in $\log x = y$.

Check: $\log_6(x-5) + \log_6 x = 2$

$\log_6({}^-4-5) + \log_6({}^-4) = 2$	Substitute the first answer ${}^-4$.
$\log_6({}^-9) + \log_6({}^-4) = 2$	This is undefined, x is less than zero.
$\log_6(9-5) + \log_6 9 = 2$	Substitute the second answer 9.
$\log_6 4 + \log_6 9 = 2$	
$\log_6(4)(9) = 2$	Multiplication property.
$\log_6 36 = 2$	
$6^2 = 36$	Write in exponential form.
$36 = 36$	

Practice problems:

1. $\log_4 x = 2\log_4 3$

2. $2\log_3 x = 2 + \log_3(x-2)$

3. Use change of base formula to find $(\log_3 4)(\log_4 3)$.

In finance, the **value of a sum of money** with compounded interest increases at a rate proportional to the original value. We use an exponential function to determine the growth of an investment accumulating compounded interest. The formula for calculating the value of an investment after a given compounding period is

$$A(t) = A_0(1 + \frac{i}{n})^{nt}.$$

A_0 is the principle, the original value of the investment. The rate of interest is *i*, the time in years is *t*, and the number of times the interest is compounded per year is *n*.

We can solve the compound interest formula for any of the variables by utilizing the properties of exponents and logarithms.

Examples:

1. Determine how long it will take \$100 to amount to \$1000 at 8% interest compounded 4 times annually.

In this problem we are given the principle (A_0 = 100), the final value (A(t) = 1000), the interest rate (*i* = .08), and the number of compounding periods per year (*n* = 4). Thus, we solve the compound interest formula for *t*. Solving for *t* involves the use of logarithms, the inverse function of exponents. To simplify calculations, we use the natural logarithm, *ln*.

$$A(t) = A_0(1 + \frac{i}{n})^{nt}$$

$$\frac{A(t)}{A_0} = (1+\frac{i}{n})^{nt}$$

$$\ln\frac{A(t)}{A_0} = \ln(1+\frac{i}{n})^{nt}$$ Take the ln of both sides.

$$\ln\frac{A(t)}{A_0} = (nt)\ln(1+\frac{i}{n})$$

Use the properties of logarithms with exponents.

$$\ln\frac{1000}{100} = (4t)\ln(1+\frac{0.08}{4})$$

Substitute and solve for time (t).

$$t = \frac{\ln 10}{4(\ln 1.02)} = 29.07\,years$$

2. Find the principle (A_0) that yields \$500 with an interest rate of 7.5% compounded semiannually for 20 years.

In this problem A(t) = 500, the interest rate (i) is 0.075, n = 2, and t = 20. To find the principle value, we solve for A_0.

$$A(t) = A_0(1+\frac{i}{n})^{nt}$$

$$500 = A_0(1+\frac{0.075}{2})^{2(20)}$$

Substitute and solve for A_0.

$$A_0 = \frac{500}{1.0375^{40}} = \$114.67$$

An **exponential function** is a function defined by the equation $y = ab^x$, where a is the starting value, b is the growth factor, and x tells how many times to multiply by the growth factor.

Example: $y = 100(1.5)^x$

x	y
0	100
1	150
2	225
3	337.5
4	506.25

This is an **exponential** or multiplicative pattern of growth.

Competency 9.0 **The teacher understands trigonometric and circular functions, analyzes their algebraic and graphical properties, and uses them to model and solve problems.**

Given 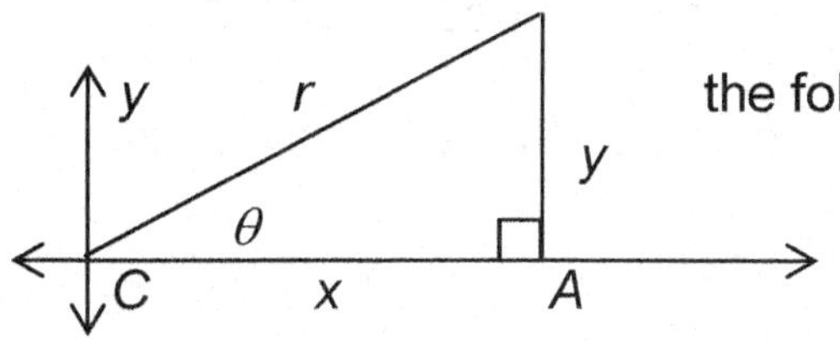 the following can be found.

Trigonometric Functions:

$\sin\theta = \frac{y}{r}$ $\csc\theta = \frac{r}{y}$

$\cos\theta = \frac{x}{r}$ $\sec\theta = \frac{r}{x}$

$\tan\theta = \frac{y}{x}$ $\cot\theta = \frac{x}{y}$

Sample problem:

1. Prove that $\sec\theta = \frac{1}{\cos\theta}$.

$\sec\theta = \frac{1}{\frac{x}{r}}$ Substitution definition of cosine.

$\sec\theta = \frac{1\times r}{\frac{x}{r}\times r}$ Multiply by $\frac{r}{r}$.

$\sec\theta = \frac{r}{x}$ Substitution.

$\sec\theta = \sec\theta$ Substitute definition of $\frac{r}{x}$.

$\sec\theta = \frac{1}{\cos\theta}$ Substitute.

2. Prove that $\sin^2 + \cos^2 = 1$.

$\left(\frac{y}{r}\right)^2 + \left(\frac{x}{r}\right)^2 = 1$ Substitute definitions of sin and cos.

$\frac{y^2 + x^2}{r^2} = 1$ $x^2 + y^2 = r^2$ Pythagorean formula.

$\frac{r^2}{r^2}=1$ Simplify.

$1=1$ Substitute.

$\sin^2\theta+\cos^2\theta=1$

Practice problems: Prove each identity.

1. $\cot\theta=\frac{\cos\theta}{\sin\theta}$
2. $1+\cot^2\theta=\csc^2\theta$

In order to **solve a right triangle** using trigonometric functions it is helpful to identify the given parts and label them. Usually more than one trigonometric function may be appropriately applied.

Some items to know about right triangles:

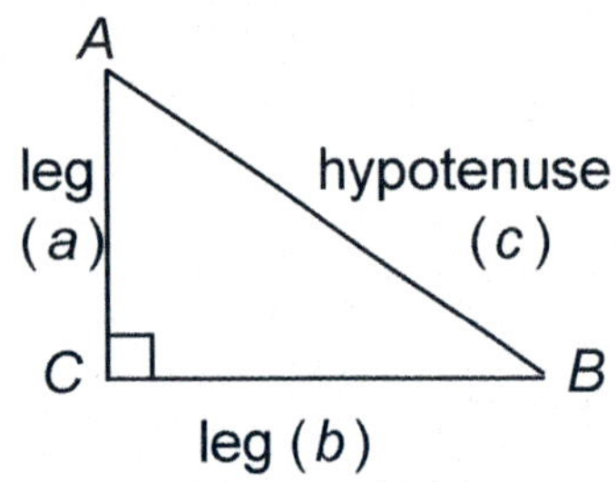

Given angle A, the side labeled leg (a) Is adjacent angle A. And the side labeled leg (b) is opposite to angle A.

Sample problem:

1. Find the missing side.

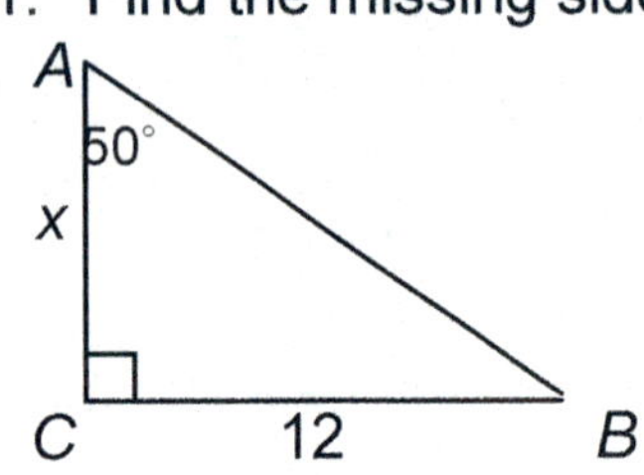

1. Identify the known values. Angle $A=50$ degrees and the side opposite the given angle is 12. The missing side is the adjacent leg.
2. The information suggests the use of the tangent function
3. Write the function.
4. Substitute.
5. Solve.

$\tan A=\frac{\text{opposite}}{\text{adjacent}}$

$\tan 50=\frac{12}{x}$

$1.192=\frac{12}{x}$

$x(1.192)=12$

$x=10.069$

Remember that since angle A and angle B are complimentary, then angle $B = 90 - 50$ or 40 degrees.

Using this information we could have solved for the same side only this time it is the leg opposite from angle B.

$\tan B = \dfrac{\text{opposite}}{\text{adjacent}}$	1. Write the formula.
$\tan 40 = \dfrac{x}{12}$	2. Substitute.
$12(.839) = x$	3. Solve.
$10.069 \approx x$	

Now that the two sides of the triangle are known, the third side can be found using the Pythagorean Theorem.

The trigonometric functions sine, cosine, and tangent are periodic functions. The values of periodic functions repeat on regular intervals. Period, amplitude, and phase shift are key properties of periodic functions that can be determined by observation of the graph.

The **period** of a function is the smallest domain containing the complete cycle of the function. For example, the period of a sine or cosine function is the distance between the peaks of the graph.

The **amplitude** of a function is half the distance between the maximum and minimum values of the function.

Phase shift is the amount of horizontal displacement of a function from its original position.

On the following page there is a generic sine/cosine graph with a labeled period and amplitude.

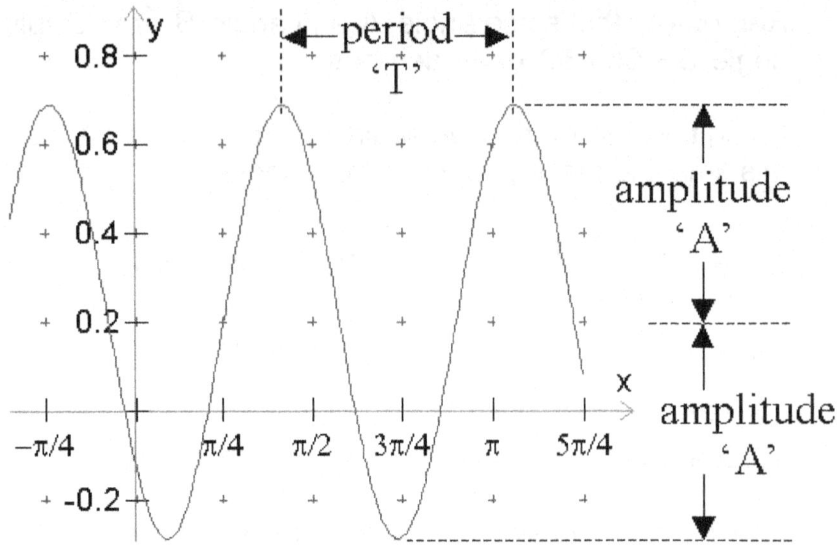

Properties of the graphs of basic trigonometric functions.

Function	Period	Amplitude
y = sin x	2π radians	1
y = cos x	2π radians	1
y = tan x	π radians	undefined

Below are the graphs of the basic trigonometric functions, (a) y = sin x; (b) y = cos x; and (c) y= tan x.

A)

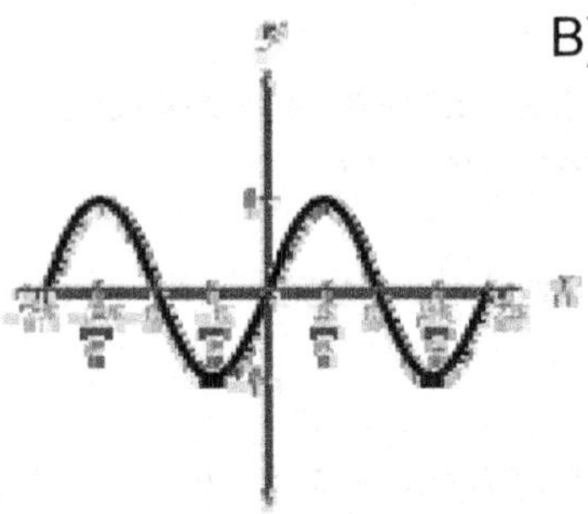

B)

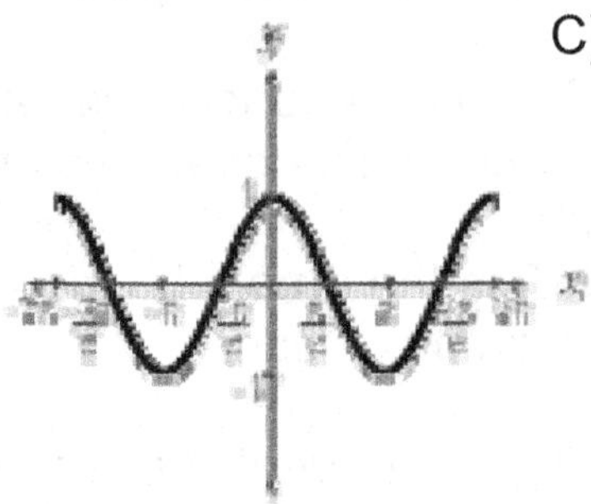

C)

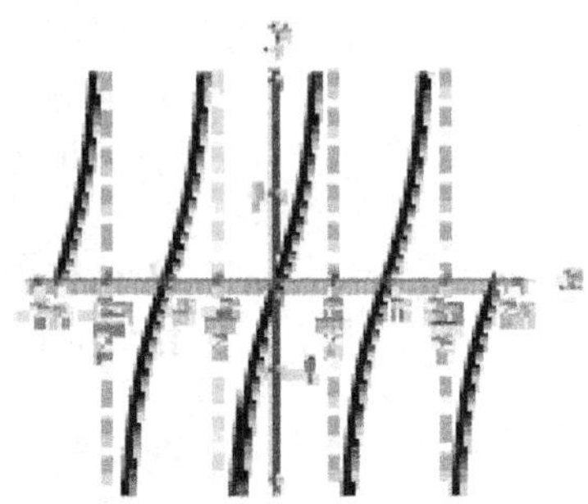

Note that the phase shift of trigonometric graphs is the horizontal distance displacement of the curve from these basic functions.

Unlike **trigonometric identities** that are true for all values of the defined variable, trigonometric equations are true for some, but not all, of the values of the variable. Most often trigonometric equations are solved for values between 0 and 360 degrees or 0 and 2π radians.

Some algebraic operation, such as squaring both sides of an equation, will give you extraneous answers. You must remember to check all solutions to be sure that they work.

Sample problems:

1. Solve: $\cos x = 1 - \sin x$ if $0 \le x < 360$ degrees.

$\cos^2 x = (1 - \sin x)^2$		1. square both sides
$1 - \sin^2 x = 1 - 2\sin x + \sin^2 x$		2. substitute
$0 = {}^-2\sin x + 2\sin^2 x$		3. set=to 0
$0 = 2\sin x({}^-1 + \sin x)$		4. factor
$2\sin x = 0$	${}^-1 + \sin x = 0$	5. set each factor=0
$\sin x = 0$	$\sin x = 1$	6. solve for $\sin x$
$x = 0$ or 180	$x = 90$	7. find value of sin at x

The solutions appear to be 0, 90 and 180. Remember to check each solution and you will find that 180 does not give you a true equation. Therefore, the only solutions are 0 and 90 degrees.

2. Solve: $\cos^2 x = \sin^2 x$ if $0 \le x < 2\pi$

$\cos^2 x = 1 - \cos^2 x$	1. substitute
$2\cos^2 x = 1$	2. simplify
$\cos^2 x = \frac{1}{2}$	3. divide by 2
$\sqrt{\cos^2 x} = \pm\sqrt{\frac{1}{2}}$	4. take square root
$\cos x = \frac{\pm\sqrt{2}}{2}$	5. rationalize denominator
$x = \frac{\pi}{4}, \frac{3\pi}{4}, \frac{5\pi}{4}, \frac{7\pi}{4}$	

There are two methods that may be used to **prove trigonometric identities**. One method is to choose one side of the equation and manipulate it until it equals the other side. The other method is to replace expressions on both sides of the equation with equivalent expressions until both sides are equal.

The Reciprocal Identities

$\sin x = \frac{1}{\csc x}$ $\quad\sin x \csc x = 1\quad$ $\csc x = \frac{1}{\sin x}$

$\cos x = \frac{1}{\sec x}$ $\quad\cos x \sec x = 1\quad$ $\sec x = \frac{1}{\cos x}$

$\tan x = \frac{1}{\cot x}$ $\quad\tan x \cot x = 1\quad$ $\cot x = \frac{1}{\tan x}$

$\tan x = \frac{\sin x}{\cos x}$ $\quad\cot x = \frac{\cos x}{\sin x}$

The Pythagorean Identities

$\sin^2 x + \cos^2 x = 1$ $\quad 1 + \tan^2 x = \sec^2 x$ $\quad 1 + \cot^2 x = \csc^2 x$

Sample problems:

1. Prove that $\cot x + \tan x = (\csc x)(\sec x)$.

$$\frac{\cos x}{\sin x} + \frac{\sin x}{\cos x}$$

Reciprocal identities.

$$\frac{\cos^2 x + \sin^2 x}{\sin x \cos x}$$

Common denominator.

$$\frac{1}{\sin x \cos x}$$

Pythagorean identity.

$$\frac{1}{\sin x} \times \frac{1}{\cos x}$$

$$\csc x(\sec x) = \csc x(\sec x)$$

Reciprocal identity, therefore,

$$\cot x + \tan x = \csc x(\sec x)$$

2. Prove that $\dfrac{\cos^2\theta}{1+2\sin\theta+\sin^2\theta}=\dfrac{\sec\theta-\tan\theta}{\sec\theta+\tan\theta}$.

$$\frac{1-\sin^2\theta}{(1+\sin\theta)(1+\sin\theta)}=\frac{\sec\theta-\tan\theta}{\sec\theta+\tan\theta}$$

Pythagorean identity factor denominator.

$$\frac{1-\sin^2\theta}{(1+\sin\theta)(1+\sin\theta)}=\frac{\dfrac{1}{\cos\theta}-\dfrac{\sin\theta}{\cos\theta}}{\dfrac{1}{\cos\theta}+\dfrac{\sin\theta}{\cos\theta}}$$

Reciprocal identities.

$$\frac{(1-\sin\theta)(1+\sin\theta)}{(1+\sin\theta)(1+\sin\theta)}=\frac{\dfrac{1-\sin\theta}{\cos\theta}(\cos\theta)}{\dfrac{1+\sin\theta}{\cos\theta}(\cos\theta)}$$

Factor $1-\sin^2\theta$.

Multiply by $\dfrac{\cos\theta}{\cos\theta}$.

$$\frac{1-\sin\theta}{1+\sin\theta}=\frac{1-\sin\theta}{1+\sin\theta}$$

Simplify.

$$\frac{\cos^2\theta}{1+2\sin\theta+\sin^2\theta}=\frac{\sec\theta-\tan\theta}{\sec\theta+\tan\theta}$$

It is easiest to **graph trigonometric functions** when using a calculator by making a table of values.

DEGREES

	0	30	45	60	90	120	135	150	180	210	225	240	270	300	315	330	360
sin	0	.5	.71	.87	1	.87	.71	.5	0	-.5	-.71	-.87	-1	-.87	-.71	-.5	0
cos	1	.87	.71	.5	0	-.5	-.71	-.87	-1	-.87	-.71	-.5	0	.5	.71	.87	1
tan	0	.58	1	1.7	--	-1.7	-1	-.58	0	.58	1	1.7	--	-1.7	-1	-.58	0
	0	$\frac{\pi}{6}$	$\frac{\pi}{4}$	$\frac{\pi}{3}$	$\frac{\pi}{2}$	$\frac{2\pi}{3}$	$\frac{3\pi}{4}$	$\frac{5\pi}{6}$	π	$\frac{7\pi}{6}$	$\frac{5\pi}{4}$	$\frac{4\pi}{3}$	$\frac{3\pi}{2}$	$\frac{5\pi}{3}$	$\frac{7\pi}{4}$	$\frac{11\pi}{6}$	2π

RADIANS

Remember the graph always ranges from $+1$ to $^-1$ for sine and cosine functions unless noted as the coefficient of the function in the equation. For example, $y=3\cos x$ has an amplitude of 3 units from the center line (0). Its maximum and minimum points would be at $+3$ and $^-3$.

Tangent is not defined at the values 90 and 270 degrees or $\frac{\pi}{2}$ and $\frac{3\pi}{2}$. Therefore, vertical asymptotes are drawn at those values.

The inverse functions can be graphed in the same manner using a calculator to create a table of values.

Mathematicians use trigonometric functions to model and solve problems involving **naturally occurring periodic phenomena**. Examples of periodic phenomena found in nature include all forms of radiation (ultraviolet rays, visible light, microwaves, etc.), sound waves and pendulums. Additionally, trigonometric functions often approximate fluctuations in temperature, employment and consumer behavior (business models).The following are examples of the use of the sine function to model problems.

Example 1:

Consider the average monthly temperatures for a hypothetical location.

Month	Avg. Temp. (F)
Jan	40
March	48
May	65
July	81
Sept	80
Nov	60

Note that the graph of the average temperatures resembles the graph of a trigonometric function with a period of one year. We can use the periodic nature of seasonal temperature fluctuation to predict weather patterns.

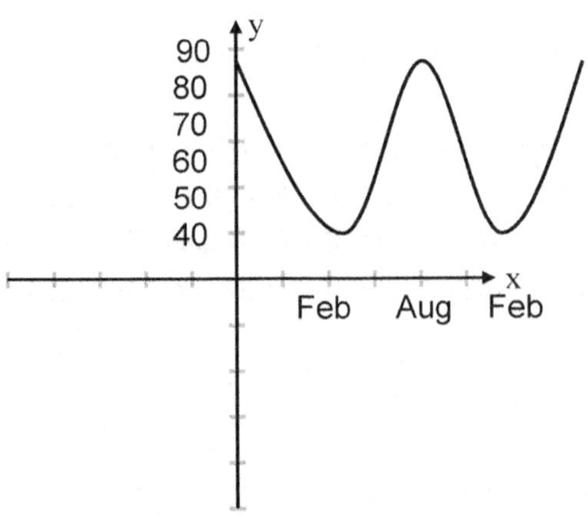

Example 2:

The general form of the sine function that is useful in modeling problems involving time is:

$$y = A[\sin \omega(t - \alpha)] + C = A[\sin(\omega t - \alpha\omega) + C$$

where A is the amplitude, C is the vertical offset, ω is the angular frequency ($\frac{2\pi}{P}$, P is period), t is the time in years and α is the horizontal offset or phase shift.

Consider the following situation. An economist at a temporary employment agency reports that the demand for temporary employment (as measured by thousands of applications per week) varies according to the following model, where t is the time in years, starting with January, 2002.

Demand (d) = 4.7sin(0.75t + 0.4) + 7.7

From this model, we can calculate the characteristics of the sine function and interpret the cyclical nature of demand for temporary employment. Applying the general sine function formula to the model:

A = 4.7
C = 7.7
ω = 0.75

$\omega\alpha$ = 0.4 → α = 0.4/0.75 = 0.53

P (period) = $\frac{2\pi}{\omega} = 8.4$

Thus, we interpret the model as follows.

The demand for temporary employment fluctuates in cycles of 8.4 years (period) about a baseline of 7,700 job applications per week (vertical offset). The peak of each cycle is approximately 12,400 job applications per week and the low point of each cycle is 3,000 (vertical offset $\pm$ amplitude).

Competency 10.0 The teacher understands and solves problems using differential and integral calculus.

The limit of a function is the y value that the graph approaches as the x values approach a certain number. To find a limit there are two points to remember.

1. Factor the expression completely and cancel all common factors in fractions.
2. Substitute the number to which the variable is approaching. In most cases this produces the value of the limit.

If the variable in the limit is approaching ∞, factor and simplify first; then examine the result. If the result does not involve a fraction with the variable in the denominator, the limit is usually also equal to ∞. If the variable is in the denominator of the fraction, the denominator is getting larger which makes the entire fraction smaller. In other words the limit is zero.

Examples:

1. $\lim_{x \to {}^-3} \frac{x^2+5x+6}{x+3} + 4x$ Factor the numerator.

$\lim_{x \to {}^-3} \frac{(x+3)(x+2)}{(x+3)} + 4x$ Cancel the common factors.

$\lim_{x \to {}^-3} (x+2) + 4x$ Substitute $^-3$ for x.

$({}^-3+2)+4({}^-3)$ Simplify.

${}^-1 + {}^-12$

${}^-13$

2. $\lim_{x \to \infty} \frac{2x^2}{x^5}$ Cancel the common factors.

$\lim_{x \to \infty} \frac{2}{x^3}$ Since the denominator is getting larger, the entire fraction is getting smaller.

$\frac{2}{\infty^3}$ The fraction is getting close to zero.

0

Practice problems:

1. $\lim_{x \to \pi} 5x^2 + \sin x$

2. $\lim_{x \to {}^-4} \frac{x^2+9x+20}{x+4}$

After simplifying an expression to evaluate a limit, substitute the value that the variable approaches. If the substitution results in either $0/0$ or ∞/∞, use L'Hopital's rule to find the limit.

L'Hopital's rule states that you can find such limits by taking the derivative of the numerator and the derivative of the denominator, and then finding the limit of the resulting quotient.

Examples:

1. $\lim_{x\to\infty} \frac{3x-1}{x^2+2x+3}$ — No factoring is possible.

$\frac{3\infty-1}{\infty^2+2\infty+3}$ — Substitute ∞ for x.

$\frac{\infty}{\infty}$ — Since a constant times infinity is still a large number, $3(\infty)=\infty$.

$\lim_{x\to\infty} \frac{3}{2x+2}$ — To find the limit, take the derivative of the numerator and denominator.

$\frac{3}{2(\infty)+2}$ — Substitute ∞ for x again.

$\frac{3}{\infty}$

0 — Since the denominator is a very large number, the fraction is getting smaller. Thus the limit is zero.

2. $\lim_{x\to 1} \frac{\ln x}{x-1}$ — Substitute 1 for x.

$\frac{\ln 1}{1-1}$ — The $\ln 1 = 0$

$\frac{0}{0}$ — To find the limit, take the derivative of the numerator and denominator.

$\lim_{x\to 1} \frac{\frac{1}{x}}{1}$ — Substitute 1 for x again.

$$\frac{1}{\frac{1}{1}}$$

Simplify. The limit is one.

1

Practice problems:

1. $\lim\limits_{x \to \infty} \dfrac{x^2 - 3}{x}$

2. $\lim\limits_{x \to \frac{\pi}{2}} \dfrac{\cos x}{x - \frac{\pi}{2}}$

The **difference quotient** is the average rate of change over an interval. For a function f, the **difference quotient** is represented by the formula:

$$\frac{f(x+h)-f(x)}{h}.$$

This formula computes the slope of the secant line through two points on the graph of f. These are the points with x-coordinates x and $x+h$.

Example: Find the difference quotient for the function $f(x) = 2x^2 + 3x - 5$.

$$\frac{f(x+h)-f(x)}{h} = \frac{2(x+h)^2 + 3(x+h) - 5 - (2x^2 + 3x - 5)}{h}$$

$$= \frac{2(x^2 + 2hx + h^2) + 3x + 3h - 5 - 2x^2 - 3x + 5}{h}$$

$$= \frac{2x^2 + 4hx + 2h^2 + 3x + 3h - 5 - 2x^2 - 3x + 5}{h}$$

$$= \frac{4hx + 2h^2 + 3h}{h}$$

$$= 4x + 2h + 3$$

The **derivative** is the slope of a tangent line to a graph $f(x)$, and is usually denoted $f'(x)$. This is also referred to as the instantaneous rate of change.

The derivative of $f(x)$ at $x = a$ is given by taking the limit of the average rates of change (computed by the difference quotient) as h approaches 0.

$$f'(a) = \lim_{h \to 0} \frac{f(a+h) - f(a)}{h}$$

Example: Suppose a company's annual profit (in millions of dollars) is represented by the above function $f(x) = 2x^2 + 3x - 5$ and x represents the number of years in the interval. Compute the rate at which the annual profit was changing over a period of 2 years.

$$f'(a) = \lim_{h \to 0} \frac{f(a+h) - f(a)}{h}$$

$$= f'(2) = \lim_{h \to 0} \frac{f(2+h) - f(2)}{h}$$

Using the difference quotient we computed above, $4x + 2h + 3$, we get

$$f'(2) = \lim_{h \to 0} (4(2) + 2h + 3)$$
$$= 8 + 3$$
$$= 11.$$

We have, therefore, determined that the annual profit for the company has increased at the average rate of $11 million per year over the two-year period.

Substituting an x value into a function produces a corresponding y value. The coordinates of the point (x, y), where y is the largest of all the y values, is said to be a **maximum point**. The coordinates of the point (x, y), where y is the smallest of all the y values, is said to be a **minimum point**. To find these points, only a few x values must be tested. First, find all of the x values that make the derivative either zero or undefined. Substitute these values into the original function to obtain the corresponding y values. Compare the y values. The largest y value is a maximum; the smallest y value is a minimum. If the question asks for the maxima or minima on an interval, be certain to also find the y values that correspond to the numbers at either end of the interval.

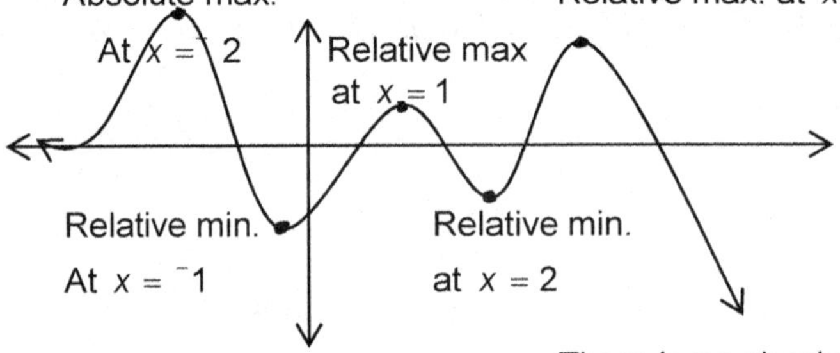

Example:

Find the maxima and minima of $f(x) = 2x^4 - 4x^2$ at the interval $({}^-2, 1)$.

$f'(x) = 8x^3 - 8x$	Take the derivative first. Find all the x values (critical values)
$8x^3 - 8x = 0$	that make the derivative zero
$8x(x^2 - 1) = 0$	or undefined. In this case,
$8x(x-1)(x+1) = 0$	there are no x values that make
$x = 0,\ x = 1,\ \text{or } x = {}^-1$	the derivative undefined.
$f(0) = 2(0)^4 - 4(0)^2 = 0$	Substitute the critical values into the original function. Also,
$f(1) = 2(1)^4 - 4(1)^2 = {}^-2$	plug in the endpoint of the
$f({}^-1) = 2({}^-1)^4 - 4({}^-1)^2 = {}^-2$	interval. Note that 1 is
$f({}^-2) = 2({}^-2)^4 - 4({}^-2)^2 = 16$	a critical point and an endpoint.

The maximum is at $({}^-2, 16)$ and there are minima at $(1, {}^-2)$ and $({}^-1, {}^-2)$. (0,0) is neither the maximum or minimum on $({}^-2, 1)$ but it is still considered a relative extra point.

A function is said to be **increasing** if it is rising from left to right and **decreasing** if it is falling from left to right. Lines with positive slopes are increasing, and lines with negative slopes are decreasing. If the function in question is something other than a line, simply refer to the slopes of the tangent lines as the test for increasing or decreasing. Take the derivative of the function and plug in an x value to get the slope of the tangent line; a positive slope means the function is increasing and a negative slope means it is decreasing. If an interval for x values is given, just pick any point between the two values to substitute.

Sample tangent line on $(^-2,0)$

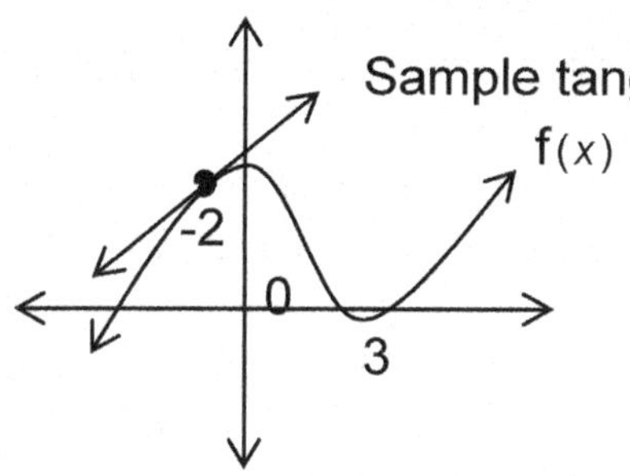

On the interval $(^-2,0)$, $f(x)$ is increasing. The tangent lines on this part of the graph have positive slopes.

Example:

The growth of a certain bacteria is given by $f(x)=x+\frac{1}{x}$. Determine if the rate of growth is increasing or decreasing on the time interval $(^-1,0)$.

$$f'(x)=1+\frac{^-1}{x^2}$$

To test for increasing or decreasing, find the slope of the tangent line by taking the derivative.

$$f'\left(\frac{^-1}{2}\right)=1+\frac{^-1}{(^-1/2)^2}$$

Pick any point on $(^-1,0)$ and substitute into the derivative.

$$f'\left(\frac{^-1}{2}\right)=1+\frac{^-1}{1/4}$$
$$=1-4$$
$$=^-3$$

The slope of the tangent line at $x=\frac{^-1}{2}$ is $^-3$. The exact value of the slope is not important. The important fact is that the slope is negative.

The first derivative reveals whether a curve is rising or falling (increasing or decreasing) from the left to the right. In much the same way, the second derivative relates whether the curve is concave up or concave down. Curves which are concave up are said to "collect water;" curves which are concave down are said to "dump water." To find the intervals where a curve is concave up or concave down, follow the following steps.

1. Take the second derivative (i.e. the derivative of the first derivative).
2. Find the critical x values.
   -Set the second derivative equal to zero and solve for critical x values.
   -Find the x values that make the second derivative undefined (i.e. make the denominator of the second derivative equal to zero).
   Such values may not always exist.
3. Pick sample values which are both less than and greater than each of the critical values.
4. Substitute each of these sample values into the second derivative and determine whether the result is positive or negative.
   -If the sample value yields a positive number for the second derivative, the curve is concave up on the interval where the sample value originated.
   -If the sample value yields a negative number for the second derivative, the curve is concave down on the interval where the sample value originated.

Example:
Find the intervals where the curve is concave up and concave down for $\mathrm{f}(x) = x^4 - 4x^3 + 16x - 16$.

$\mathrm{f}'(x) = 4x^3 - 12x^2 + 16$ Take the second derivative.

$\mathrm{f}''(x) = 12x^2 - 24x$ Find the critical values by setting the second derivative equal to zero.

$12x^2 - 24x = 0$

$12x(x-2) = 0$ There are no values that make the second derivative undefined.

$x = 0$ or $x = 2$

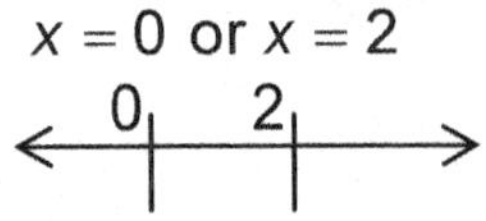

Set up a number line with the critical values.

Sample values: $^{-}1,\ 1,\ 3$

$f''(^{-}1) = 12(^{-}1)^2 - 24(^{-}1) = 36$
$f''(1) = 12(1)^2 - 24(1) = {}^{-}12$
$f''(3) = 12(3)^2 - 24(3) = 36$

Pick sample values in each of the 3 intervals. If the sample value produces a negative number, the function is concave down.

If the value produces a positive number, the curve is concave up. If the value produces a zero, the function is linear.

A **point of inflection** is a point where a curve changes from being concave up to concave down or vice versa. To find these points, follow the steps for finding the intervals where a curve is concave up or concave down. A critical value is part of an inflection point if the curve is concave up on one side of the value and concave down on the other. The critical value is the x coordinate of the inflection point. To get the y coordinate, plug the critical value into the **original** function.

<u>Example:</u> Find the inflection points of $f(x) = 2x - \tan x$ where $\frac{^{-}\pi}{2} < x < \frac{\pi}{2}$.

$(x) = 2x - \tan x \qquad \frac{^{-}\pi}{2} < x < \frac{\pi}{2}$

Note the restriction on x.

$f'(x) = 2 - \sec^2 x$

Take the second derivative.

$f''(x) = 0 - 2 \bullet \sec x \bullet (\sec x \tan x)$

Use the Power rule.

$= {}^{-}2 \bullet \frac{1}{\cos x} \bullet \frac{1}{\cos x} \bullet \frac{\sin x}{\cos x}$

The derivative of $\sec x$ is $(\sec x \tan x)$.

$f''(x) = \frac{^{-}2\sin x}{\cos^3 x}$

Find critical values by solving for the second derivative equal to zero.

$0 = \frac{^{-}2\sin x}{\cos^3 x}$

No x values on $\left(\frac{^{-}\pi}{2}, \frac{\pi}{2}\right)$ make the denominator zero.

$^{-}2\sin x = 0$
$\sin x = 0$
$x = 0$

Pick sample values on each side of the critical value $x = 0$.

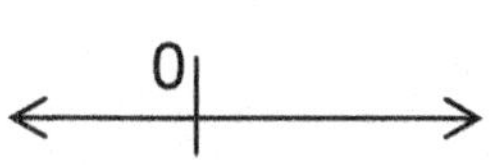

Sample values: $x = \frac{^-\pi}{4}$ and $x = \frac{\pi}{4}$

$$f''\left(\frac{^-\pi}{4}\right) = \frac{^-2\sin(^-\pi/4)}{\cos^3(\pi/4)} = \frac{^-2(^-\sqrt{2}/2)}{(\sqrt{2}/2)^3} = \frac{\sqrt{2}}{(\sqrt{8}/8)} = \frac{8\sqrt{2}}{\sqrt{8}} = \frac{8\sqrt{2}}{\sqrt{8}} \cdot \frac{\sqrt{8}}{\sqrt{8}}$$

$$= \frac{8\sqrt{16}}{8} = 4$$

$$f''\left(\frac{\pi}{4}\right) = \frac{^-2\sin(\pi/4)}{\cos^3(\pi/4)} = \frac{^-2(\sqrt{2}/2)}{(\sqrt{2}/2)^3} = \frac{^-\sqrt{2}}{(\sqrt{8}/8)} = \frac{^-8\sqrt{2}}{\sqrt{8}} = -4$$

The second derivative is positive on $(0,\infty)$ and negative on $(^-\infty,0)$. So the curve changes concavity at $x = 0$. Use the original equation to find the y value that inflection occurs at.

$f(0) = 2(0) - \tan 0 = 0 - 0 = 0$ The inflection point is (0,0).

Extreme value problems are also known as max-min problems. Extreme value problems require using the first derivative to find values which either maximize or minimize some quantity such as area, profit, or volume. Follow these steps to solve an extreme value problem.

1. Write an equation for the quantity to be maximized or minimized.
2. Use the other information in the problem to write secondary equations.
3. Use the secondary equations for substitutions, and rewrite the original equation in terms of only one variable.
4. Find the derivative of the primary equation (step 1) and the critical values of this derivative.
5. Substitute these critical values into the primary equation. The value which produces either the largest or smallest value is used to find the solution.

Example:

A manufacturer wishes to construct an open box from the piece of metal shown below by cutting squares from each corner and folding up the sides. The square piece of metal is 12 feet on a side. What are the dimensions of the squares to be cut out which will maximize the volume?

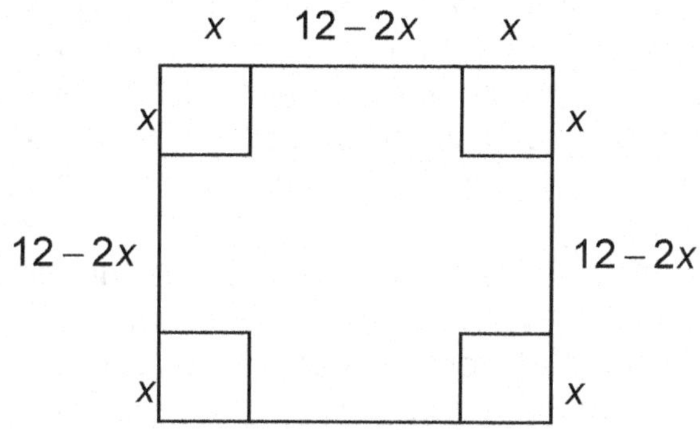

Volume $= lwh$ — Primary equation.

$l = 12 - 2x$

$w = 12 - 2x$ — Secondary equations.

$h = x$

$V = (12 - 2x)(12 - 2x)(x)$ — Make substitutions.

$V = (144x - 48x^2 + 4x^3)$ — Take the derivative.

$$\frac{dV}{dx} = 144 - 96x + 12x^2$$

Find critical values by setting the derivative equal to zero.

$0 = 12(x^2 - 8x + 12)$

$0 = 12(x - 6)(x - 2)$

$x = 6$ and $x = 2$

Substitute critical values into volume equation.

$V = 144(6) - 48(6)^2 + 4(6)^3$

$V = 0$ ft^3 when $x = 6$

$V = 144(2) - 48(2)^2 + 4(2)^3$

$V = 128$ ft^3 when $x = 2$

Therefore, the manufacturer can maximize the volume if the squares to be cut out are 2 feet by 2 feet ($x = 2$).

If a particle (or a car, a bullet, etc.) is moving along a line, then the **distance that the particle travels** can be expressed by a function in terms of time.

1. The first derivative of the distance function will provide the velocity function for the particle. Substituting a value for time into this expression will provide the instantaneous velocity of the particle at the time. Velocity is the rate of change of the distance traveled by the particle. Taking the absolute value of the derivative provides the speed of the particle. A positive value for the velocity indicates that the particle is moving forward, and a negative value indicates the particle is moving backwards.
2. The second derivative of the distance function (which would also be the first derivative of the velocity function) provides the acceleration function. The acceleration of the particle is the rate of change of the velocity. If a value for time produces a positive acceleration, the particle is speeding up; if it produces a negative value, the particle is slowing down. If the acceleration is zero, the particle is moving at a constant speed.

To find the time when a particle stops, set the first derivative (i.e. the velocity function) equal to zero and solve for time. This time value is also the instant when the particle changes direction.

Example:

The motion of a particle moving along a line is according to the equation:

$s(t) = 20 + 3t - 5t^2$ where s is in meters and t is in seconds. Find the position, velocity, and acceleration of a particle at $t = 2$ seconds.

$s(2) = 20 + 3(2) - 5(2)^2$ $= 6$ meters	Plug $t = 2$ into the original equation to find the position.
$s'(t) = v(t) = 3 - 10t$	The derivative of the first function gives the velocity.
$v(2) = 3 - 10(2) = {}^-17$ m/s	Plug $t = 2$ into the velocity function to find the velocity. $^-17$ m/s indicates the particle is moving backwards.
$s''(t) = a(t) = {}^-10$	The second derivation of position gives the acceleration.
$a(2) = {}^-10$ m/s²	Substitute $t = 2$, yields an acceleration of $^-10$ m/s², which indicates the particle is slowing down.

An **integral** is almost the same thing as an antiderivative, the only difference is the notation.

$\int_{-2}^{1} 2x dx$ is the integral form of the antiderivative of $2x$. The numbers at the top and bottom of the integral sign (1 and $^-2$) are the numbers used to find the exact value of this integral. If these numbers are used the integral is said to be *definite* and does not have an unknown constant c in the answer.

The fundamental theorem of calculus states that an integral such as the one above is equal to the antiderivative of the function inside (here $2x$) evaluated from $x = {}^-2$ to $x = 1$. To do this, follow these steps.

1. Take the antiderivative of the function inside the integral.
2. Plug in the upper number (here $x = 1$) and plug in the lower number (here $x = {}^-2$), giving two expressions.
3. Subtract the second expression from the first to achieve the integral value.

Examples:

1. $\int_{-2}^{1} 2x dx = x^2\Big]_{-2}^{1}$ — Take the antiderivative.

   $\int_{-2}^{1} 2x dx = 1^2 - ({}^-2)^2$ — Substitute in $x = 1$ and $x = {}^-2$ and subtract the results.

   $\int_{-2}^{1} 2x dx = 1 - 4 = {}^-3$ — The integral has the value $^-3$.

2. $\int_{0}^{\pi/2} \cos x dx = \sin x\Big]_{0}^{\pi/2}$ — The antiderivative of $\cos x$ is $\sin x$.

   $\int_{0}^{\frac{\pi}{2}} \cos x dx = \sin\frac{\pi}{2} - \sin 0$ — Substitute in $x = \frac{\pi}{2}$ and $x = 0$. Subtract the results.

   $\int_{0}^{\frac{\pi}{2}} \cos x dx = 1 - 0 = 1$ — The integral has the value 1.

The **derivative of a distance function** provides a velocity function, and the derivative of a velocity function provides an acceleration function. Therefore taking the antiderivative of an acceleration function yields the velocity function, and the antiderivative of the velocity function yields the distance function.

Example:

A particle moves along the x axis with acceleration $a(t) = 3t - 1$ cm/sec/sec. At time, $t = 4$, the particle is moving to the left at 3 cm per second. Find the velocity of the particle at time $t = 2$ seconds.

$a(t) = 3t - 1$ — Before taking the antiderivative, make sure the correct coefficients are present.

$a(t) = 3 \bullet \frac{1}{2} \bullet 2t - 1$

$v(t) = \frac{3}{2}t^2 - 1 \bullet t + c$ — $v(t)$ is the antiderivative of $a(t)$.

$v(4) = \frac{3}{2}(4)^2 - 1(4) + c = {}^-3$ — Use the given information that $v(4) = {}^-3$ to find c.

$24 - 4 + c = {}^-3$

$20 + c = {}^-3$

$c = {}^-23$ — The constant is ${}^-23$.

$v(t) = \frac{3}{2}t^2 - 1t + {}^-23$ — Rewrite $v(t)$ using $c = {}^-23$.

$v(2) = \frac{3}{2}2^2 - 1(2) + {}^-23$ — Solve $v(t)$ at t=2.

$v(2) = 6 + 2 + {}^-23 = {}^-15$ — the velocity at $t = 2$ is -15 cm/sec.

Practice problem:

A particle moves along a line with acceleration $a(t) = 5t + 2$. The velocity after 2 seconds is ${}^-10$ m/sec.

1. Find the initial velocity.
2. Find the velocity at $t = 4$.

To find the **distance function**, take the antiderivative of the velocity function. And to find the velocity function, find the antiderivative of the acceleration function. Use the information in the problem to solve for the constants that result from taking the antiderivatives.

Example:

A particle moves along the x axis with acceleration $a(t) = 6t - 6$. The initial velocity is 0 m/sec and the initial position is 8 cm to the right of the origin. Find the velocity and position functions.

$v(0) = 0$
$s(0) = 8$ — Interpret the given information.

$a(t) = 6t - 6$ — Put in the coefficients needed to take the antiderivative.

$a(t) = 6 \bullet \frac{1}{2} \bullet 2t - 6$ — Take the antiderivative of $a(t)$ to get $v(t)$.

$v(t) = \frac{6}{2}t^2 - 6t + c$

$v\,v(0) = 3(0)^2 - 6(0) = c = 0$ — Use $v(0) = 0$ to solve for c.
$0 - 0 + c = 0$
$c = 0$ — $c = 0$

$v(t) = 3t^2 - 6t + 0$ — Rewrite $v(t)$ using $c = 0$.

$v(t) = 3t^2 - 6\frac{1}{2} \bullet 2t$ — Put in the coefficients needed to take the antiderivative.

$s(t) = t^3 - \frac{6}{2}t^2 + c$ — Take the antiderivative of $v(t)$ to get $s(t) \rightarrow$ the distance function.

$s(0) = 0^3 - 3(0)^2 + c = 8$ — Use $s(0) = 8$ to solve for c.
$c = 8$
$s(t) = t^3 - 3t^2 + 8$

DOMAIN III. GEOMETRY AND MEASUREMENT

Competency 11.0 The teacher understands measurement as a process.

Use appropriate problem solving strategies to find the solution.

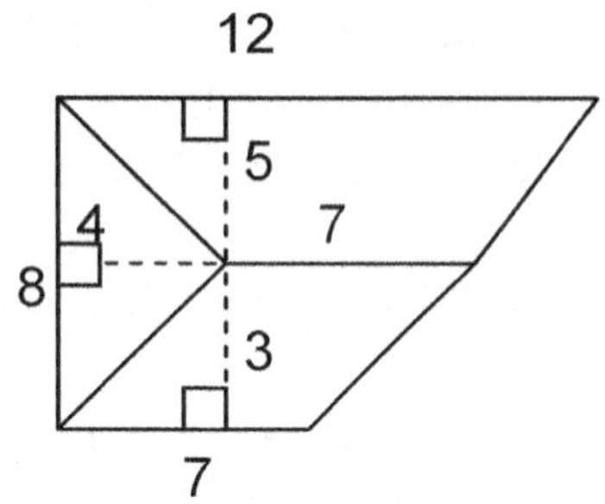

1. Find the area of the given figure.
2. Cut the figure into familiar shapes.
3. Identify what type figures are given and write the appropriate formulas.

Area of figure 1 (triangle)

$A = \frac{1}{2}bh$

$A = \frac{1}{2}(8)(4)$

$A = 16$ sq. ft

Area of figure 2 (parallelogram)

$A = bh$

$A = (7)(3)$

$A = 21$ sq. ft

Area of figure 3 (trapezoid)

$A = \frac{1}{2}h(a+b)$

$A = \frac{1}{2}(5)(12+7)$

$A = 47.5$ sq. ft

Now find the total area by adding the area of all figures.

Total area $= 16 + 21 + 47.5$
Total area $= 84.5$ square ft

Given the figure below, find the area by **dividing the polygon into smaller shapes**.

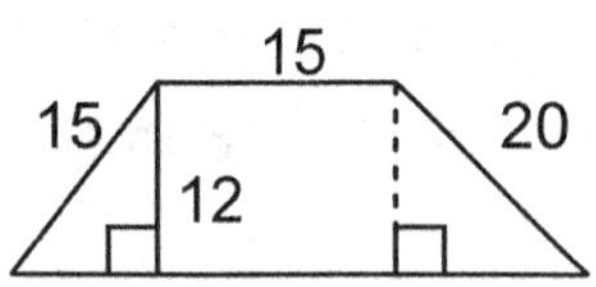

1. divide the figure into two triangles and a rectangle.
2. find the missing lengths.
3. find the area of each part.
4. find the sum of all areas.

Find the base of both right triangles using the Pythagorean Formula:

$$\begin{aligned} a^2 + b^2 &= c^2 \\ a^2 + 12^2 &= 15^2 \\ a^2 &= 225 - 144 \\ a^2 &= 81 \\ a &= 9 \end{aligned} \qquad \begin{aligned} a^2 + b^2 &= c^2 \\ a^2 + 12^2 &= 20^2 \\ a^2 &= 400 - 144 \\ a^2 &= 256 \\ a &= 16 \end{aligned}$$

Area of triangle 1	Area of triangle 2	Area of rectangle
$A = \frac{1}{2}bh$	$A = \frac{1}{2}bh$	$A = LW$
$A = \frac{1}{2}(9)(12)$	$A = \frac{1}{2}(16)(12)$	$A = (15)(12)$
$A = 54$ sq. units	$A = 96$ sq. units	$A = 180$ sq. units

Find the sum of all three figures.

$$54 + 96 + 180 = 330 \text{ square units}$$

Polygons are similar if and only if there is a one-to-one correspondence between their vertices such that the corresponding angles are congruent and the lengths of corresponding sides are proportional.

Given the rectangles below, compare the area and perimeter.

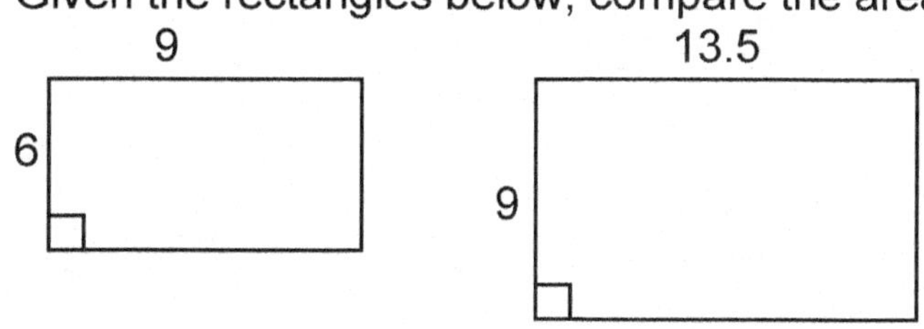

$A = LW$	$A = LW$	1. write formula
$A = (6)(9)$	$A = (9)(13.5)$	2. substitute known values
$A = 54$ sq. units	$A = 121.5$ sq. units	3. compute
$P = 2(L + W)$	$P = 2(L + W)$	1. write formula
$P = 2(6 + 9)$	$P = 2(9 + 13.5)$	2. substitute known values
$P = 30$ units	$P = 45$ units	3. compute

Notice that the areas relate to each other in the following manner:

Ratio of sides $9/13.5 = 2/3$

Multiply the first area by the square of the reciprocal $(3/2)^2$ to get the second area.

$$54 \times (3/2)^2 = 121.5$$

The perimeters relate to each other in the following manner:

$$\text{Ratio of sides} \quad 9/13.5 = 2/3$$

Multiply the perimeter of the first by the reciprocal of the ratio to get the perimeter of the second.

$$30 \times 3/2 = 45$$

Examining the **change in area or volume of a given figure** requires first to find the existing area given the original dimensions and then finding the new area given the increased dimensions.

Sample problem:

Given the rectangle below determine the change in area if the length is increased by 5 and the width is increased by 7.

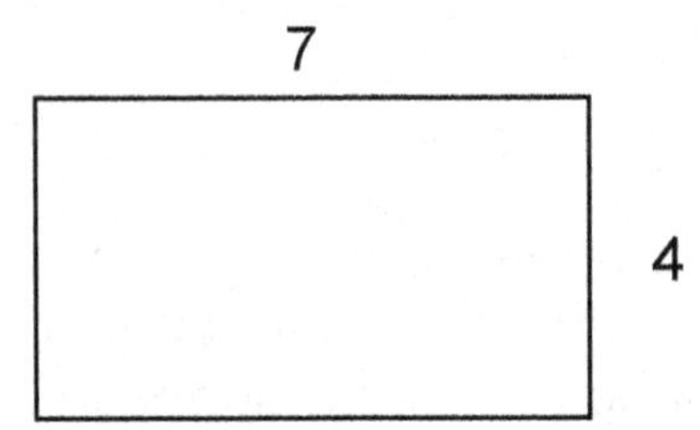

Draw and label a sketch of the new rectangle.

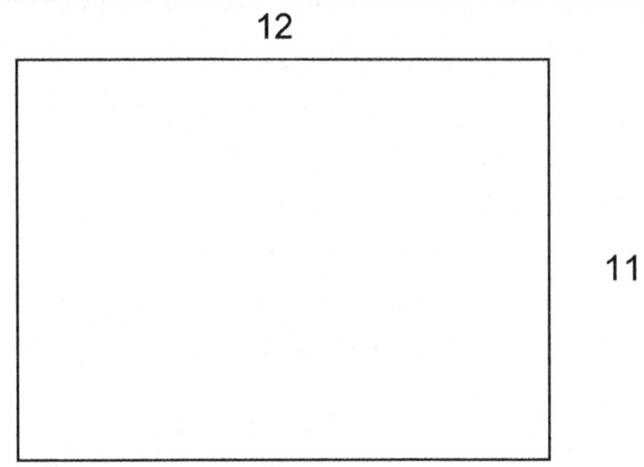

Find the areas.

Area of original = LW
= (7)(4)
= 28 units^2

Area of enlarged shape = LW
= (12)(11)
= 132 units^2

The change in area is 132 – 28 = 104 units^2.

FIGURE	LATERAL AREA	TOTAL AREA	VOLUME
Right prism	sum of area of lateral faces (rectangles)	lateral area plus 2 times the area of base	area of base times height
regular pyramid	sum of area of lateral faces (triangles)	lateral area plus area of base	1/3 times the area of the base times the height

Find the total area of the given figure:

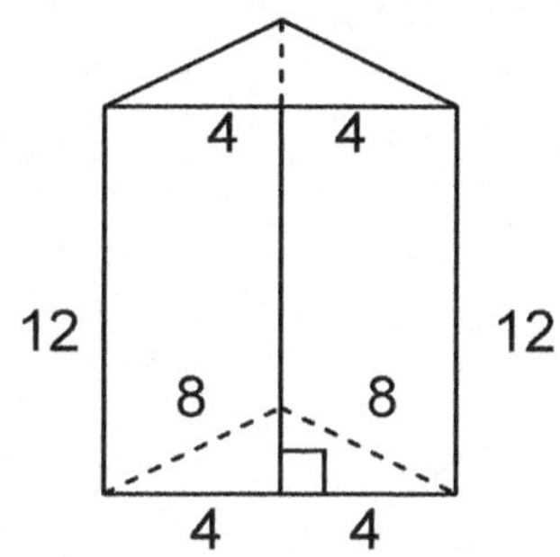

1. Since this is a triangular prism, first find the bases.
2. Find the area of each rectangular lateral face.
3. Add the areas together.

$A=\frac{1}{2}bh$	$A=LW$	1. write formula
$8^2=4^2+h^2$ $h=6.928$		2. find the height of the base triangle
$A=\frac{1}{2}(8)(6.928)$	$A=(8)(12)$	3. substitute known values
$A=27.713$ sq. units	$A=96$ sq. units	4. compute

$$\text{Total Area} = 2(27.713)+3(96)$$
$$= 343.426 \text{ sq. units}$$

FIGURE	VOLUME	TOTAL SURFACE AREA	LATERAL AREA
Right Cylinder	$\pi r^2 h$	$2\pi rh + 2\pi r^2$	$2\pi rh$
Right Cone	$\frac{\pi r^2 h}{3}$	$\pi r\sqrt{r^2+h^2} + \pi r^2$	$\pi r\sqrt{r^2+h^2}$

Note: $\sqrt{r^2+h^2}$ is equal to the slant height of the cone.

Sample problem:

1. A water company is trying to decide whether to use traditional cylindrical paper cups or to offer conical paper cups since both cost the same. The traditional cups are 8 cm wide and 14 cm high. The conical cups are 12 cm wide and 19 cm high. The company will use the cup that holds the most water.

1. Draw and label a sketch of each.

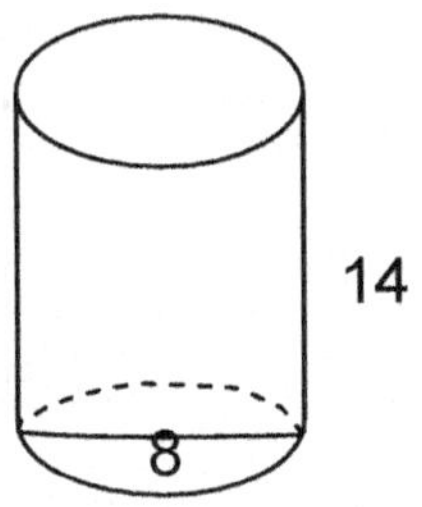

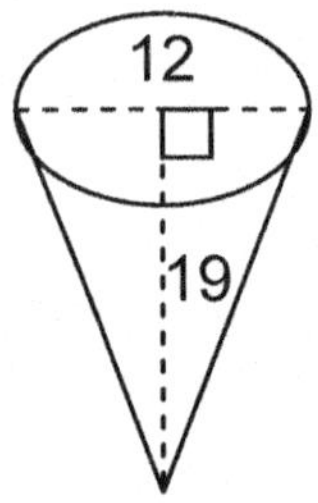

$V = \pi r^2 h$	$V = \frac{\pi r^2 h}{3}$	1. write formula
$V = \pi (4)^2 (14)$	$V = \frac{1}{3}\pi (6)^2 (19)$	2. substitute
$V = 703.717 \text{ cm}^3$	$V = 716.283 \text{ cm}^3$	3. solve

The choice should be the conical cup since its volume is more.

FIGURE	VOLUME	TOTAL SURFACE AREA
Sphere	$\frac{4}{3}\pi r^3$	$4\pi r^2$

Sample problem:

1. How much material is needed to make a basketball that has a diameter of 15 inches? How much air is needed to fill the basketball?

Draw and label a sketch:

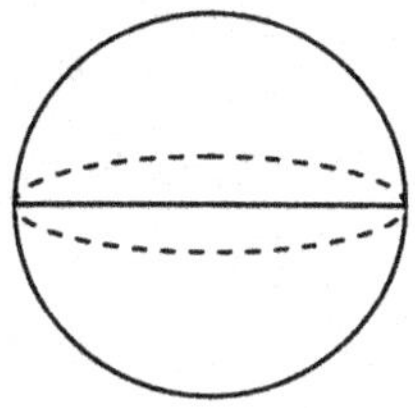

D=15 inches

Total surface area	Volume	
$TSA = 4\pi r^2$	$V = \frac{4}{3}\pi r^3$	1. write formula
$= 4\pi(7.5)^2$	$= \frac{4}{3}\pi(7.5)^3$	2. substitute
$= 706.9 \text{ in}^2$	$= 1767.1 \text{ in}^3$	3. solve

Use the formulas to find the volume and surface area.

FIGURE	VOLUME	TOTAL SURFACE AREA
Right Cylinder	$\pi r^2 h$	$2\pi rh + 2\pi r^2$
Right Cone	$\frac{\pi r^2 h}{3}$	$\pi r\sqrt{r^2 + h^2} + \pi r^2$
Sphere	$\frac{4}{3}\pi r^3$	$4\pi r^2$
Rectangular Solid	LWH	$2LW + 2WH + 2LH$

Note: $\sqrt{r^2 + h^2}$ is equal to the slant height of the cone.

Sample problem:

1. Given the figure below, find the volume and surface area.

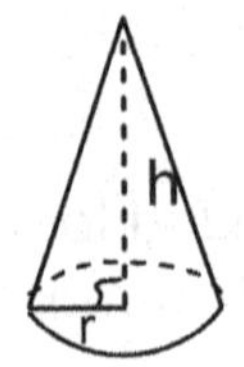

$r = 5$ in $\qquad h = 6.2$ in

Volume $= \dfrac{\pi r^2 h}{3}$	First write the formula.
$\dfrac{1}{3}\pi(5^2)(6.2)$	Then substitute.
162.3 cubic inches	Finally solve the problem.

Surface area = $\pi r\sqrt{r^2 + h^2} + \pi r^2$	First write the formula.
$\pi 5\sqrt{5^2 + 6.2^2} + \pi 5^2$	Then substitute.
203.6 square inches	Compute.

Note: volume is always given in cubic units and area is always given in square units.

Cut the **compound shape** into smaller, more familiar shapes and then compute the total area by adding the areas of the smaller parts.

Sample problem:
Find the area of the given shape.

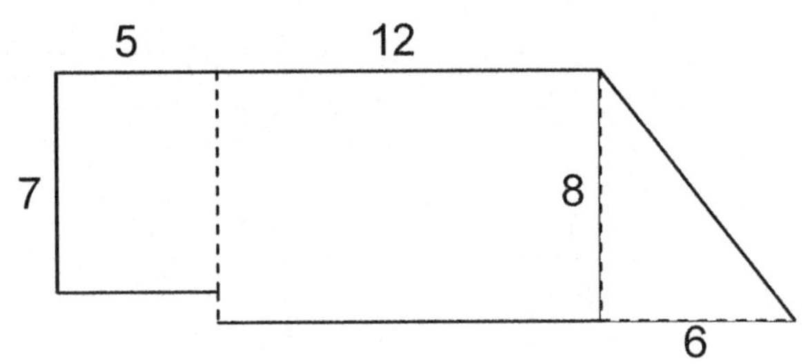

1. Using a dotted line we have cut the shape into smaller parts that are familiar.
2. Use the appropriate formula for each shape and find the sum of all areas.

Area 1 = LW	Area 2 = LW	Area 3 = ½bh
= (5)(7)	= (12)(8)	= ½(6)(8)
= 35 units^2	= 96 units^2	= 24 units^2

Total area = Area 1 + Area 2 + Area 3
= 35 + 96 + 24
= 155 units^2

FIGURE	AREA FORMULA	PERIMETER FORMULA
Rectangle	LW	$2(L+W)$
Triangle	$\frac{1}{2}bh$	$a+b+c$
Parallelogram	bh	sum of lengths of sides
Trapezoid	$\frac{1}{2}h(a+b)$	sum of lengths of sides

Sample problems:

1. Find the area and perimeter of a rectangle if its length is 12 inches and its diagonal is 15 inches.

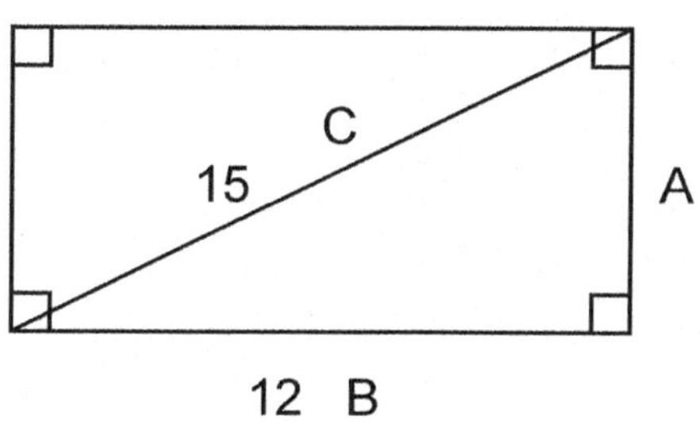

1. Draw and label sketch.

2. Since the height is still needed use Pythagorean formula to find missing leg of the triangle.

$$A^2+B^2=C^2$$
$$A^2+12^2=15^2$$
$$A^2=15^2-12^2$$
$$A^2=81$$
$$A=9$$

Now use this information to find the area and perimeter.

$\text{A}=LW$	$\text{P}=2(L+W)$	1. write formula
$\text{A}=(12)(9)$	$\text{P}=2(12+9)$	2. substitute
$\text{A}=108\text{ in}^2$	$\text{P}=42$ inches	3. solve

Examining the change in area or volume of a given figure requires first to find the existing area given the original dimensions and then finding the new area given the increased dimensions.

Sample problem:

Given the rectangle below determine the change in area if the length is increased by 5 and the width is increased by 7.

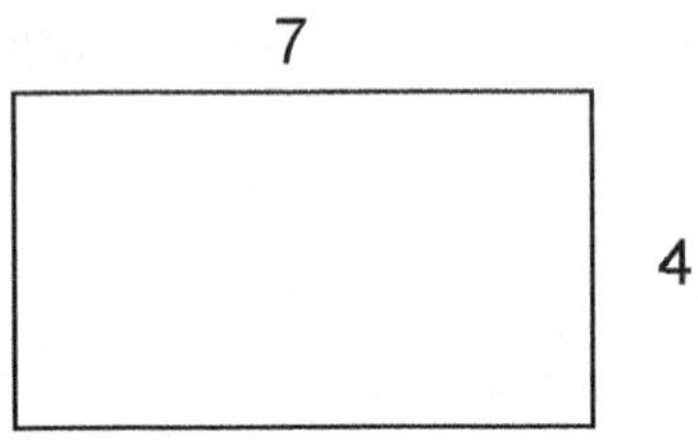

Draw and label a sketch of the new rectangle.

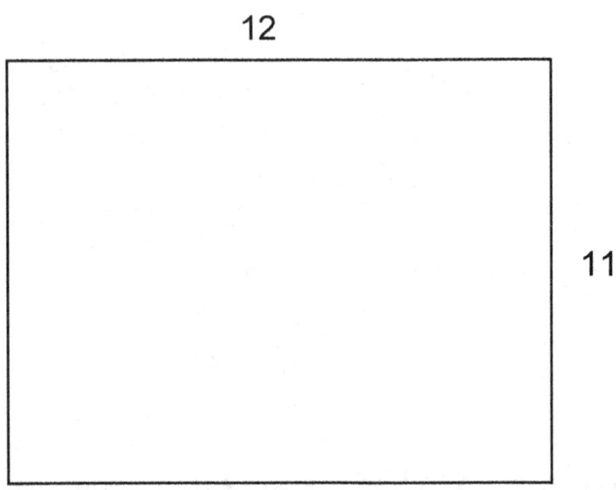

Find the areas.

Area of original = LW
= (7)(4)
= 28 units^2

Area of enlarged shape = LW
= (12)(11)
= 132 units^2

The change in area is 132 – 28 = 104 units^2.

Pythagorean theorem states that the square of the length of the hypotenuse is equal to the sum of the squares of the lengths of the legs. Symbolically, this is stated as:

$$c^2 = a^2 + b^2$$

Given the right triangle below, find the missing side.

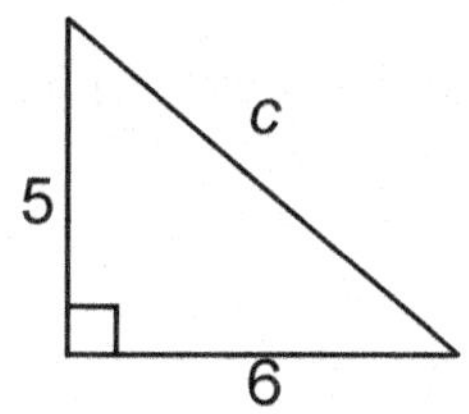

$c^2 = a^2 + b^2$	1. write formula
$c^2 = 5^2 + 6^2$	2. substitute known values
$c^2 = 61$	3. take square root
$c = \sqrt{61}$ or 7.81	4. solve

The **definite integral** is defined as the limit approached by the *n*th upper and lower Riemann sums as $n \to \infty$. The **Riemann sum** is the sum of the areas of all rectangles approximating the area under the graph of a function.

Example: For $f(x) = x^2$, find the values of the Riemann Sums over the interval [0, 1] using *n* subintervals of equal width evaluated at the midpoint of each subinterval. Find the limit of the Riemann Sums.

$$\int_0^1 x^2 dx$$

Let *n* be a positive integer and let $\lim_{n\to\infty} \frac{n(n+1)(2n+1)}{6n^3} = \frac{1}{3}$.

Take the interval [0,1] and subdivide it into *n* subintervals each of length $\frac{1}{n}$.

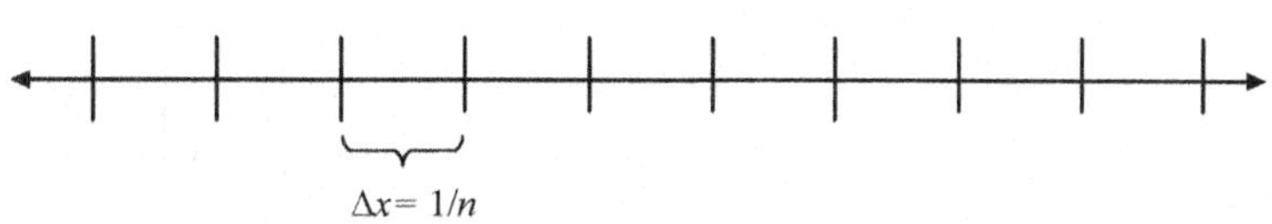

Let $a_i = \dfrac{i}{n}$; the endpoints of the ith subinterval are

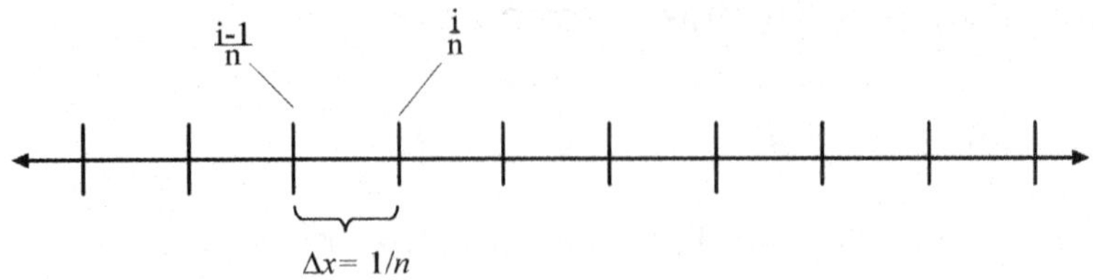

Let $x_i = \dfrac{i}{n}$ be the right-hand endpoint. Draw a line of length

$f(x_i) = \left[\dfrac{i}{n}\right]^2$ at the right-hand endpoint.

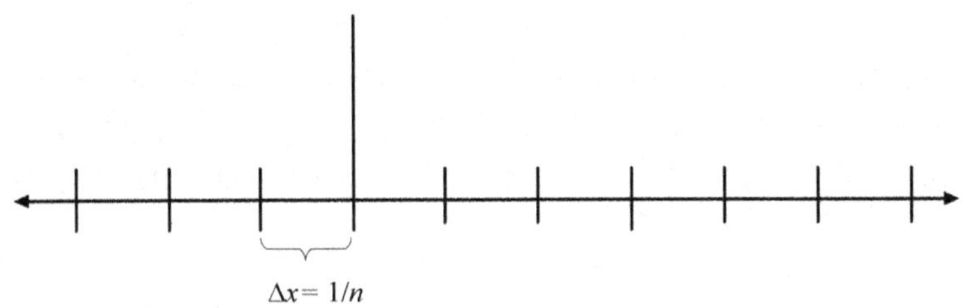

Draw a rectangle.

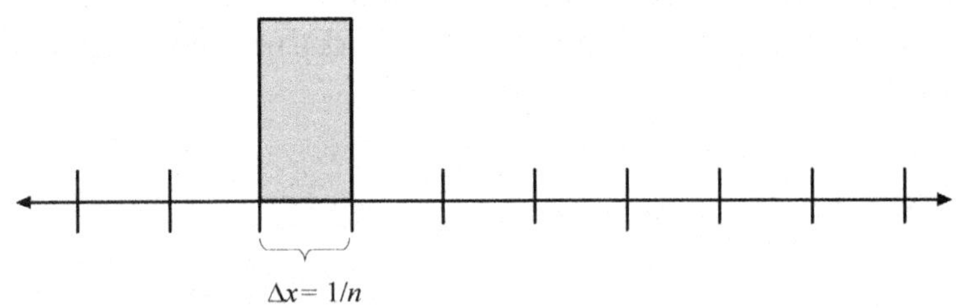

The area of this rectangle is $f(x)\Delta x = \left[\dfrac{i}{n}\right]^2 \dfrac{1}{n} = \dfrac{i^2}{n^3}$.

Now draw all 9 rectangles.

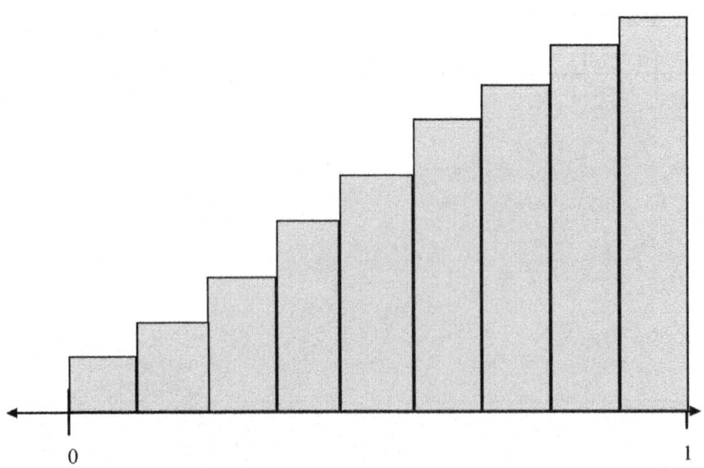

The sum of the area of these rectangles is:

$$\sum_{i=1}^{n} f(x_i)\Delta x = \sum_{i=1}^{n} \frac{i^2}{n^3} = \frac{1}{n^3}\sum_{i=1}^{n} i^2 = \frac{n(n+1)(2n+1)}{6n^3}.$$

Finally, to evaluate the integral, we take the limit

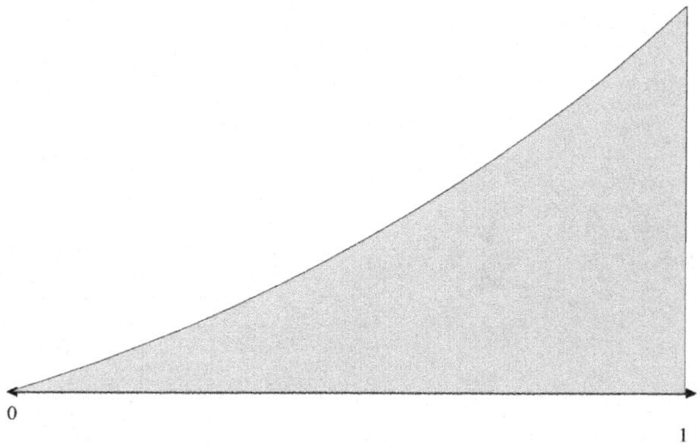

$$\lim_{n\to\infty} \frac{n(n+1)(2n+1)}{6n^3} = \frac{1}{3}.$$

Finding the **area between two curves** is much the same as finding the area under one curve. But instead of finding the roots of the functions, you need to find the *x* values which produce the same number from both functions (set the functions equal and solve). Use these numbers and the given boundaries to write the intervals. On each interval you must pick sample values to determine which function is "on top" of the other. Find the integral of each function. For each interval, subtract the "bottom" integral from the "top" integral. Use the interval numbers to evaluate each of these differences. Add the evaluated integrals to get the total area between the curves.

Example:

Find the area of the regions bounded by the two functions on the indicated intervals.

$f(x) = x + 2$ and $g(x) = x^2$ $\quad \left[{}^-2,3\right]$ — Set the functions equal and solve.

$x + 2 = x^2$
$0 = (x-2)(x+1)$

$x = 2$ or $x = {}^-1$

Use the solutions and the boundary numbers to write the intervals.

$({}^-2, {}^-1)$ $({}^-1, 2)$ $(2, 3)$

$$f({}^-3/2) = \left(\frac{{}^-3}{2}\right) + 2 = \frac{1}{2}$$

Pick sample values on the integral and valuate each function as that number.

$$g({}^-3/2) = \left(\frac{{}^-3}{2}\right)^2 = \frac{9}{4}$$

g(x) is "on top" on $\left[{}^-2, {}^-1\right]$.

$$f(0) = 2$$

f(x) is "on top" on $\left[{}^-1, 2\right]$.

$$g(0) = 0$$

$$f(5/2) = \frac{5}{2} + 2 = \frac{9}{2}$$

g(x) is "on top" on [2,3].

$$g(5/2) = \left(\frac{5}{2}\right)^2 = \frac{25}{4}$$

$$\int f(x)dx = \int (x+2)dx$$
$$\int f(x)dx = \int xdx + 2\int dx$$
$$\int f(x)dx = \frac{1}{1+1}x^{1+1} + 2x$$
$$\int f(x)dx = \frac{1}{2}x^2 + 2x$$
$$\int g(x)dx = \int x^2 dx$$
$$\int g(x)dx = \frac{1}{2+1}x^{2+1} = \frac{1}{3}x^3$$
$$\text{Area } 1 = \int g(x)dx - \int f(x)dx$$

g(x) is "on top" on $\left[{}^-2, {}^-1\right]$.

$$\text{Area } 1 = \frac{1}{3}x^3 - \left(\frac{1}{2}x^2 + 2x\right)\Big]_{-2}^{-1}$$

$$\text{Area } 1=\left[\frac{1}{3}(^-1)^3-\left(\frac{1}{2}(^-1)^2+2(^-1)\right)\right]-\left[\frac{1}{3}(^-2)^3-\left(\frac{1}{2}(^-2)^2+2(^-2)\right)\right]$$

$$\text{Area } 1=\left[\frac{^-1}{3}-\left(\frac{^-3}{2}\right)\right]-\left[\frac{^-8}{3}-(^-2)\right]$$

$$\text{Area } 1=\left(\frac{7}{6}\right)-\left(\frac{^-2}{3}\right)=\frac{11}{6}$$

$$\text{Area } 2=\int f(x)dx-\int g(x)dx$$

f(x) is "on top" on $\left[^-1,2\right]$.

$$\text{Area } 2=\frac{1}{2}x^2+2x-\frac{1}{3}x^3\Big]_{^-1}^{2}$$

$$\text{Area } 2=\left(\frac{1}{2}(2)^2+2(2)-\frac{1}{3}(2)^3\right)-\left(\frac{1}{2}(^-1)^2+2(^-1)-\frac{1}{3}(^-1)^3\right)$$

$$\text{Area } 2=\left(\frac{10}{3}\right)-\left(\frac{1}{2}-2+\frac{1}{3}\right)$$

$$\text{Area } 2=\frac{27}{6}$$

$$\text{Area } 3=\int g(x)dx-\int f(x)dx$$

g(x) is "on top" on [2,3].

$$\text{Area } 3=\frac{1}{3}x^3-\left(\frac{1}{2}x^2+2x\right)\Big]_2^3$$

$$\text{Area } 3=\left[\frac{1}{3}(3)^3-\left(\frac{1}{2}(3^2)+2(3)\right)\right]-\left[\frac{1}{3}(2)^3-\left(\frac{1}{2}(2)^2+2(2)\right)\right]$$

$$\text{Area } 3=\left(\frac{^-3}{2}\right)-\left(\frac{^-10}{3}\right)=\frac{11}{6}$$

$$\text{Total area}=\frac{11}{6}+\frac{27}{6}+\frac{11}{6}=\frac{49}{6}=8\frac{1}{6}$$

Taking the integral of a function and evaluating it from one x value to another provides the total **area under the curve** (i.e. between the curve and the x axis). Remember, though, that regions above the x axis have "positive" area and regions below the x axis have "negative" area. You must account for these positive and negative values when finding the area under curves. Follow these steps.

1. Determine the x values that will serve as the left and right boundaries of the region.
2. Find all x values between the boundaries that are either solutions to the function or are values which are not in the domain of the function. These numbers are the interval numbers.
3. Integrate the function.
4. Evaluate the integral once for each of the intervals using the boundary numbers.
5. If any of the intervals evaluates to a negative number, make it positive (the negative simply tells you that the region is below the x axis).
6. Add the value of each integral to arrive at the area under the curve.

Example:

Find the area under the following function on the given intervals.
$f(x) = \sin x\ ;\ (0, 2\pi)$

$\sin x = 0$ — Find any roots to f(x) on $(0, 2\pi)$.

$x = \pi$

$(0, \pi)\quad (\pi, 2\pi)$ — Determine the intervals using the boundary numbers and the roots.

$\int \sin x dx = {}^{-}\cos x$ — Integrate f(x). We can ignore the constant c because we have numbers to use to evaluate the integral.

$${}^{-}\cos x\Big]_{x=0}^{x=\pi} = {}^{-}\cos\pi - ({}^{-}\cos 0)$$

$${}^{-}\cos x\Big]_{x=0}^{x=\pi} = {}^{-}(-1) + (1) = 2$$

$${}^{-}\cos x\Big]_{x=\pi}^{x=2\pi} = {}^{-}\cos 2\pi - ({}^{-}\cos\pi)$$

$${}^{-}\cos x\Big]_{x=\pi}^{x=2\pi} = {}^{-}1 + ({}^{-}1) = {}^{-}2$$

The ${}^{-}2$ means that for $(\pi, 2\pi)$, the region is below the x axis, but the area is still 2.

$\text{Area} = 2 + 2 = 4$ — Add the 2 integrals together to get the area.

Competency 12.0 The teacher understands geometries, in particular Euclidean geometry, as axiomatic systems.

The 3 undefined terms of geometry are point, line, and plane.

A plane is a flat surface that extends forever in two dimensions. It has no ends or edges. It has no thickness to it. It is usually drawn as a parallelogram that can be named either by 3 non-colinear points (3 points that are not on the same line) on the plane or by placing a letter in the corner of the plane that is not used elsewhere in the diagram.

A line extends forever in one dimension. It is determined and named by 2 points that are on the line. The line consists of every point that is between those 2 points as well as the points that are on the "straight" extension each way. A line is drawn as a line segment with arrows facing opposite directions on each end to indicate that the line continues in both directions forever.

A point is a position in space, on a line, or on a plane. It has no thickness and no width. Only 1 line can go through any 2 points. A point is represented by a dot named by a single letter.

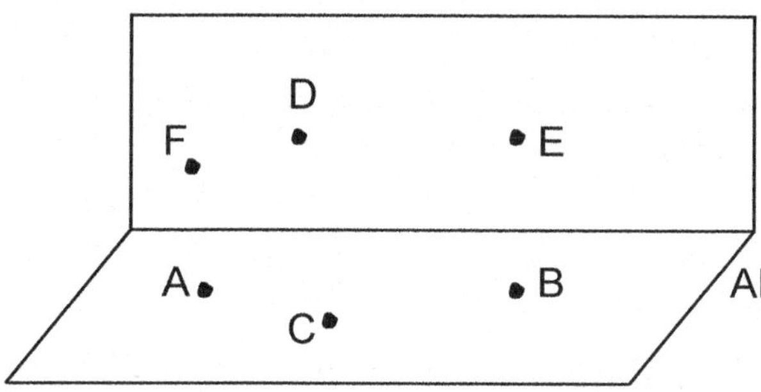

DEF is a plane.

ABC is another plane.

This line can be named by any two points on the line.

It could be named $\overleftrightarrow{GH}$, $\overleftrightarrow{HI}$, $\overleftrightarrow{GI}$, $\overleftrightarrow{IG}$, $\overleftrightarrow{IH}$, or $\overleftrightarrow{HG}$. Any 2 points (letters) on the line can be used and their order is not important in naming a line.

In the above diagrams, A, B, C, D, E, F, G, H, and I are all locations of individual points.

A ray is not an undefined term. A ray consists of all the points on a line starting at one given point and extending in only one of the two opposite directions along the line. The ray is named by naming 2 points on the ray. The first point must be the endpoint of the ray, while the second point can be any other point along the ray. The symbol for a ray is a ray above the 2 letters used to name it. The endpoint of the ray MUST be the first letter.

```
J•----•-------------•------>
      K             L
```

This ray could be named $\overrightarrow{JK}$ or $\overrightarrow{JL}$. It can not be called $\overrightarrow{KJ}$ or $\overrightarrow{LJ}$ or $\overrightarrow{LK}$ or $\overrightarrow{KL}$ because none of those names start with the endpoint, J.

The **distance between 2 points on a number line** is equal to the absolute value of the difference of the two numbers associated with the points.

If one point is located at "a" and the other point is at "b", then the distance between them is found by this formula:

$$\text{distance} = |a-b| \text{ or } |b-a|$$

If one point is located at $^{-}3$ and another point is located at 5, the distance between them is found by:

$$\text{distance} = |a-b| = |(^{-}3)-5| = |^{-}8| = 8$$

The only **undefined terms** are point, line and plane.

Definitions are explanations of all mathematical terms except those that are undefined.

Postulates are mathematical statements that are accepted as true statements without providing a proof.

Theorems are mathematical statements that can be proven to be true based on postulates, definitions, algebraic properties, given information, and previously proved theorems.

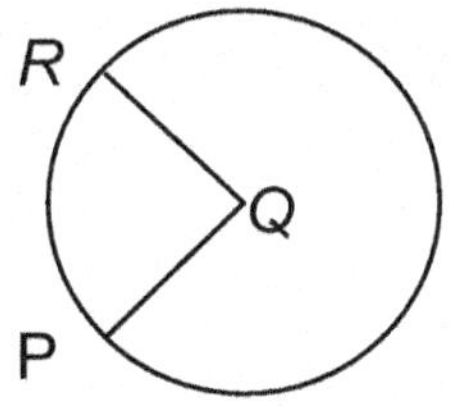

$$\frac{\measuredangle PQR}{360^\circ} = \frac{\text{length of arc } RP}{\text{circumference of } \bigcirc Q} = \frac{\text{area of sector } PQR}{\text{area of } \bigcirc Q}$$

While an arc has a measure associated to the degree measure of a central angle, it also has a length which is a fraction of the circumference of the circle.

For each central angle and its associated arc, there is a sector of the circle which resembles a pie piece. The area of such a sector is a fraction of the area of the circle.

The fractions used for the area of a sector and length of its associated arc are both equal to the ratio of the central angle to 360°.

Examples:

1.

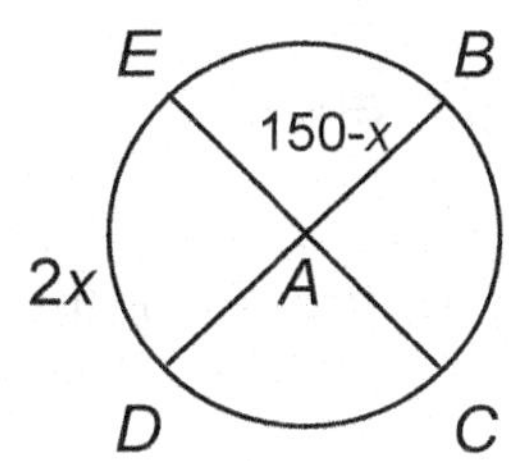

$\bigcirc A$ has a radius of 4 cm. What is the length of arc *ED*?

$$2x + 150 - x = 180$$
$$x + 150 = 180$$
$$x = 30$$

Arc *BE* and arc *DE* make a semicircle.

Arc $ED = 2(30) = 60^\circ$

The ratio 60° to 360° is equal to the ratio of arch length *ED* to the circumference of $\bigcirc A$.

$$\frac{60}{360} = \frac{\text{arc length } ED}{2\pi 4}$$
$$\frac{1}{6} = \frac{\text{arc length}}{8\pi}$$

Cross multiply and solve for the arc length.

$$\frac{8\pi}{6} = \text{arc length}$$

$$\text{arc length } ED = \frac{4\pi}{3} \text{ cm}.$$

2.

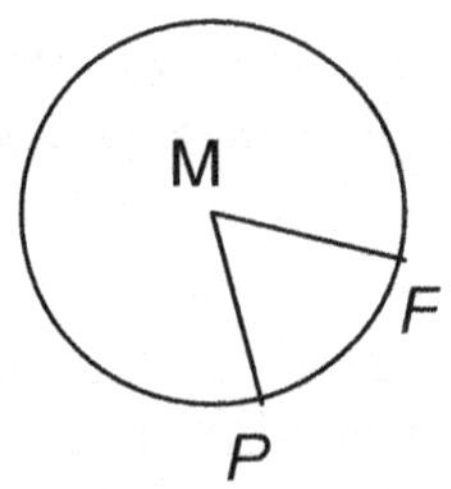

The radius of $\bigcirc$M is 3 cm. The length of arc *PF* is 2π cm.

What is the area of sector PMF?

Circumference of $\bigcirc M = 2\pi(3) = 6\pi$
Area of $\bigcirc M = \pi(3)^2 = 9\pi$

Find the circumference and area of the circle.

$$\frac{\text{area of } PMF}{9\pi} = \frac{2\pi}{6\pi}$$

$$\frac{\text{area of } PMF}{9\pi} = \frac{1}{3}$$

$$\text{area of } PMF = \frac{9\pi}{3}$$

The ratio of the sector area to the circle area is the same as the arc length to the circumference.

$$\text{area of } PMF = 3\pi$$

Solve for the area of the sector.

The **distance between two parallel lines**, such as line AB and line CD as shown below is the line segment RS, the perpendicular between the two parallels.

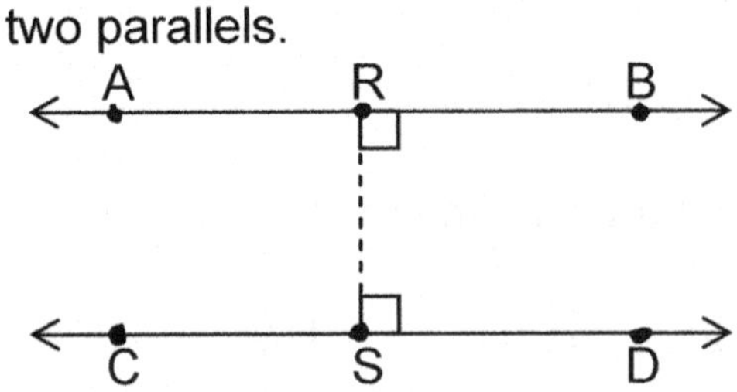

Sample Problem:

Given the geometric figure below, find the distance between the two parallel sides *AB* and *CD*.

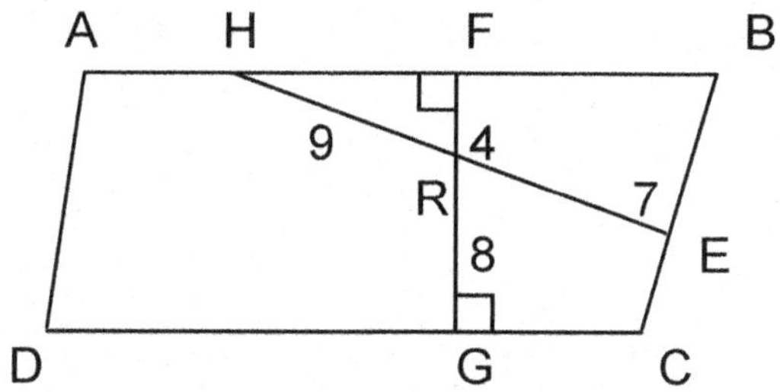

The distance *FG* is 12 units.

Two triangles **can be proven congruent** by comparing pairs of appropriate congruent corresponding parts.

SSS POSTULATE

If three sides of one triangle are congruent to three sides of another triangle, then the two triangles are congruent.

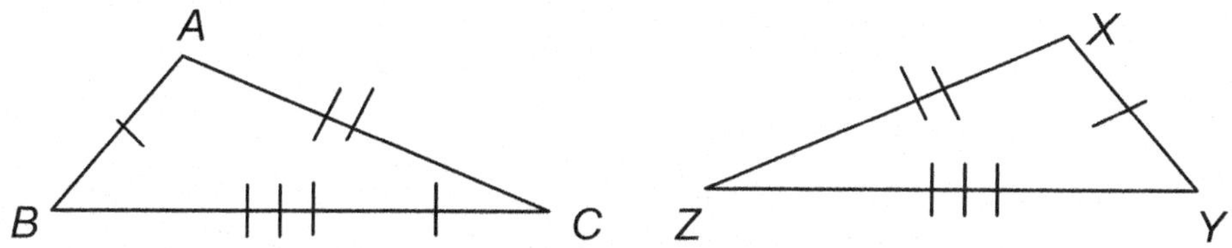

Since $AB \cong XY$, $BC \cong YZ$ and $AC \cong XZ$, then $\Delta ABC \cong \Delta XYZ$.

<u>Example</u>: Given isosceles triangle *ABC* with *D* the midpoint of base *AC*, prove the two triangles formed by *AD* are congruent.

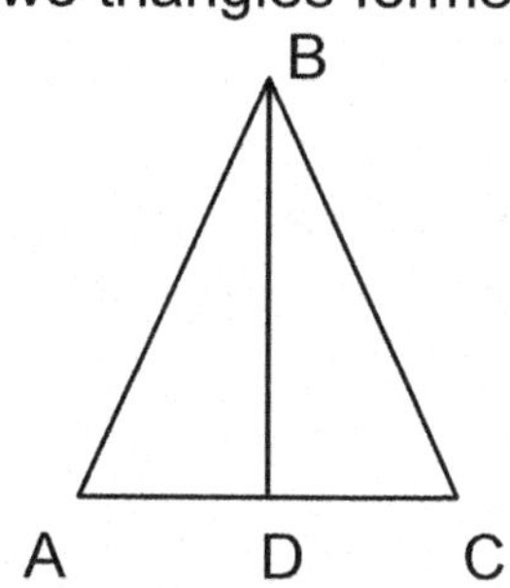

Proof:

1. Isosceles triangle *ABC*, *D* midpoint of base *AC*	Given
2. $AB \cong BC$	An isosceles Δ has two congruent sides
3. $AD \cong DC$	Midpoint divides a line into two equal parts
4. $BD \cong BD$	Reflexive
5. $\Delta ABD \cong \Delta BCD$	SSS

SAS POSTULATE

If two sides and the included angle of one triangle are congruent to two sides and the included angle of another triangle, then the two triangles are congruent.

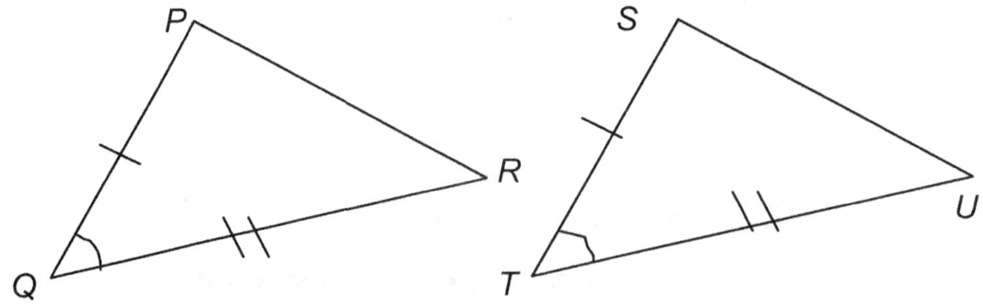

Example:

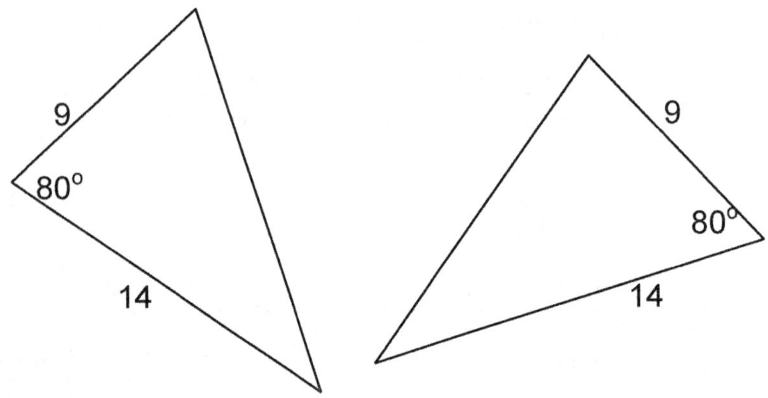

The two triangles are congruent by SAS.

ASA POSTULATE

If two angles and the included side of one triangle are congruent to two angles and the included side of another triangle, the triangles are congruent.

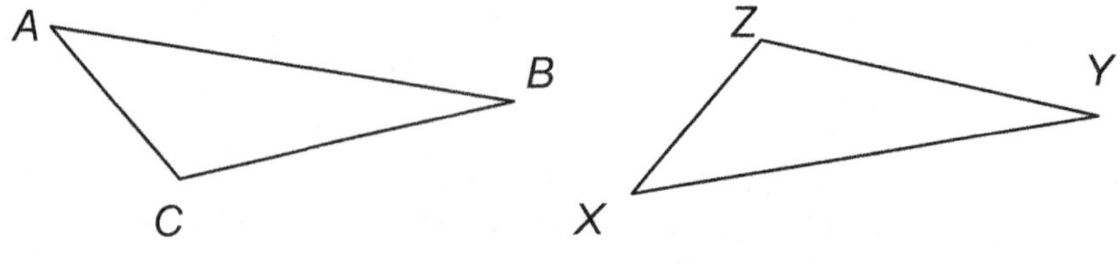

$\angle A \cong \angle X$, $\angle B \cong \angle Y$, $AB \cong XY$ then $\Delta ABC \cong \Delta XYZ$ by ASA

Example 1: Given two right triangles with one leg of each measuring 6 cm and the adjacent angle 37°, prove the triangles are congruent.

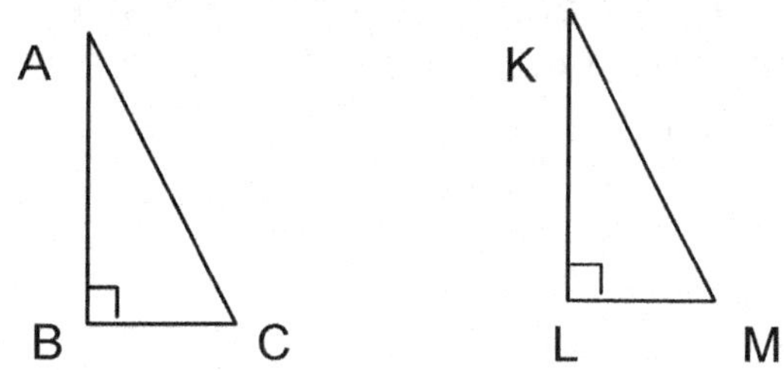

1. Right triangles *ABC* and *KLM* — Given
   $AB = KL = 6$ cm
   $\angle A = \angle K = 37°$
2. $AB \cong KL$ — Figures with the same measure are congruent.
   $\angle A \cong \angle K$
3. $\angle B \cong \angle L$ — All right angles are congruent.
4. $\Delta ABC \cong \Delta KLM$ — ASA

Example 2:
What method would you use to prove the triangles congruent?

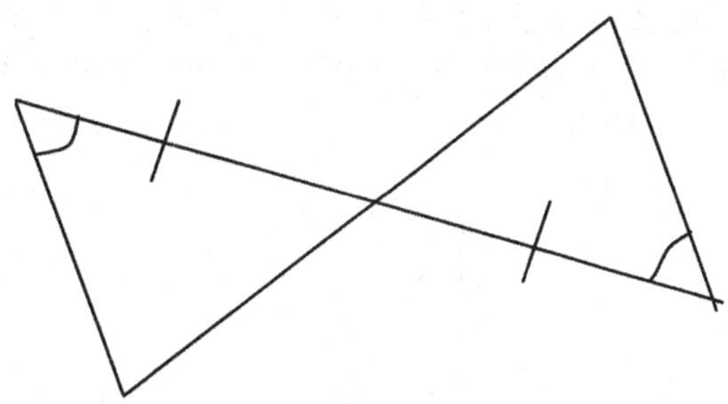

ASA because vertical angles are congruent.

AAS THEOREM

If two angles and a non-included side of one triangle are congruent to the corresponding parts of another triangle, then the triangles are congruent.

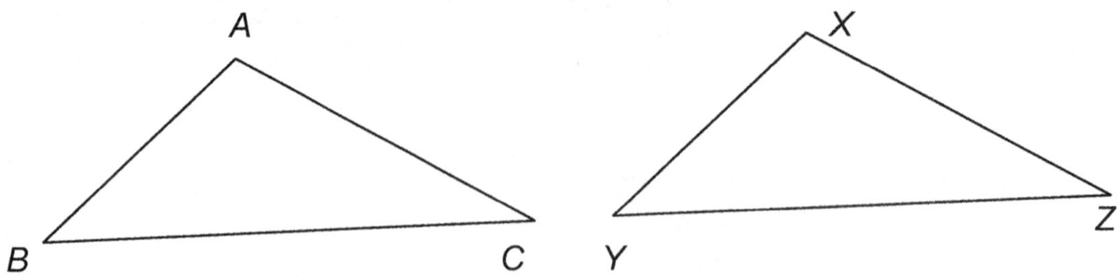

$\angle B \cong \angle Y$, $\angle C \cong \angle Z$, $AC \cong XZ$, then $\Delta ABC \cong \Delta XYZ$ by AAS.

We can derive this theorem because if two angles of the triangles are congruent, then the third angle must also be congruent. Therefore, we can uses the ASA postulate.

HL THEOREM

If the hypotenuse and a leg of one right triangle are congruent to the corresponding parts of another right triangle, the triangles are congruent.

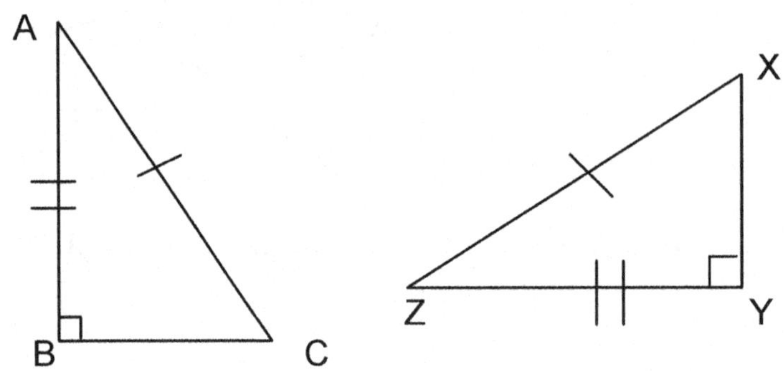

Since $\angle B$ and $\angle Y$ are right angles and $AC \cong XZ$ (hypotenuse of each triangle), $AB \cong YZ$ (corresponding leg of each triangle), then $\Delta ABC \cong \Delta XYZ$ by HL.

Example: What method would you use to prove the triangles congruent?

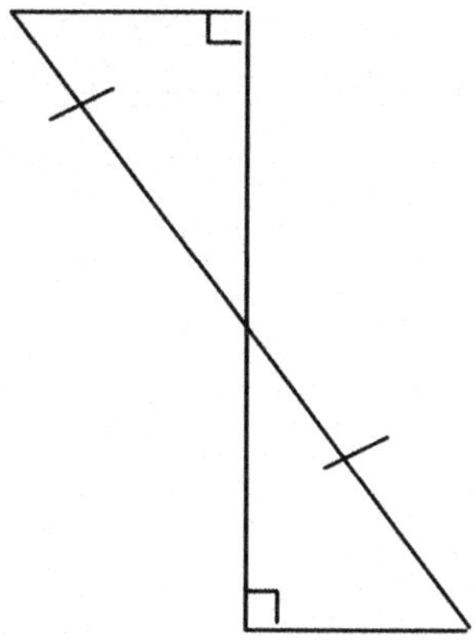

AAS

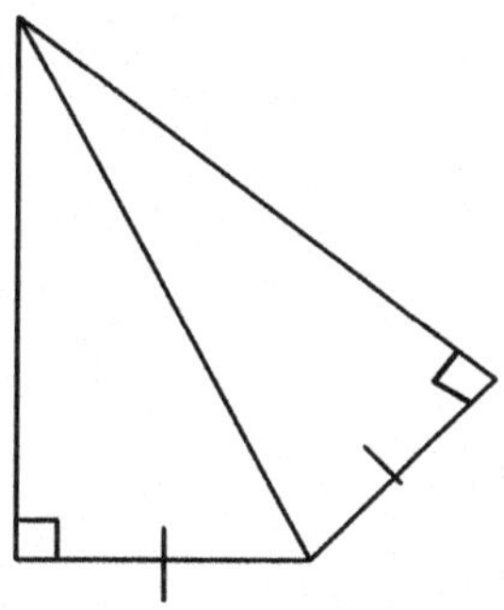

HL

Angle Bisector Theorem Proof

For a triangle **ABC** let the angle bisector of angle **A** intersect side **BC** at a point **D**. The angle bisector theorem states that the ratio of the length of the line segment **BD** to the length of segment **DC** is equal to the ratio of the length of side **AB** to the length of side **AC**.

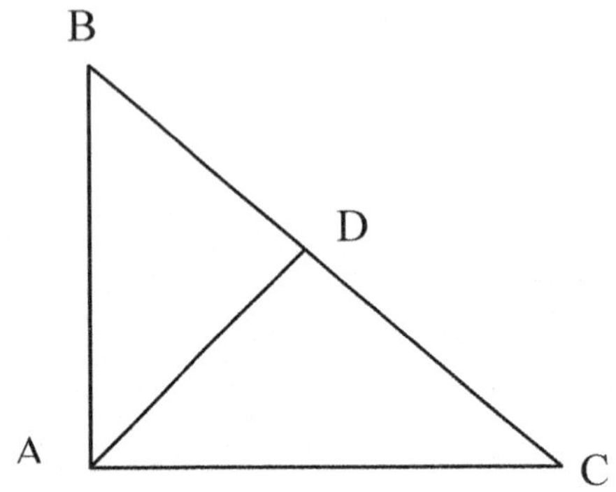

A **geometric construction** is a drawing made using only a compass and straightedge. A construction consists of only segments, arcs, and points. The easiest construction to make is to duplicate a given line segment. Given segment *AB*, construct a segment equal in length to segment AB by following these steps.

The easiest construction to make is to duplicate a given line segment. Given segment *AB*, construct a segment equal in length to segment AB by following these steps.

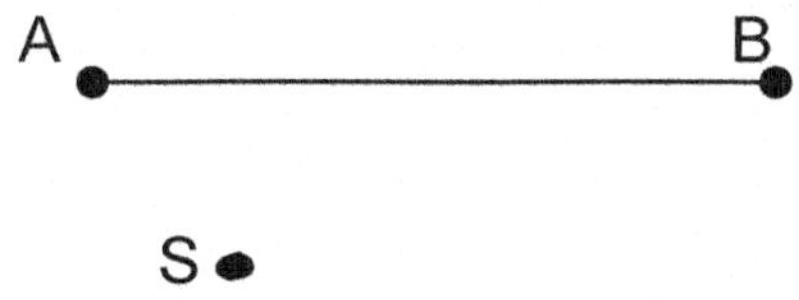

1. Place a point anywhere in the plane to anchor the duplicate segment. Call this point S.

2. Open the compass to match the length of segment *AB*. Keeping the compass rigid, swing an arc from S.

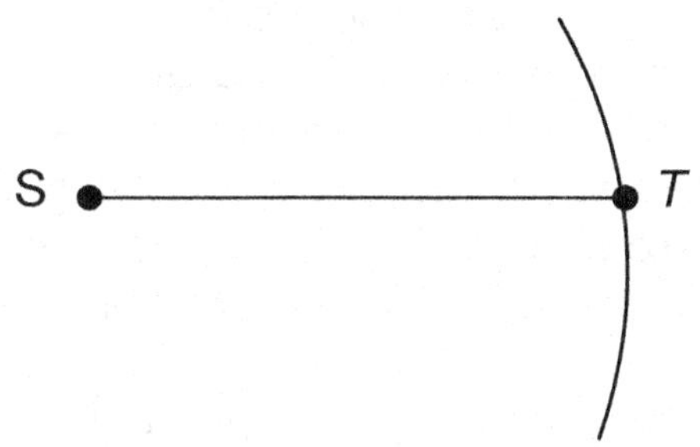

3. Draw a segment from S to any point on the arc. This segment will be the same length as *AB*.

Samples:

Construct segments congruent to the given segments.

1. 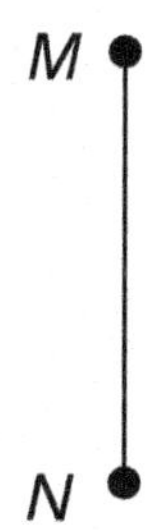

2. 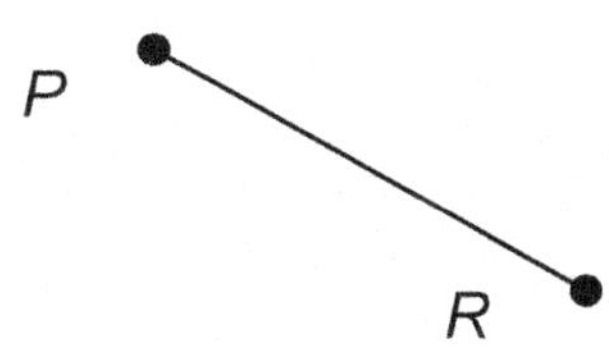

To construct an angle congruent to a given angle such as angle TAP follow these steps.

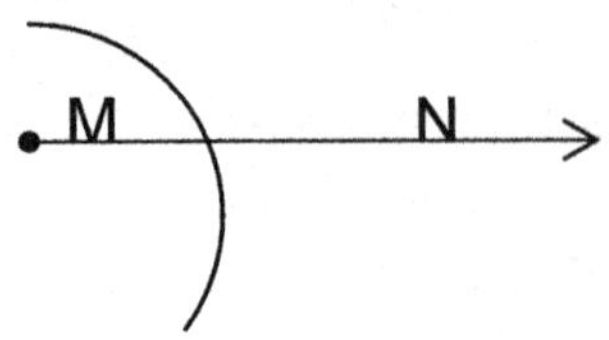

1. Draw ray *MN* using a straightedge. This ray will be one side of the duplicate angle.

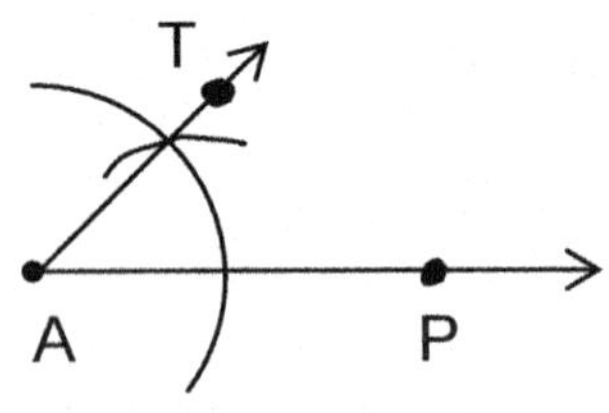

2. Using the compass, draw an arc of any radius with its central at the A. vertex Draw an arc of the same radius with center M.

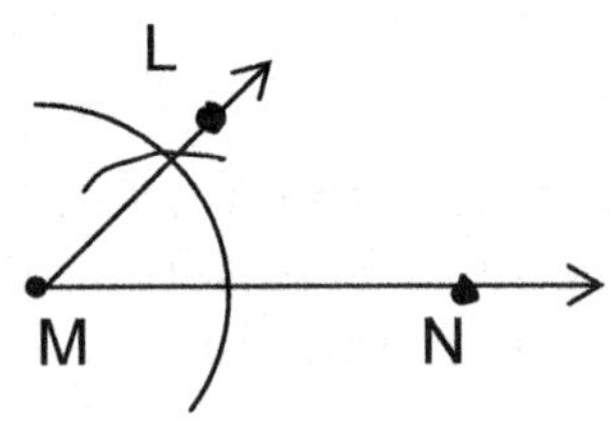

3. Use the point where the arc intercepts ray *AP* to draw another arc that intercepts the intersection of the arc and ray *AT*. Swing an arc of the same radius from the intersection point on ray *MN*.

4. Connect M and the point of intersection of the two arcs to form angle *LMN* which will be congruent to angle *TAP*.

Euclid wrote a set of 13 books around 330 B.C. called the Elements. He outlined ten axioms and then deduced 465 theorems. Euclidean geometry is based on the undefined concept of the point, line and plane.

The fifth of Euclid's axioms (referred to as the parallel postulate) was not as readily accepted as the other nine axioms. Many mathematicians throughout the years have attempted to prove that this axiom is not necessary because it could be proved by the other nine. Among the many who attempted to prove this was Carl Friedrich Gauss. His works led to the development of hyperbolic geometry. Elliptical or Reimannian geometry was suggested by G.F. Berhard Riemann. He based his work on the theory of surfaces and used models as physical interpretations of the undefined terms that satisfy the axioms.

The chart below lists the fifth axiom (parallel postulate) as it is given in each of the three geometries.

EUCLIDEAN	ELLIPTICAL	HYPERBOLIC
Given a line and a point not on that line, one and only one line can be drawn through the given point parallel to the given line.	Given a line and a point not on that line, no line can be drawn through the given point parallel to the given line.	Given a line and a point not on that line, two or more lines can be drawn through the point parallel to the given line.

Competency 13.0 The teacher understands the results, uses, and applications of Euclidean geometry.

In order to determine if a figure is convex and then determine if it is regular, it is necessary to apply the definition of convex first.

Convex polygons: polygons in which no line containing the side of the polygon contains a point on the interior of the polygon.

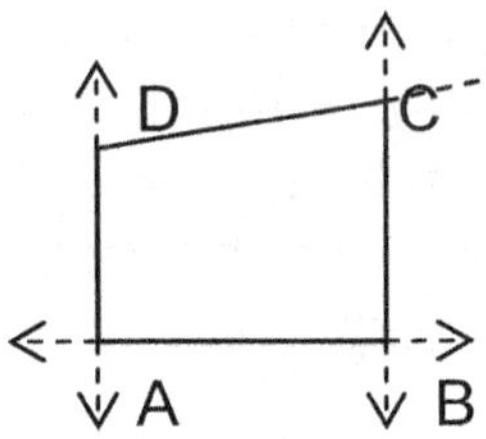

Convex

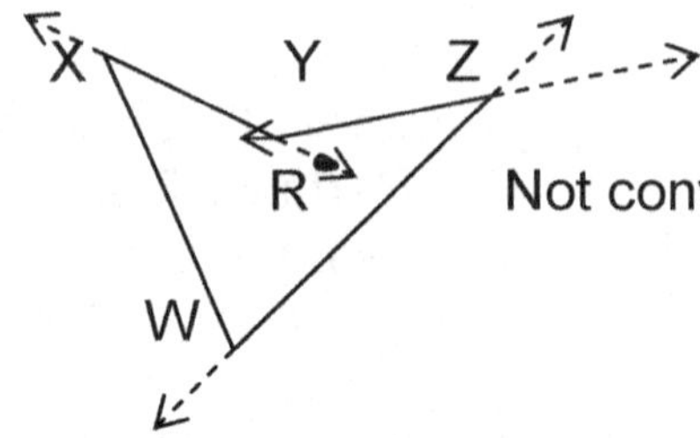

Not convex

Regular polygons: convex polygons in which all sides are congruent and all angles are congruent (in other words, a regular polygon must be both equilateral and equiangular).

A **polygon** is a simple closed figure composed of line segments. In a **regular polygon** all sides are the same length and all angles are the same measure.

The sum of the measures of the **interior angles** of a polygon can be determined using the following formula, where *n* represents the number of angles in the polygon.

$$\text{Sum of } \angle s = 180(n - 2)$$

The measure of each angle of a regular polygon can be found by dividing the sum of the measures by the number of angles.

$$\text{Measure of } \angle = \frac{180(n-2)}{n}$$

Example: Find the measure of each angle of a regular octagon. Since an octagon has eight sides, each angle equals:

$$\frac{180(8-2)}{8} = \frac{180(6)}{8} = 135^\circ$$

The sum of the measures of the **exterior angles** of a polygon, taken one angle at each vertex, equals 360°.

The measure of each exterior angle of a regular polygon can be determined using the following formula, where n represents the number of angles in the polygon.

$$\text{Measure of exterior} \angle \text{ of regular polygon} = 180 - \frac{180(n-2)}{n}$$

$$\text{or, more simply} = \frac{360}{n}$$

<u>Example</u>: Find the measure of the interior and exterior angles of a regular pentagon.

Since a pentagon has five sides, each exterior angle measures:

$$\frac{360}{5} = 72^\circ$$

Since each exterior angles is supplementary to its interior angle, the interior angle measures 180 - 72 or 108°.

A **quadrilateral** is a polygon with four sides.
The sum of the measures of the angles of a quadrilateral is 360°.

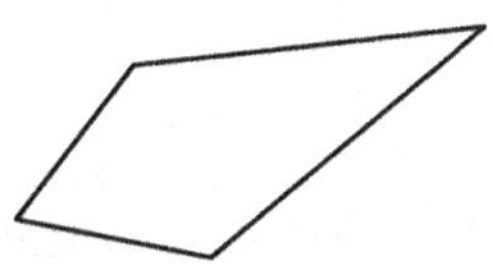

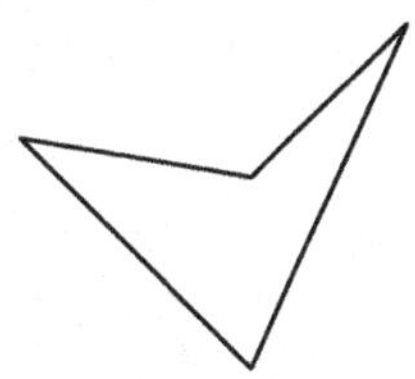

A **trapezoid** is a quadrilateral with exactly <u>one</u> pair of parallel sides.

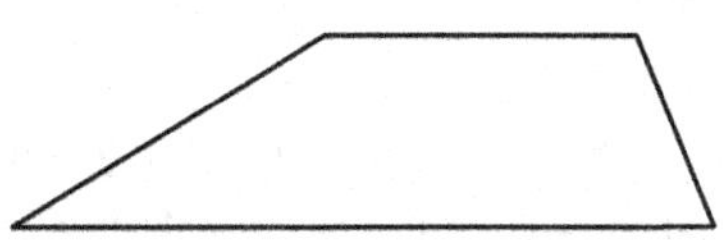

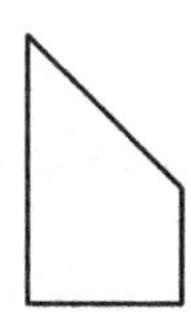

In an **isosceles trapezoid**, the non-parallel sides are congruent.

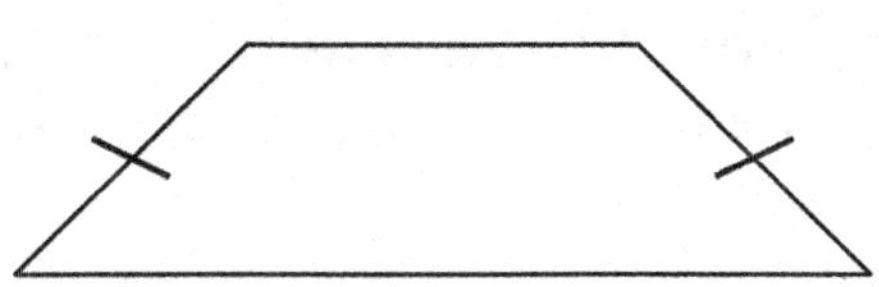

A **parallelogram** is a quadrilateral with <u>two</u> pairs of parallel sides.

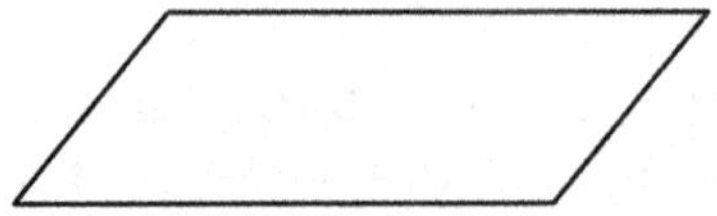

A **rectangle** is a parallelogram with a right angle.

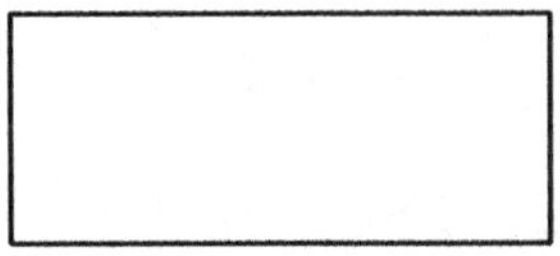

A **rhombus** is a parallelogram with all sides equal length.

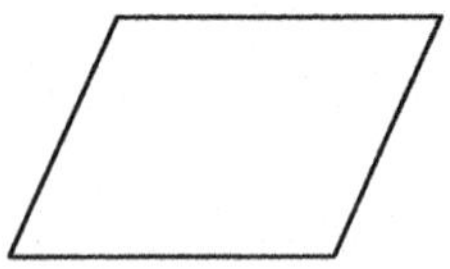

A **square** is a rectangle with all sides equal length.

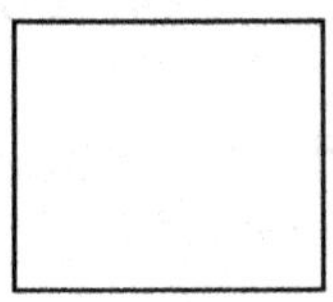

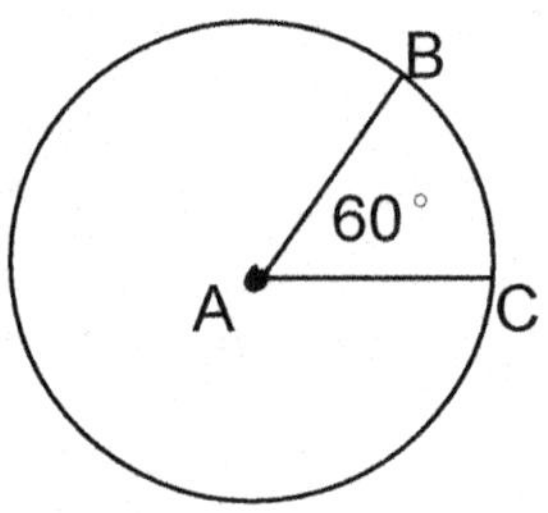

Central angle $BAC = 60^\circ$
Minor arc $BC = 60^\circ$
Major arc $BC = 360 - 60 = 300^\circ$

If you draw **two radii** in a circle, the angle they form with the center as the vertex is a central angle. The piece of the circle "inside" the angle is an arc. Just like a central angle, an arc can have any degree measure from 0 to 360. The measure of an arc is equal to the measure of the central angle which forms the arc. Since a diameter forms a semicircle and the measure of a straight angle like a diameter is 180°, the measure of a semicircle is also 180°.

Given two points on a circle, there are two different arcs which the two points form. Except in the case of semicircles, one of the two arcs will always be greater than 180° and the other will be less than 180°. The arc less than 180° is a minor arc and the arc greater than 180° is a major arc.

Examples:

1.

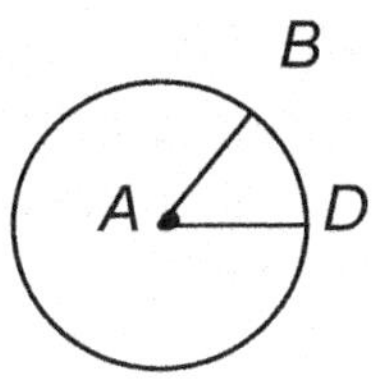

$m\measuredangle BAD = 45^\circ$
What is the measure of the major arc *BD?*

$\measuredangle BAD =$ minor arc *BD*
$45^\circ =$ minor arc *BD*

The measure of the central angle is the same as the measure of the arc it forms.

$360 - 45 =$ major arc *BD*
$315^\circ =$ major arc *BD*

A major and minor arc always add to 360°.

2.

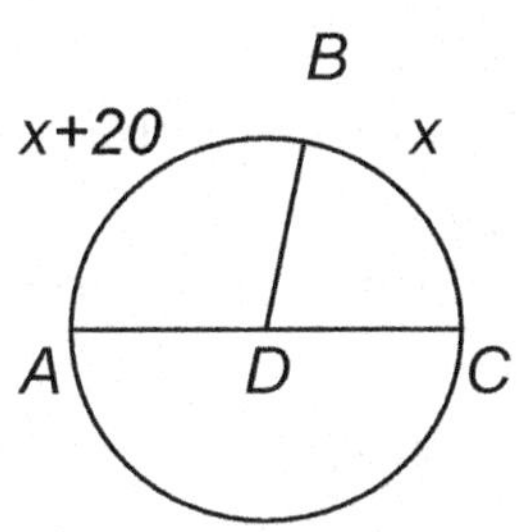

$\overline{AC}$ is a diameter of circle *D*.
What is the measure of $\measuredangle BDC$?

$$m\measuredangle ADB + m\measuredangle BDC = 180^\circ$$
$$x + 20 + x = 180$$
$$2x + 20 = 180$$
$$2x = 160$$
$$x = 80$$

A diameter forms a semicircle which has a measure of 180°.

minor arc $BC = 80^\circ$
$m\measuredangle BDC = 80^\circ$

A central angle has the same measure as the arc it forms.

A **tangent line** intersects a circle in exactly one point. If a radius is drawn to that point, the radius will be perpendicular to the tangent.

A **chord** is a segment with endpoints on the circle. If a radius or diameter is perpendicular to a chord, the radius will cut the chord into two equal parts.

If **two chords** in the same circle have the same length, the two chords will have arcs that are the same length, and the two chords will be equidistant from the center of the circle. Distance from the center to a chord is measured by finding the length of a segment from the center perpendicular to the chord.

Examples:

1.

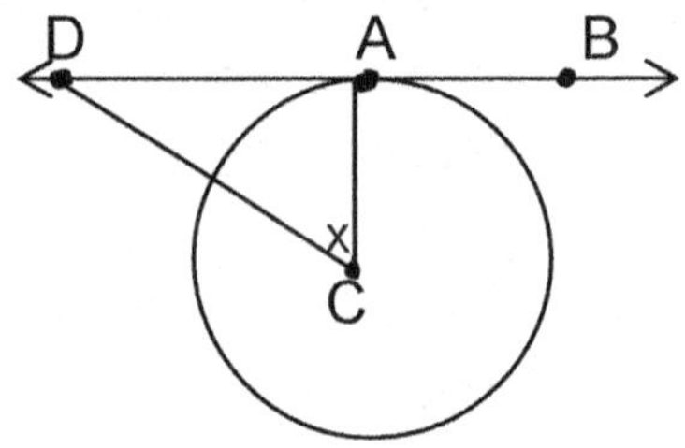

$\overleftrightarrow{DB}$ is tangent to $\odot C$ at A.

$m\measuredangle ADC = 40^\circ$. Find x.

$\overline{AC} \perp \overleftrightarrow{DB}$	A radius is $\perp$ to a tangent at the point of tangency.
$m\measuredangle DAC = 90^\circ$	Two segments that are $\perp$ form a 90° angle.
$40 + 90 + x = 180$	The sum of the angles of a triangle is 180°.
$x = 50^\circ$	Solve for x.

2.

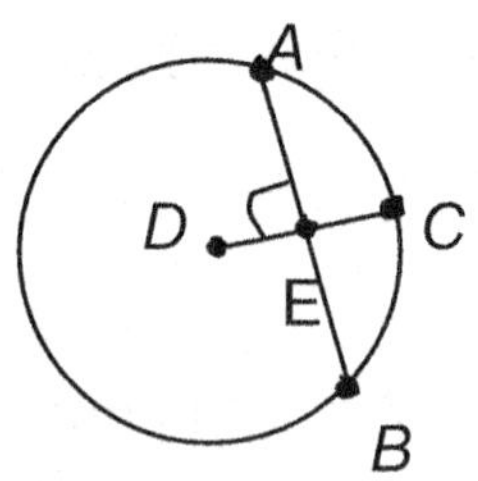

x $\overline{CD}$ is a radius and $\overline{CD} \perp$ chord $\overline{AB}$. $\overline{AB} = 10$. Find x.

$$x = \frac{1}{2}(10)$$

$x = 5$ If a radius is $\perp$ to a chord, the radius bisects the chord.

Angles with their vertices on the circle:

An inscribed angle is an angle whose vertex is on the circle. Such an angle could be formed by two chords, two diameters, two secants, or a secant and a tangent. An inscribed angle has one arc of the circle in its interior. The measure of the inscribed angle is one-half the measure of this intercepted arc. If two inscribed angles intercept the same arc, the two angles are congruent (i.e. their measures are equal). If an inscribed angle intercepts an entire semicircle, the angle is a right angle.

Angles with their vertices in a circle's interior:

When two chords intersect inside a circle, two sets of vertical angles are formed.

Each set of vertical angles intercepts two arcs which are across from each other. The measure of an angle formed by two chords in a circle is equal to one-half the sum of the angle intercepted by the angle and the arc intercepted by its vertical angle.

Angles with their vertices in a circle's exterior:

If an angle has its vertex outside of the circle and each side of the circle intersects the circle, then the angle contains two different arcs. The measure of the angle is equal to one-half the difference of the two arcs.

Examples:

1.

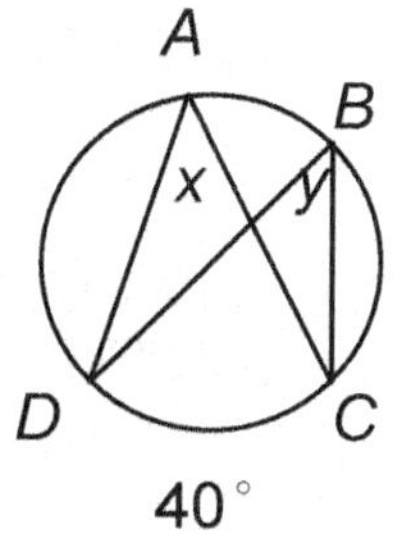

Find *x and y.*
arc $DC = 40^\circ$

$m\measuredangle DAC = \frac{1}{2}(40) = 20^\circ$

$m\measuredangle DBC = \frac{1}{2}(40) = 20^\circ$

$x = 20^\circ$ and $y = 20^\circ$

$\measuredangle DAC$ and $\measuredangle DBC$ are both inscribed angles, so each one has a measure equal to one-half the measure of arc *DC*.

Intersecting chords:

If two chords intersect inside a circle, each chord is divided into two smaller segments. The product of the lengths of the two segments formed from one chord equals the product of the lengths of the two segments formed from the other chord.

Intersecting tangent segments:

If two tangent segments intersect outside of a circle, the two segments have the same length.

Intersecting secant segments:

If two secant segments intersect outside a circle, a portion of each segment will lie inside the circle and a portion (called the exterior segment) will lie outside the circle. The product of the length of one secant segment and the length of its exterior segment equals the product of the length of the other secant segment and the length of its exterior segment.

Tangent segments intersecting secant segments:

If a tangent segment and a secant segment intersect outside a circle, the square of the length of the tangent segment equals the product of the length of the secant segment and its exterior segment.

Examples:

1.

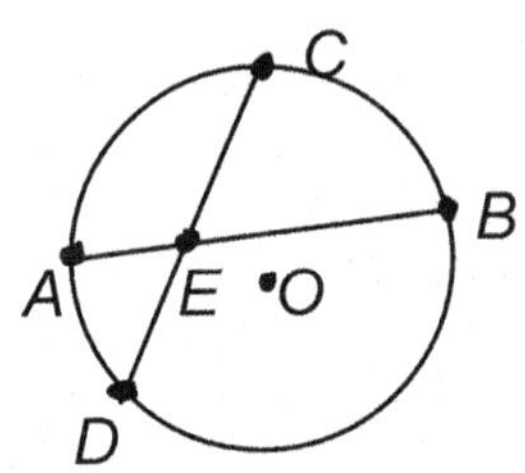

$\overline{AB}$ and $\overline{CD}$ are chords.
$CE=10$, $ED=x$, $AE=5$, $EB=4$

$(AE)(EB) = (CE)(ED)$	Since the chords intersect in the circle,
$5(4) = 10x$	the products of the segment pieces are
$20 = 10x$	equal.
$x = 2$	Solve for x.

2.

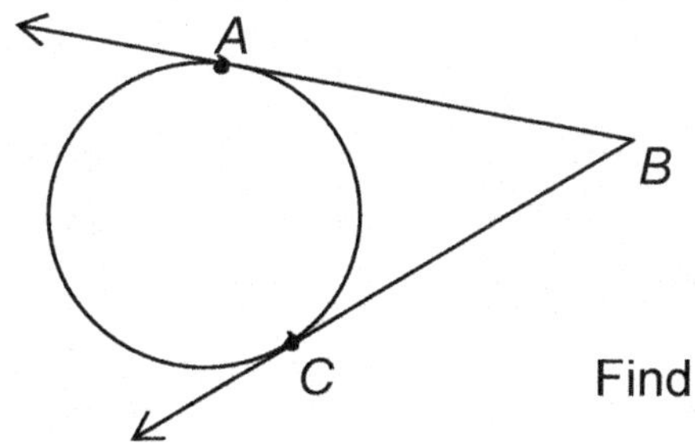

$\overline{AB}$ and $\overline{CD}$ are chords.
$\overline{AB} = x^2 + x - 2$
$\overline{BC} = x^2 - 3x + 5$
Find the length of $\overline{AB}$ and $\overline{BC}$.

$\overline{AB} = x^2 + x - 2$
$\overline{BC} = x^2 - 3x + 5$ — Given

$\overline{AB} = \overline{BC}$ — Intersecting tangents are equal.

$x^2 + x - 2 = x^2 - 3x + 5$ — Set the expressions equal to each other and solve.

$4x = 7$
$x = 1.75$ — Substitute and solve.

$(1.75)^2 + 1.75 - 2 = \overline{AB}$
$\overline{AB} = \overline{BC} = 2.81$

Congruent figures have the same size and shape. If one is placed above the other, it will fit exactly. Congruent lines have the same length. Congruent angles have equal measures.
The symbol for congruent is $\cong$.

Polygons (pentagons) *ABCDE* and *VWXYZ* are congruent. They are exactly the same size and shape.

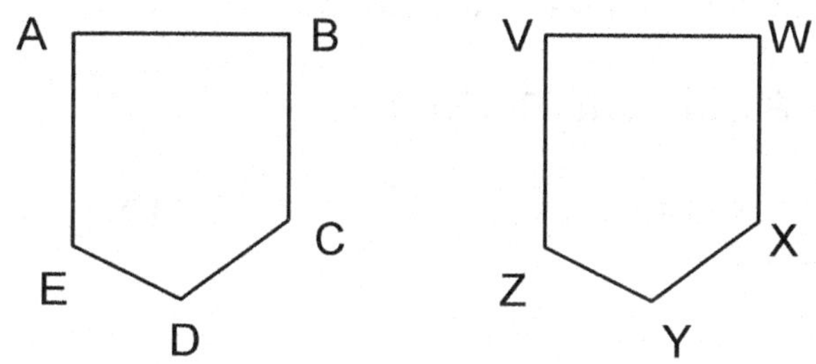

$ABCDE \cong VWXYZ$

Corresponding parts are those congruent angles and congruent sides, that is:

corresponding angles	corresponding sides
$\angle A \leftrightarrow \angle V$	$AB \leftrightarrow VW$
$\angle B \leftrightarrow \angle W$	$BC \leftrightarrow WX$
$\angle C \leftrightarrow \angle X$	$CD \leftrightarrow XY$
$\angle D \leftrightarrow \angle Y$	$DE \leftrightarrow YZ$
$\angle E \leftrightarrow \angle Z$	$AE \leftrightarrow VZ$

Two figures that have the **same shape** are **similar**. Two polygons are similar if corresponding angles are congruent and corresponding sides are in proportion. Corresponding parts of similar polygons are proportional.

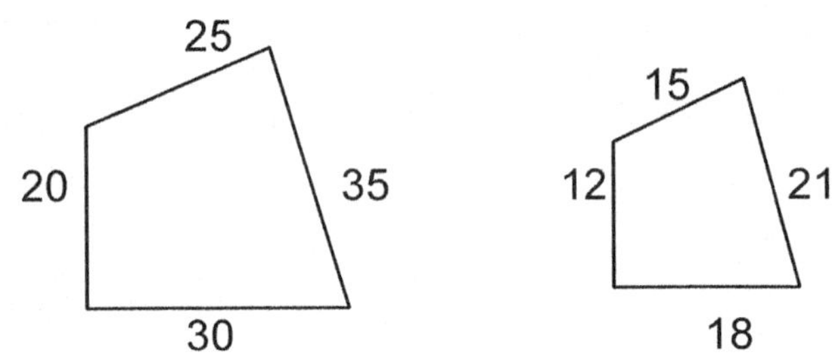

SIMILAR TRIANGLES

AA Similarity Postulate

If two angles of one triangle are congruent to two angles of another triangle, then the triangles are similar.

SAS Similarity Theorem

If an angle of one triangle is congruent to an angle of another triangle and the sides adjacent to those angles are in proportion, then the triangles are similar.

SSS Similarity Theorem

If the sides of two triangles are in proportion, then the triangles are similar.

The strategy for solving problems of this nature should be to identify the given shapes and choose the correct formulas. Subtract the smaller cut out shape from the larger shape.

Sample problems:

1. Find the area of one side of the metal in the circular flat washer shown below:

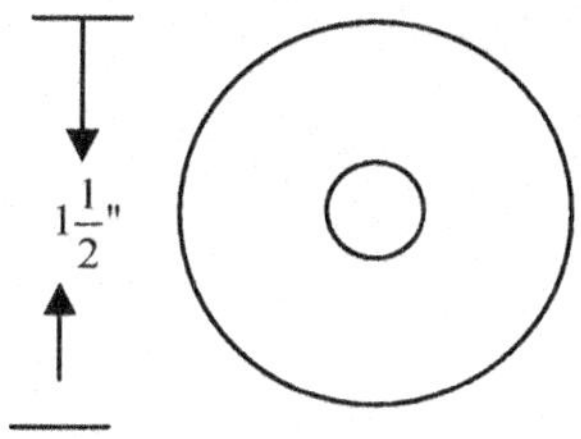

1. the shapes are both circles.
2. use the formula A = πr^2 for both.

(Inside diameter is 3/8")

Area of larger circle	Area of smaller circle
$A = \pi r^2$	$A = \pi r^2$
$A = \pi(.75^2)$	$A = \pi(.1875^2)$
$A = 1.76625 \text{ in}^2$	$A = .1104466 \text{ in}^2$

Area of metal washer = larger area - smaller area

$= 1.76625 \text{ in}^2 - .1104466 \text{ in}^2$

$= 1.6558034 \text{ in}^2$

2. You have decided to fertilize your lawn. The shapes and dimensions of your lot, house, pool and garden are given in the diagram below. The shaded area will not be fertilized. If each bag of fertilizer costs \$7.95 and covers 4,500 square feet, find the total number of bags needed and the total cost of the fertilizer.

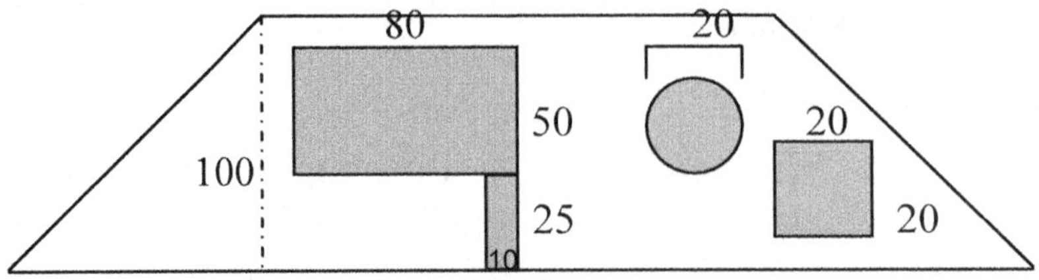

Area of Lot	Area of House	Area of Driveway
$A = \frac{1}{2}h(b_1 + b_2)$	$A = LW$	$A = LW$
$A = \frac{1}{2}(100)(180 + 160)$	$A = (80)(50)$	$A = (10)(25)$
$A = 17,000$ sq ft	$A = 4,000$ sq ft	$A = 250$ sq ft

Area of Pool	Area of Garden
$A = \pi r^2$	$A = s^2$
$A = \pi(10)^2$	$A = (20)^2$
$A = 314.159$ sq. ft.	$A = 400$ sq. ft.

Total area to fertilize = Lot area - (House + Driveway + Pool + Garden)

= 17,000 - (4,000 + 250 + 314.159 + 400)
= 12,035.841 sq ft

Number of bags needed = Total area to fertilize / 4,500 sq.ft. bag

$= 12,035.841 / 4,500$
= 2.67 bags

Since we cannot purchase 2.67 bags we must purchase 3 full bags.

Total cost = Number of bags * \$7.95
= 3 * \$7.95
= \$23.85

The union of all points on a simple closed surface and all points in its interior form a space figure called a **solid**. The five regular solids, or **polyhedra**, are the cube, tetrahedron, octahedron, icosahedron, and dodecahedron. A **net** is a two-dimensional figure that can be cut out and folded up to make a three-dimensional solid. Below are models of the five regular solids with their corresponding face polygons and nets.

Cube 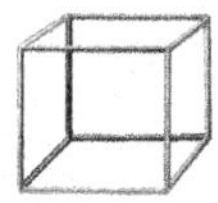 6 squares 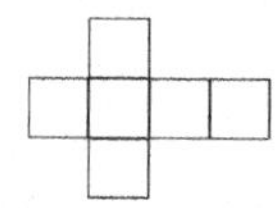

Tetrahedron 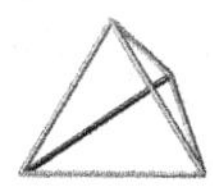 4 equilateral triangles 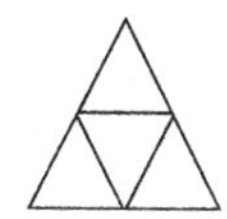

Octahedron  8 equilateral triangles 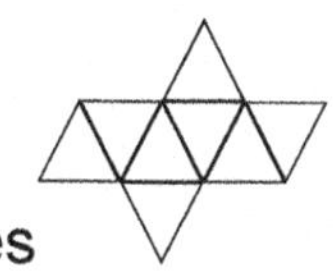

Icosahedron 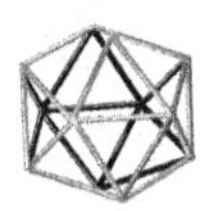 20 equilateral triangles 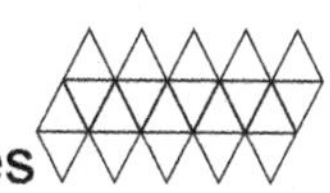

Dodecahedron 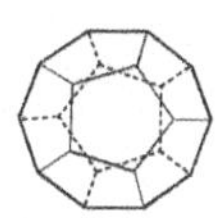 12 regular pentagons 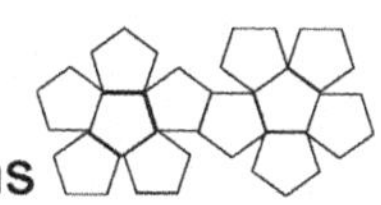

Other examples of solids:

Sphere 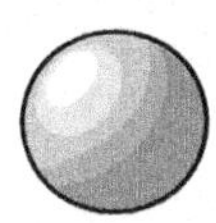

Cone 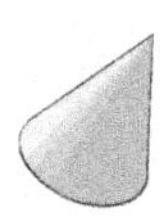

We refer to three-dimensional figures in geometry as **solids**. A solid is the union of all points on a simple closed surface and all points in its interior. A **polyhedron** is a simple closed surface formed from planar polygonal regions. Each polygonal region is called a **face** of the polyhedron. The vertices and edges of the polygonal regions are called the **vertices** and **edges** of the polyhedron.

We may form a cube from three congruent squares. However, if we tried to put four squares about a single vertex, their interior angle measures would add up to 360°; i.e., four edge-to-edge squares with a common vertex lie in a common plane and therefore cannot form a corner figure of a regular polyhedron.

There are five ways to form corner figures with congruent regular polygons:

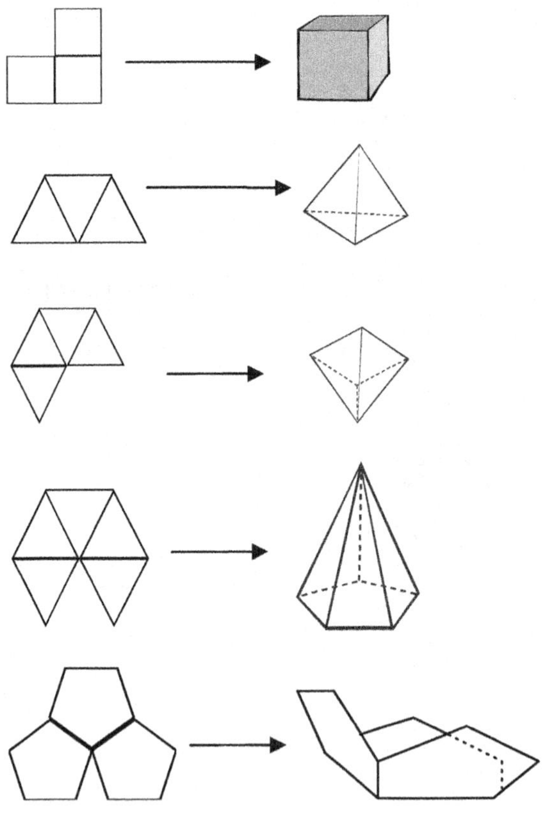

When creating a three-dimensional figure, if we know any two values of the vertices, faces, and edges, we can find the remaining value by using **Euler's Formula**: $V+F=E+2$.

For example:

We want to create a pentagonal pyramid, and we know it has six vertices and six faces. Using Euler's Formula, we compute:

$$\begin{aligned} V+F&=E+2\\ 6+6&=E+2\\ 12&=E+2\\ 10&=E \end{aligned}$$

Thus, we know that our figure should have 10 edges.

We can represent any two-dimensional geometric figure in the **Cartesian** or **rectangular coordinate system**. The Cartesian or rectangular coordinate system is formed by two perpendicular axes (coordinate axes): the X-axis and the Y-axis. If we know the dimensions of a two-dimensional, or planar, figure, we can use this coordinate system to visualize the shape of the figure.

Example: Represent an isosceles triangle with two sides of length 4.

Draw the two sides along the x- and y- axes and connect the points (vertices).

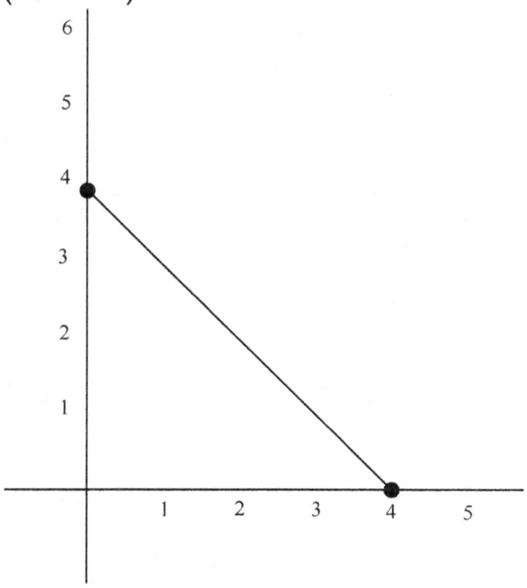

In order to represent three-dimensional figures, we need three coordinate axes (X, Y, and Z) which are all mutually perpendicular to each other. Since we cannot draw three mutually perpendicular axes on a two-dimensional surface, we use oblique representations.

Example: Represent a cube with sides of 2.
Once again, we draw three sides along the three axes to make things easier.

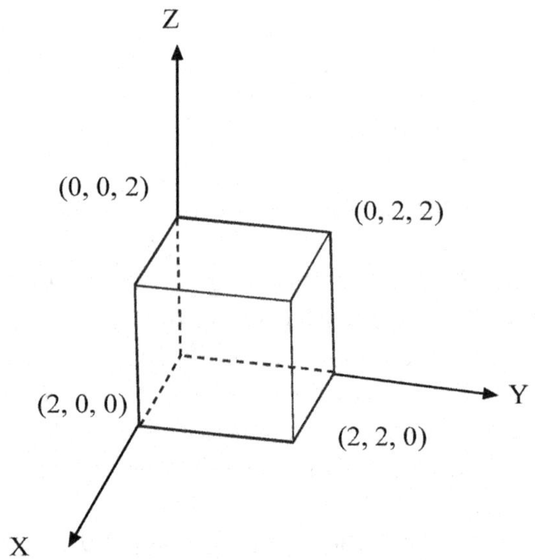

Each point has three coordinates (x, y, z).

Competency 14.0 The teacher understands coordinate, transformational, and vector geometry and their connections.

A **transformation** is a change in the position, shape, or size of a geometric figure. **Transformational geometry** is the study of manipulating objects by flipping, twisting, turning and scaling. **Symmetry** is exact similarity between two parts or halves, as if one were a mirror image of the other.

There are four basic transformational symmetries: **translation, rotation, reflection,** and **glide reflection**. The transformation of an object is called its image. If the original object was labeled with letters, such as $ABCD$, the image may be labeled with the same letters followed by a prime symbol, $A'B'C'D'$.

A **translation** is a transformation that "slides" an object a fixed distance in a given direction. The original object and its translation have the same shape and size, and they face in the same direction.

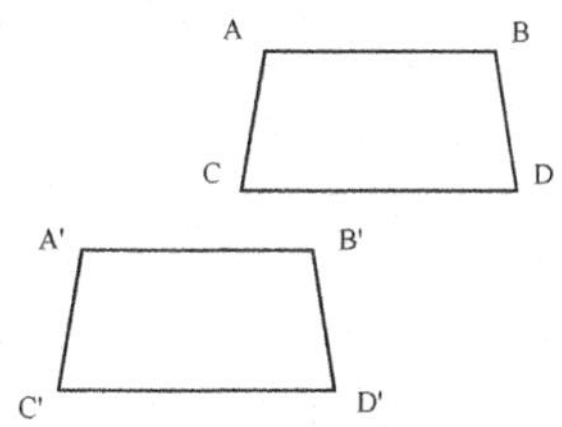

An example of a translation in architecture would be stadium seating. The seats are the same size and the same shape and face in the same direction.

A **rotation** is a transformation that turns a figure about a fixed point called the center of rotation. An object and its rotation are the same shape and size, but the figures may be turned in different directions. Rotations can occur in either a clockwise or a counterclockwise direction.

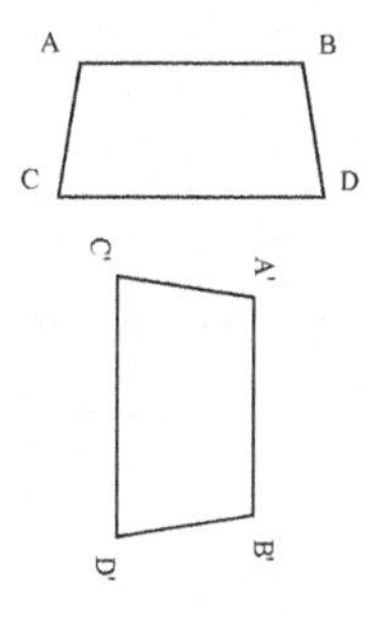

Rotations can be seen in wallpaper and art, and a Ferris wheel is an example of rotation.

An object and its **reflection** have the same shape and size, but the figures face in opposite directions.

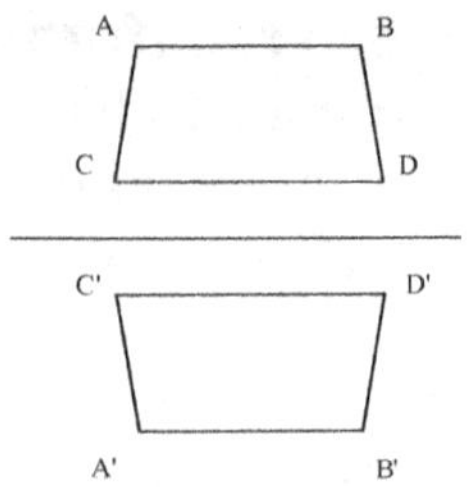

The line (where a mirror may be placed) is called the **line of reflection**. The distance from a point to the line of reflection is the same as the distance from the point's image to the line of reflection.

A **glide reflection** is a combination of a reflection and a translation.

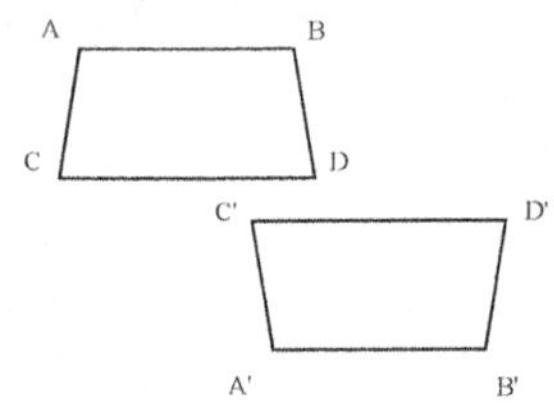

Another type of transformation is **dilation**. Dilation is a transformation that "shrinks" or "makes it bigger."

Example:

Using dilation to transform a diagram.

Starting with a triangle whose center of dilation is point P,

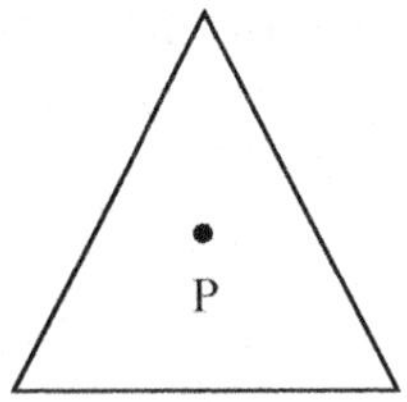

we dilate the lengths of the sides by the same factor to create a new triangle.

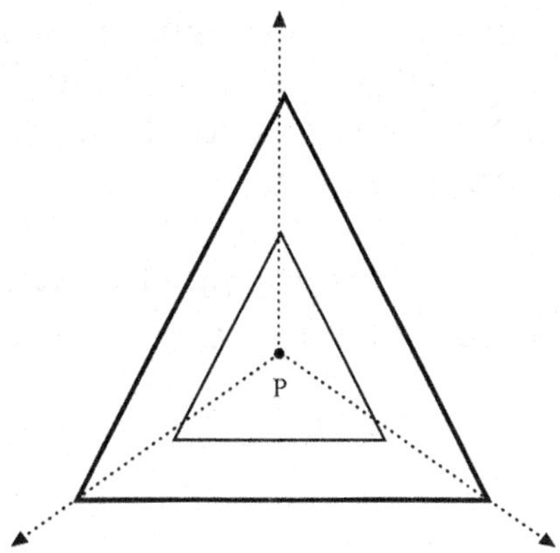

Example:

Plot the given ordered pairs on a coordinate plane and join them in the given order, then join the first and last points.

(-3, -2), (3, -2), (5, -4), (5, -6), (2, -4), (-2, -4), (-5, -6), (-5, -4)

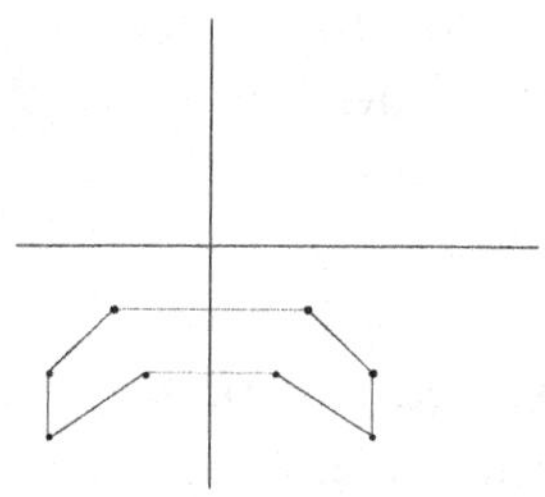

Increase all y-coordinates by 6.
(-3, 4), (3, 4), (5, 2), (5, 0), (2, 2), (-2, 2), (-5, 0), (-5, 2)

Plot the points and join them to form a second figure.

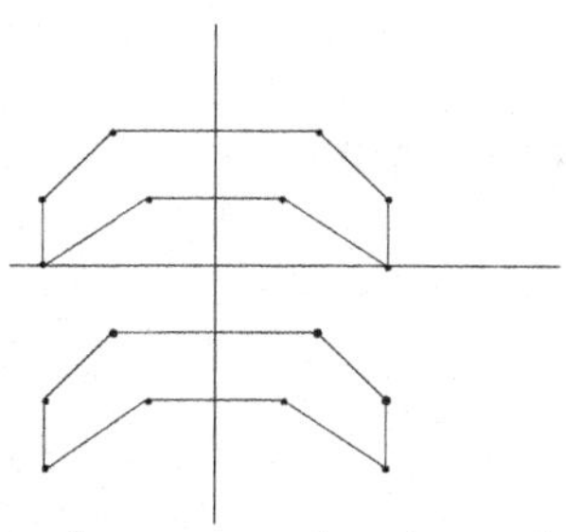

A figure on a coordinate plane can be translated by changing the ordered pairs.

A **fractal** is an endlessly repeating pattern that varies according to a set formula, a mixture of art and geometry. A fractal is any pattern that reveals greater complexity as it is enlarged. An example of a fractal is an ice crystal freezing on a glass window.

Fractals are **self-similar** and have **fractional (fractal) dimension**. Self-similar means that a fractal looks the same over all ranges of scale.

Example:

Fractional, or fractal dimension, means that the dimension of the figure is a non-integer, or fraction.

Example,

In the above figure, notice that the second triangle is composed of three miniature triangles exactly like the original. The length of any side of one of the miniature triangles could be multiplied by two to produce the entire triangle (S=2). The resulting figure consists of three separate identical miniature pieces (N=3). The formula to find the dimension of a strictly self-similar fractal is

$$D = \frac{\log N}{\log S}$$

For the above figure,

$$D = \frac{\log 3}{\log 2}$$

$$D = 1.585$$

Example:

Using transformational methods, prove the Pythagorean Theorem.

Assume you have a right triangle with legs b and a, and hypotenuse c. Construct a segment subdivided into two parts of lengths a and b.

Using several rotations, construct a square on side a and a square on side b to create two regions whose total area is $a^2 + b^2$.

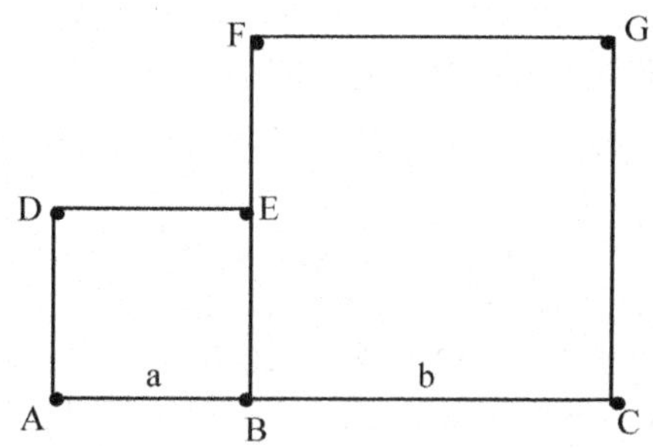

Define a translation from B to A and translate point C to get point H. Connect H to D and H to G, resulting in two right triangles.

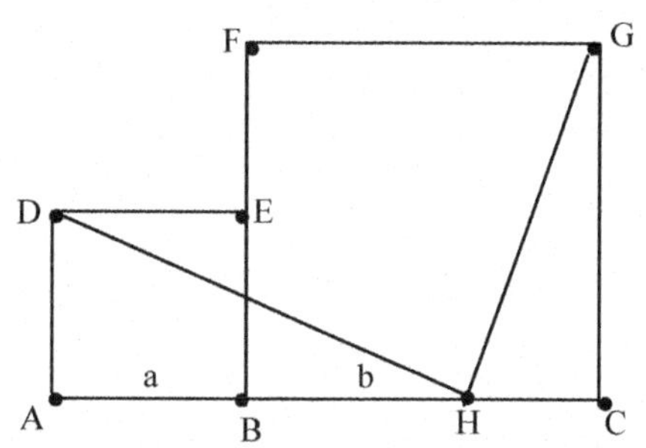

Hide segment BC and create segments BH and HC. This is so that we have well-defined triangle sides for the next step – rotating right triangle ADH 90 degrees about its top vertex, and right triangle HGC 90 degrees about its top vertex.

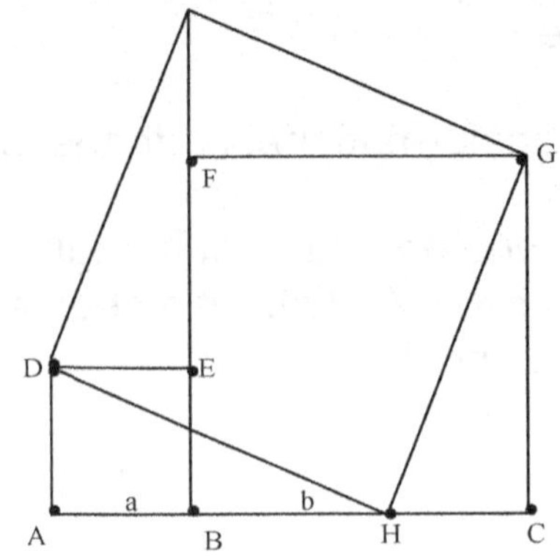

Note: Figure not drawn to scale.

Now prove that this construction yields a square (on DH) of side length c, and thus, since the area of this square is clearly equal to the sum of the areas of the original two squares, we have $a^2 + b^2 = c^2$, and the proof will be complete.

By SAS, triangle HCG must be congruent to the original right triangle, and thus its hypotenuse must be c. Also, by SAS, triangle DAH is also congruent to the original triangle, and so its hypotenuse is also c. Then, angles AHD and CHG (=ADH) must sum to 90 degrees, and the angle DHG is a right angle. Thus, you have shown that the construction yields a square on DH of side length c, and the proof is complete.

Midpoint Definition:

If a line segment has endpoints of (x_1, y_1) and (x_2, y_2), then the midpoint can be found using:

$$\left(\frac{x_1 + x_2}{2}, \frac{y_1 + y_2}{2}\right)$$

Sample problems:

1. Find the center of a circle with a diameter whose endpoints are (3,7) and $(^-4, {}^-5)$.

$$\text{Midpoint} = \left(\frac{3 + (^-4)}{2}, \frac{7 + (^-5)}{2}\right)$$

$$\text{Midpoint} = \left(\frac{^-1}{2}, 1\right)$$

2. Find the midpoint given the two points $\left(5, 8\sqrt{6}\right)$ and $\left(9, {}^-4\sqrt{6}\right)$.

$$\text{Midpoint} = \left(\frac{5 + 9}{2}, \frac{8\sqrt{6} + (^-4\sqrt{6})}{2}\right)$$

$$\text{Midpoint} = \left(7, 2\sqrt{6}\right)$$

The equation of a circle with its center at (h,k) and a radius r units is:

$$(x-h)^2+(y-k)^2=r^2$$

Sample Problem:

1. Given the equation $x^2+y^2=9$, find the center and the radius of the circle. Then graph the equation.

First, writing the equation in standard circle form gives:

$$(x-0)^2+(y-0)^2=3^2$$

therefore, the center is (0,0) and the radius is 3 units.

Sketch the circle:

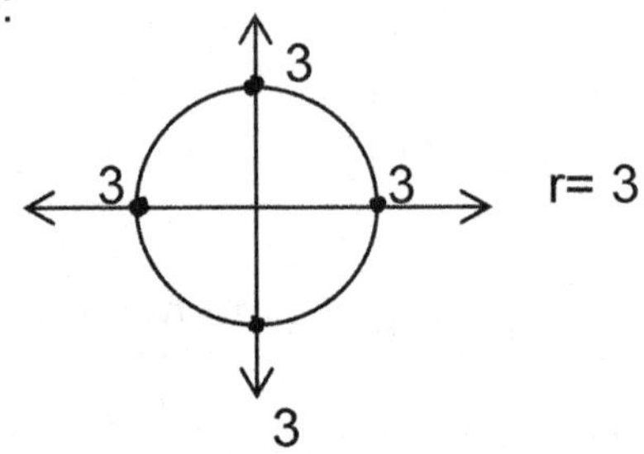

2. Given the equation $x^2+y^2-3x+8y-20=0$, find the center and the radius. Then graph the circle.

First, write the equation in standard circle form by completing the square for both variables.

$x^2+y^2-3x+8y-20=0$ 1. Complete the squares.

$(x^2-3x+9/4)+(y^2+8y+16)=20+9/4+16$
$(x-3/2)^2+(y+4)^2=153/4$

The center is $(3/2, {}^-4)$ and the radius is $\frac{\sqrt{153}}{2}$ or $\frac{3\sqrt{17}}{2}$.

Graph the circle.

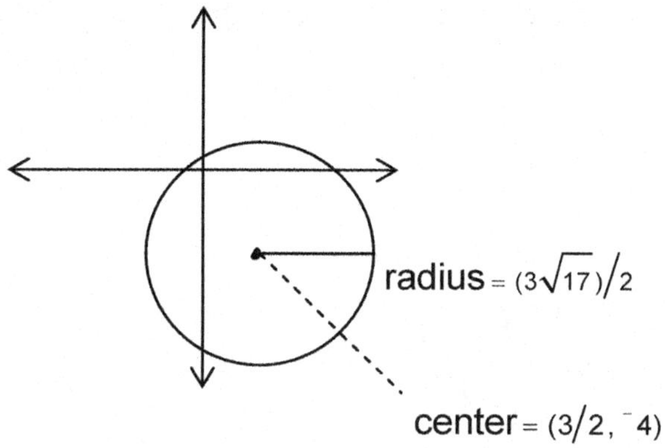

To write the equation given the center and the radius use the standard form of the equation of a circle:

$$(x-h)^2+(y-k)^2=r^2$$

Sample problems:

Given the center and radius, write the equation of the circle.

1. Center ($^-1$,4); radius 11

$(x-h)^2+(y-k)^2=r^2$	1. Write standard equation.
$(x-(^-1))^2+(y-(4))^2=11^2$	2. Substitute.
$(x+1)^2+(y-4)^2=121$	3. Simplify.

2. Center ($\sqrt{3}$, $^-1/2$); radius $=5\sqrt{2}$

$(x-h)^2+(y-k)^2=r^2$	1. Write standard equation.
$(x-\sqrt{3})^2+(y-(^-1/2))^2=(5\sqrt{2})^2$	2. Substitute.
$(x-\sqrt{3})^2+(y+1/2)^2=50$	3. Simplify.

Conic sections result from the intersection of a cone and a plane. The three main types of conics are parabolas, ellipses, and hyperbolas.

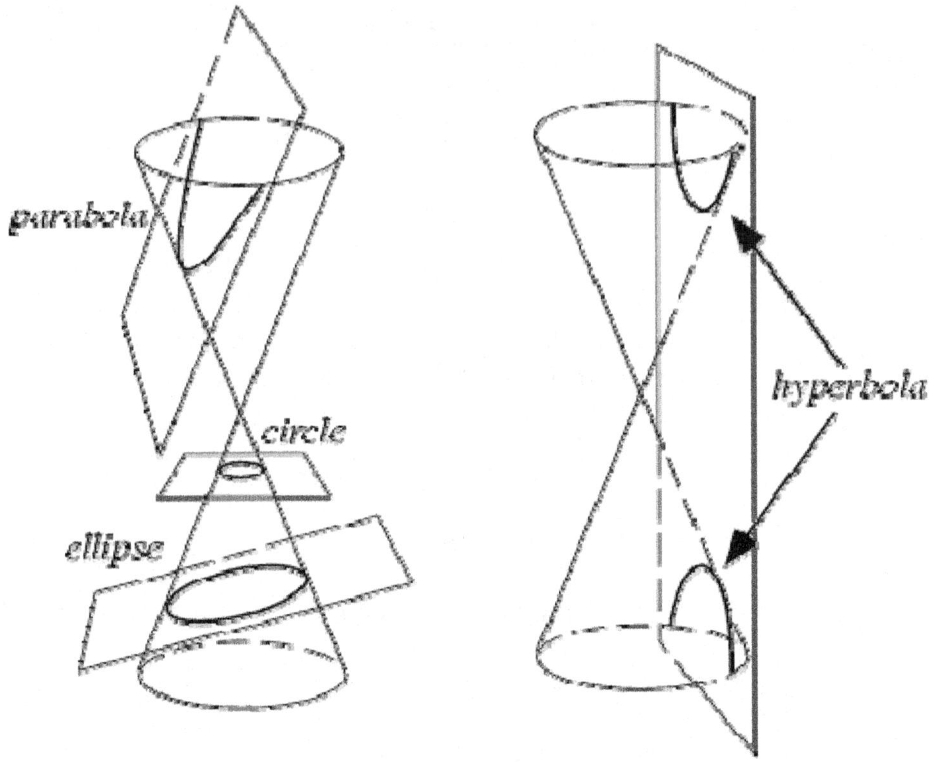

The general equation for a conic section is:

$$Ax^2 + Bxy + Cy^2 + Dx + Ey + F = 0$$

The value of $B^2 - 4AC$ determines the type of conic. If $B^2 - 4AC$ is less than zero the curve is an ellipse or a circle. If equal to zero, the curve is a parabola. If greater than zero, the curve is a hyperbola.

PARABOLAS

-A parabola is a set of all points in a plane that are equidistant from a fixed point (focus) and a line (directrix).

FORM OF EQUATION	$y = a(x-h)^2 + k$	$x = a(y-k)^2 + h$
IDENTIFICATION	x^2 term, y not squared	y^2 term, x not squared

SKETCH OF GRAPH

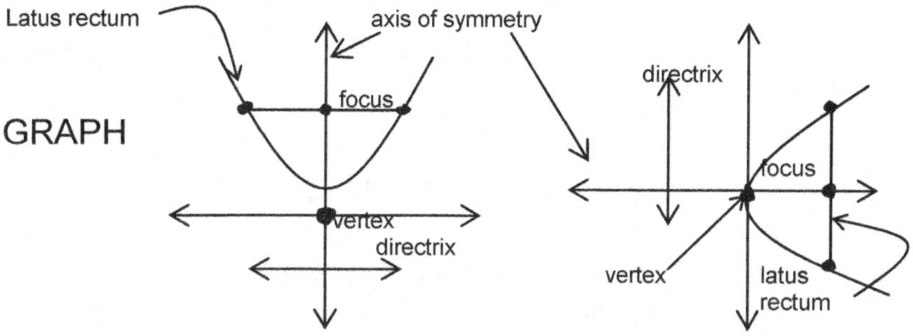

AXIS OF SYMMETRY	$x = h$	$y = k$

-A line through the vertex and focus upon which the parabola is symmetric.

VERTEX	(h,k)	(h,k)

-The point where the parabola intersects the axis of symmetry.

FOCUS	$(h, k + 1/4a)$	$(h + 1/4a, k)$
DIRECTRIX	$y = k - 1/4a$	$x = h - 1/4a$
DIRECTION OF OPENING	up if $a > 0$, down if $a < 0$	right if $a > 0$, left if $a < 0$
LENGTH OF LATUS RECTUM	$\lvert 1/a \rvert$	$\lvert 1/a \rvert$

-A chord through the focus, perpendicular to the axis of symmetry, with endpoints on the parabola.

Sample Problem:

1. Find all identifying features of $y = {}^{-}3x^2 + 6x - 1$.

First, the equation must be put into the general form $y = a(x-h)^2 + k$.

$y = {}^{-}3x^2 + 6x - 1$	1. Begin by completing the square.
$= {}^{-}3(x^2 - 2x + 1) - 1 + 3$	
$= {}^{-}3(x-1)^2 + 2$	2. Using the general form of the equation begin to identify known variables.

$a = {}^{-}3 \quad h = 1 \quad k = 2$

axis of symmetry: $x = 1$
vertex: (1,2)
focus: $(1, 1\tfrac{1}{4})$
directrix: $y = 2\tfrac{3}{4}$
direction of opening: down since $a < 0$
length of latus rectum: 1/3

ELLIPSE

FORM OF EQUATION	$\frac{(x-h)^2}{a^2} + \frac{(y-k)^2}{b^2} = 1$	$\frac{(x-h)^2}{b^2} + \frac{(y-k)^2}{a^2} = 1$
(for ellipses where $a^2 > b^2$).	where $b^2 = a^2 - c^2$	where $b^2 = a^2 - c^2$
IDENTIFICATION	horizontal major axis	vertical major axis
SKETCH	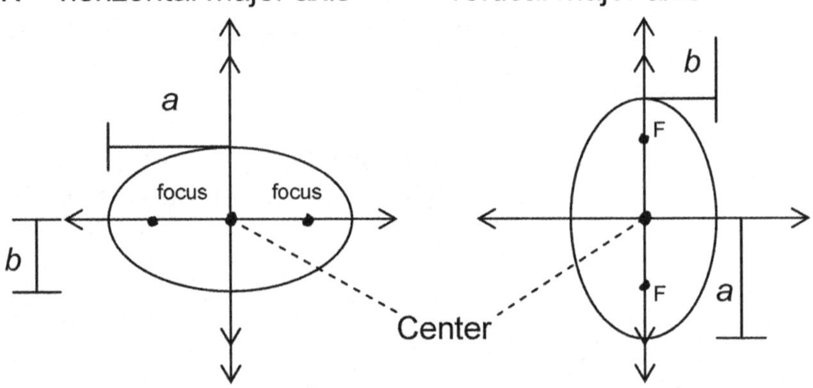	

CENTER	(h,k)	(h,k)
FOCI	$(h \pm c,k)$	$(h,k \pm c)$
MAJOR AXIS LENGTH	$2a$	$2a$
MINOR AXIS LENGTH	$2b$	$2b$

Sample Problem:

Find all identifying features of the ellipse $2x^2 + y^2 - 4x + 8y - 6 = 0$. First, begin by writing the equation in standard form for an ellipse.

$2x^2 + y^2 - 4x + 8y - 6 = 0$ — 1. Complete the square for each variable.

$2(x^2 - 2x + 1) + (y^2 + 8y + 16) = 6 + 2(1) + 16$

$2(x-1)^2 + (y+4)^2 = 24$ — 2. Divide both sides by 24.

$\frac{(x-1)^2}{12} + \frac{(y+4)^2}{24} = 1$ — 3. Now the equation is in standard form.

Identify known variables: $h = 1$ $k = {}^-4$ $a = \sqrt{24}$ or $2\sqrt{6}$
$b = \sqrt{12}$ or $2\sqrt{3}$ $c = 2\sqrt{3}$

Identification: vertical major axis
Center: $(1, {}^-4)$
Foci: $(1, {}^-4 \pm 2\sqrt{3})$
Major axis: $4\sqrt{6}$
Minor axis: $4\sqrt{3}$

HYPERBOLA

FORM OF EQUATION	$\frac{(x-h)^2}{a^2} - \frac{(y-k)^2}{b^2} = 1$ where $c^2 = a^2 + b^2$	$\frac{(y-k)^2}{a^2} - \frac{(x-h)^2}{b^2} = 1$ where $c^2 = a^2 + b^2$

IDENTIFICATION	horizontal transverse axis (y^2 is negative)	vertical transverse axis (x^2 is negative)
SKETCH	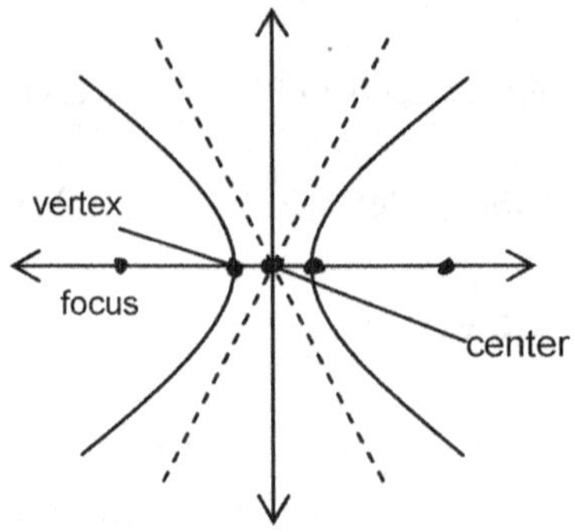	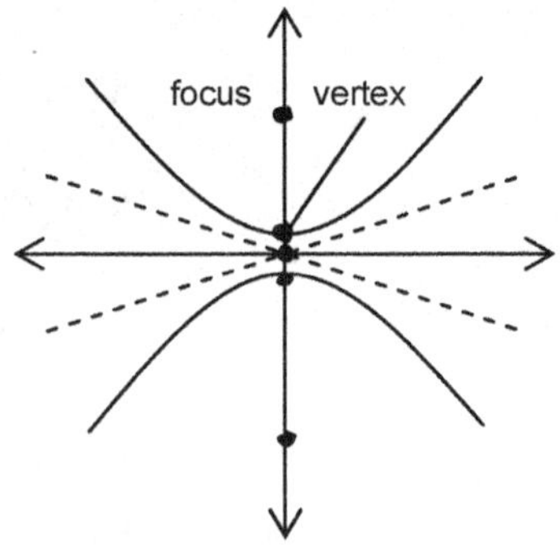
SLOPE OF ASYMPTOTES	$\pm(b/a)$	$\pm(a/b)$
TRANSVERSE AXIS (endpoints are vertices of the hyperbola and goes through the center)	$2a$ -on y axis	$2a$ -on x axis
CONJUGATE AXIS (perpendicular to transverse axis at center)	$2b$ -on y axis	$2b$ -on x axis
CENTER	(h,k)	(h,k)
FOCI	$(h \pm c,k)$	$(h,k \pm c)$
VERTICES	$(h \pm a,k)$	$(h,k \pm a)$

Sample Problem:

Find all the identifying features of a hyperbola given its equation.

$$\frac{(x+3)^2}{4} - \frac{(y-4)^2}{16} = 1$$

Identify all known variables: $h = {}^{-}3$ $k = 4$ $a = 2$ $b = 4$
$c = 2\sqrt{5}$

Slope of asymptotes: $\pm 4/2$ or ± 2
Transverse axis: 4 units long
Conjugate axis: 8 units long
Center: $({}^{-}3,4)$
Foci: $({}^{-}3 \pm 2\sqrt{5},4)$
Vertices: $({}^{-}1,4)$ and $({}^{-}5,4)$

A **transformation matrix** defines how to map points from one coordinate space into another coordinate space. The matrix used to accomplish two-dimensional transformations is described mathematically by a 3-by-3 matrix.

Example:

A point transformed by a 3-by-3 matrix

$$[x \;\; y \;\; 1] \times \begin{pmatrix} a & b & u \\ c & d & v \\ t_x & t_y & w \end{pmatrix} = [x' \;\; y' \;\; 1]$$

A 3-by-3 matrix transforms a point (x, y) into a point (x', y') by means of the following equations:

$$x' = ax + cy + t_x$$

Occasionally, it is important to reverse the addition or subtraction process and express the single vector as the sum or difference of two other vectors. It may be critically important for a pilot to understand not only the air velocity but also the ground speed and the climbing speed.

Sample problem:

A pilot is traveling at an air speed of 300 mph and a direction of 20 degrees. Find the horizontal vector (ground speed) and the vertical vector (climbing speed).

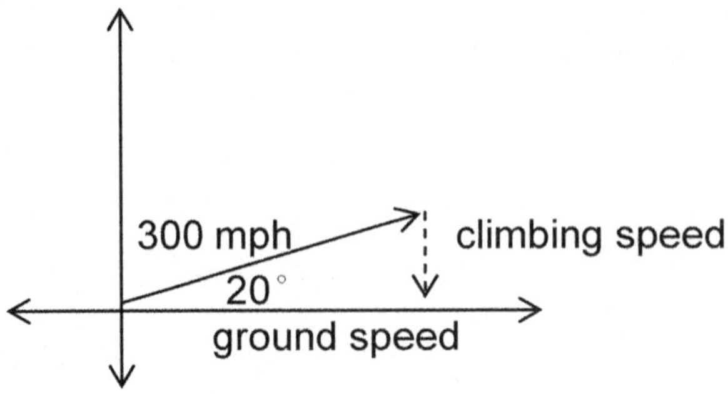

1. Draw sketch.
2. Use appropriate trigonometric ratio to calculate the component vectors.

To find the vertical vector:

$$\sin x = \frac{\text{opposite}}{\text{hypotenuse}}$$

$$\sin(20) = \frac{c}{300}$$

$$c = (.3420)(300)$$

$$c = 102.606$$

To find the horizontal vector:

$$\cos x = \frac{\text{adjacent}}{\text{hypotenuse}}$$

$$\cos(20) = \frac{g}{300}$$

$$g = (.9397)(300)$$

$$g = 281.908$$

Vectors are used to measure displacement of an object or force.
Addition of vectors:

$$(a,b)+(c,d)=(a+c,b+d)$$

Addition Properties of vectors:

$$a+b=b+a$$

$$a+(b+c)=(a+b)+c$$

$$a+0=a$$

$$a+(^{-}a)=0$$

Subtraction of vectors:

$$a-b=a+(^{-}b) \text{ therefore,}$$

$$a-b=(a_1,a_2)+(^{-}b_1,{}^{-}b_2) \text{ or}$$

$$a-b=(a_1-b_1,a_2-b_2)$$

Sample problem:

If $a=(4,{}^{-}1)$ and $b=(^{-}3,6)$, find $a+b$ and $a-b$.

Using the rule for addition of vectors:

$$(4,{}^{-}1)+(^{-}3,6)=(4+(^{-}3),{}^{-}1+6)$$
$$=(1,5)$$

Using the rule for subtraction of vectors:

$$(4,{}^{-}1)-(^{-}3,6)=(4-(^{-}3),{}^{-}1-6)$$
$$=(7,{}^{-}7)$$

The **dot product** $a{\bullet}b$:

$$a = (a_1, a_2) = a_1 i + a_2 j \quad \text{and} \quad b = (b_1, b_2) = b_1 i + b_2 j$$

$$a{\bullet}b = a_1 b_1 + a_2 b_2$$

$a{\bullet}b$ is read "a dot b". Dot products are also called scalar or inner products. When discussing dot products, it is important to remember that "a dot b" is not a vector, but a real number.

Properties of the dot product:

$a{\bullet}a = |a|^2$

$a{\bullet}b = b{\bullet}a$

$a{\bullet}(b + c) = a{\bullet}b + a{\bullet}c$

$(ca){\bullet}b = c(a{\bullet}b) = a{\bullet}(cb)$

$0{\bullet}a = 0$

Sample problems:

Find the dot product.

1. $a = (5,2), b = ({}^{-}3,6)$

$$\begin{aligned} a{\bullet}b &= (5)({}^{-}3) + (2)(6) \\ &= {}^{-}15 + 12 \\ &= {}^{-}3 \end{aligned}$$

2. $a = (5i + 3j), b = (4i - 5j)$

$$\begin{aligned} a{\bullet}b &= (5)(4) + (3)({}^{-}5) \\ &= 20 - 15 \\ &= 5 \end{aligned}$$

3. The magnitude and direction of a constant force are given by $a = 4i + 5j$. Find the amount of work done if the point of application of the force moves from the origin to the point $P(7,2)$.

The work W done by a constant force a as its point of application moves along a vector b is $W = a{\bullet}b$.

Sketch the constant force vector a and the vector b.

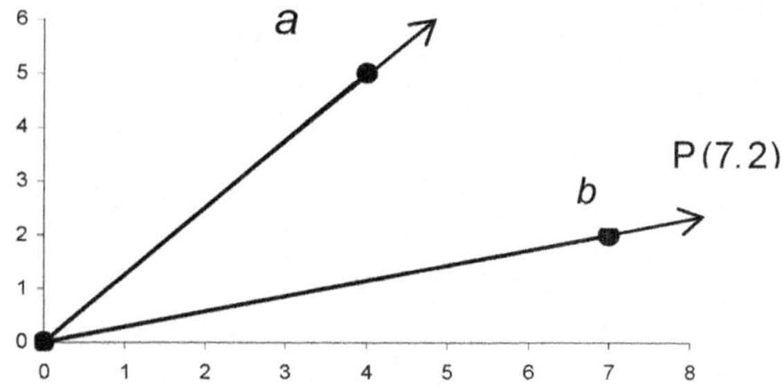

$b = (7,2) = 7i + 2j$

Use the definition of work done to solve.

$$\begin{aligned} W &= a \bullet b \\ &= (4i + 5j)(7i + 2j) \\ &= (4)(7) + (5)(2) \\ &= (28) + (10) \\ &= 38 \end{aligned}$$

Vectors are used often in navigation of ships and aircraft as well as in force and work problems.

Sample problem:

1. An airplane is flying with a heading of 60 degrees east of north at 450 mph. A wind is blowing at 37 mph from the north. Find the plane's ground speed to the nearest mph and direction to the nearest degree.

Draw a sketch.

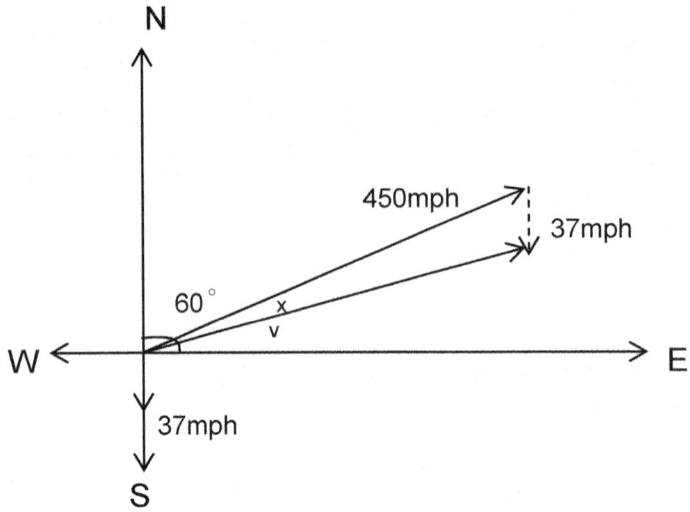

Use the law of cosines to find the ground speed.

$$a^2 = b^2 + c^2 - 2bc\cos A$$
$$|v|^2 = 450^2 + 37^2 - 2(450)(37)\cos 60$$
$$|v|^2 = 187{,}219$$
$$|v| = 432.688$$
$$|v| \approx 433 \text{ mph}$$

Use the law of sines to find the measure of x.

$$\frac{\sin A}{a} = \frac{\sin C}{c}$$
$$\frac{\sin 60}{433} = \frac{\sin x}{37}$$
$$\sin x = \frac{37(\sin 60)}{433}$$
$$x = 4.24$$
$$x \approx 4 \text{ degrees}$$

The plane's actual course is $60 + 4 = 64$ degrees east of north and the ground speed is approximately 433 miles per hour.

DOMAIN IV. PROBABILITY AND STATISTICS

Competency 15.0 The teacher understands how to use appropriate graphical and numerical techniques to explore data, characterize patterns, and describe departures from patterns.

The four main types of measurement scales used in statistical analysis are nominal, ordinal, interval, and ratio. The type of variable measured and the research questions asked determine the appropriate measurement scale. The different measurement scales have distinctive qualities and attributes.

The nominal measurement scale is the most basic measurement scale. When measuring using the nominal scale, we simply label or classify responses into categories. Examples of variables measured on the nominal scale are gender, religion, ethnicity, and marital status. The essential attribute of the nominal scale is that the classifications have no numerical or comparative value. For example, when classifying people by marital status, there is no sense in which "single" is greater or less than "married". The only measure of central tendency applicable to the nominal scale is mode and the only applicable arithmetic operation is counting.

The ordinal measurement scale is more descriptive than the nominal scale in that the ordinal scale allows comparison between categories. Examples of variables measured on the ordinal scale are movie ratings, consumer satisfaction surveys, and the rank or order of anything. While we can compare categories of responses on the ordinal scale (e.g. "highly satisfied" indicates a higher level of satisfaction than "somewhat satisfied"), we cannot determine anything about the difference between the categories. In other words, we cannot presume that the difference between two categories is the same as the distance between two other categories. Even if the responses are in numeric form (e.g. 1 = good, 2 = fair, 3 = poor), we can presume nothing about the intervals separating the groups. Ordinal scales allow greater or less-than comparisons and the applicable measures of central tendency are range and median.

The next measurement scale is the interval scale. Interval scales are numeric scales where intervals have a fixed, uniform value throughout the scale. An example of an interval scale is the Fahrenheit temperature scale. On the Fahrenheit scale, the difference between 40 degrees and 50 degrees is the same as the difference between 70 degrees and 80 degrees. The major limitation of interval scales is that there is no fixed zero point. For example, while the Fahrenheit scale has a value of zero degrees, this assignment is arbitrary because the measurement does not represent the absence of temperature. Because interval scales lack a true zero point, ratio comparison of values has no meaning. Thus, the arithmetic operations addition and subtraction are applicable to interval scales while multiplication and division are not. The measures of central tendency applicable to interval scales are mode, median, and arithmetic mean.

The final, and most informative, measurement scale is the ratio scale. The ratio scale is essentially an interval scale with a true zero point. Examples of ratio scales are measurement of length (meters, inches, etc.), monetary systems, and degrees Kelvin. The zero value of each of these scales represents the absence of length, money, and temperature, respectively. The presence of a true zero point allows proportional comparisons. For example, we can say that someone with one dollar has twice as much money as someone with fifty cents. Because ratios have meaning on ratio scales, we can apply the arithmetic operations of multiplication and division to the data sets.

There are many graphical ways in which to represent data: line plots, line graphs, scatter plots, stem and leaf plots, histograms, bar graphs, pie charts, and pictographs.

A **line plot** organizes data in numerical order along a number line. An x is placed above the number line for each occurrence of the corresponding number. Line plots allow you to see at a glance a range of data and where typical and atypical data falls. They are generally used to summarize relatively small sets of data.

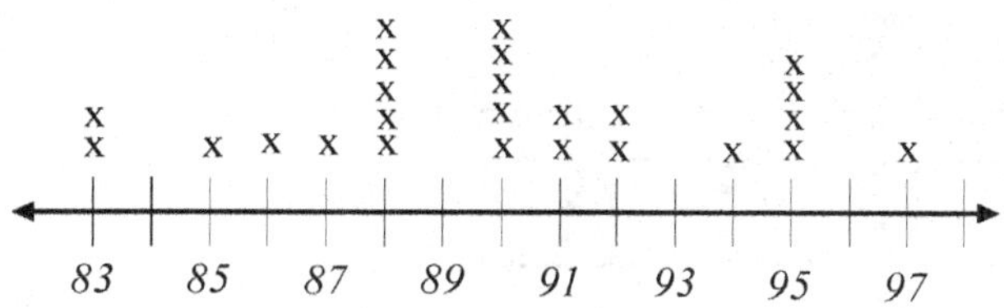

A **line graph** compares two variables, and each variable is plotted along an axis. A line graph highlights trends by drawing connecting lines between data points. They are particularly appropriate for representing data that varies continuously. Line graphs are sometimes referred to as **frequency polygons**.

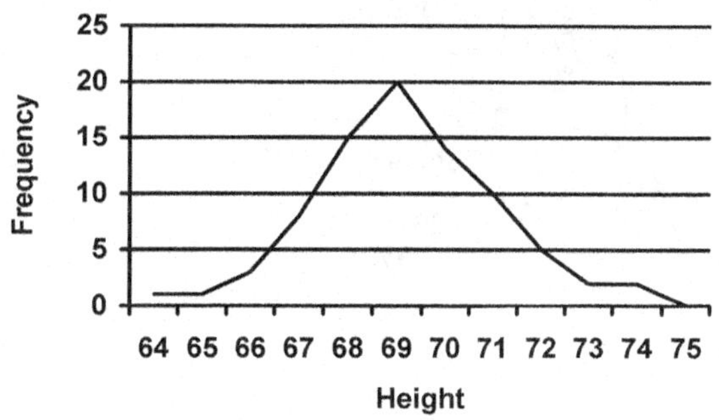

Bar graphs are similar to histograms. However, bar graphs are often used to convey information about categorical data where the horizontal scale represents a non-numeric attributes such as cities or years. Another difference is that the bars in bar graphs rarely touch. Bar graphs are also useful in comparing data about two or more similar groups of items.

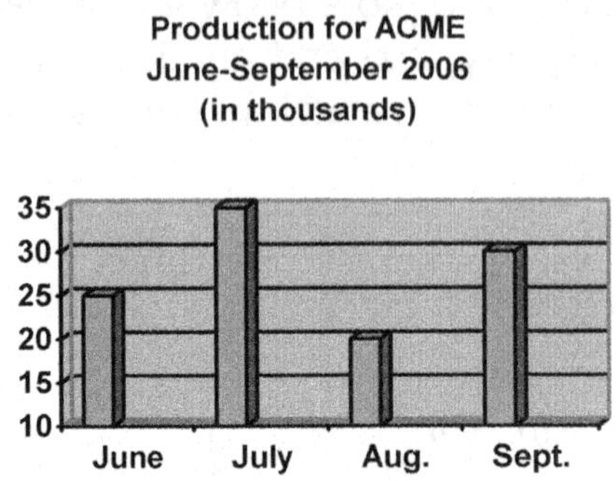

A **pie chart**, also known as a **circle graph**, is used to represent relative amounts of a whole.

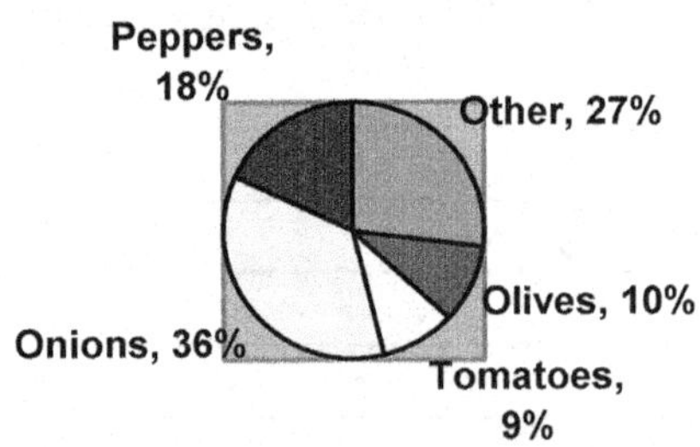

Scatter plots compare two characteristics of the same group of things or people and usually consist of a large body of data. They show how much one variable is affected by another. The relationship between the two variables is their **correlation**. The closer the data points come to making a straight line when plotted, the closer the correlation.

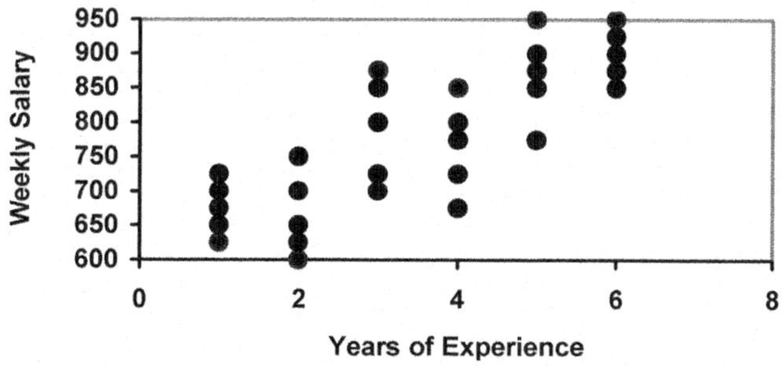

Stem and leaf plots are visually similar to line plots. The **stems** are the digits in the greatest place value of the data values, and the **leaves** are the digits in the next greatest place values. Stem and leaf plots are best suited for small sets of data and are especially useful for comparing two sets of data. The following is an example using test scores:

4	9
5	4 9
6	1 2 3 4 6 7 8 8
7	0 3 4 6 6 6 7 7 7 7 8 8 8 8
8	3 5 5 7 8
9	0 0 3 4 5
10	0 0

Histograms are used to summarize information from large sets of data that can be naturally grouped into intervals. The vertical axis indicates **frequency** (the number of times any particular data value occurs), and the horizontal axis indicates data values or ranges of data values. The number of data values in any interval is the **frequency of the interval**.

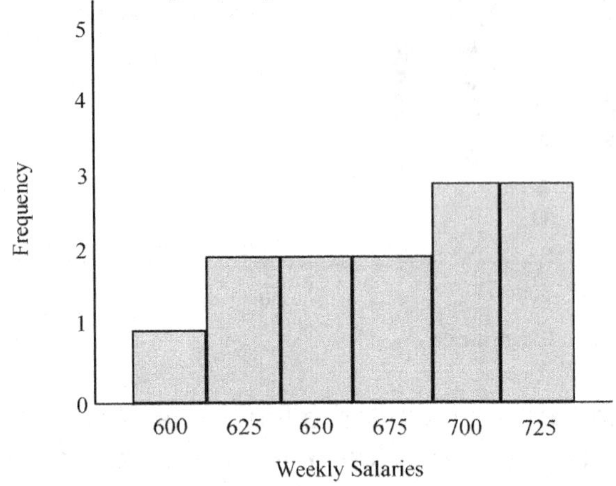

A **pictograph** uses small figures or icons to represent data. Pictographs are used to summarize relative amounts, trends, and data sets. They are useful in comparing quantities.

Monarch Butterfly Migration to the U.S.
in millions

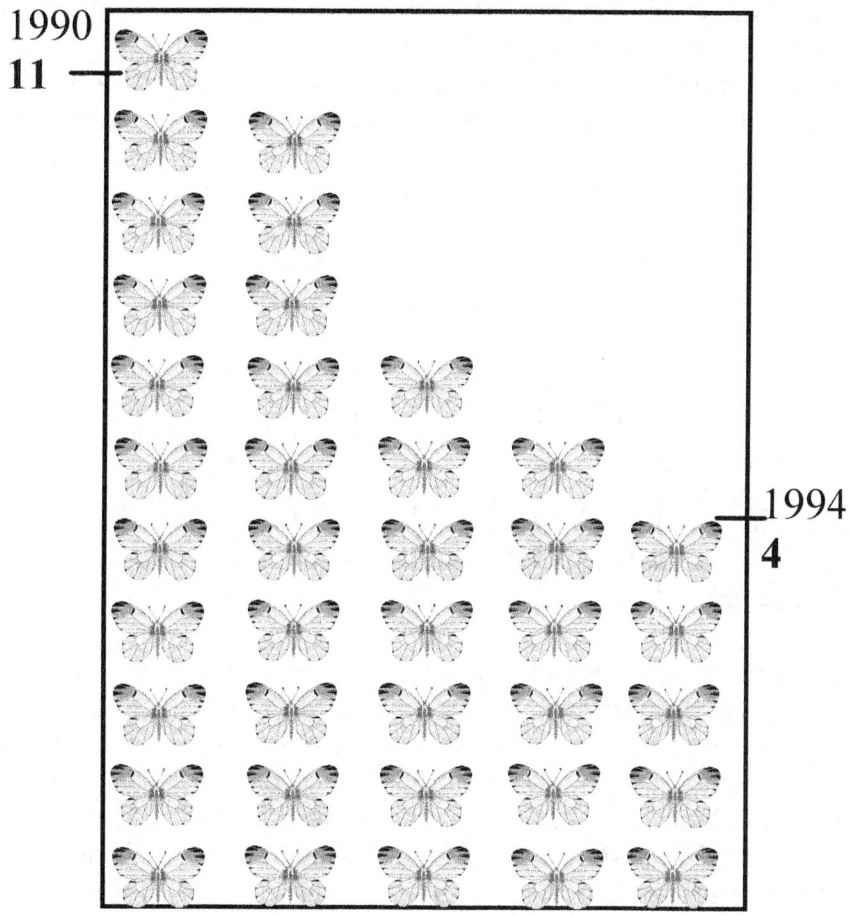

The data in this graph is not accurate. It is for illustration purposes only.

Percentiles divide data into 100 equal parts. A person whose score falls in the 65th percentile has outperformed 65 percent of all those who took the test. This does not mean that the score was 65 percent out of 100 nor does it mean that 65 percent of the questions answered were correct. It means that the grade was higher than 65 percent of all those who took the test.

Stanine "standard nine" scores combine the understandability of percentages with the properties of the normal curve of probability. Stanines divide the bell curve into nine sections, the largest of which stretches from the 40th to the 60th percentile and is the "Fifth Stanine" (the average of taking into account error possibilities).

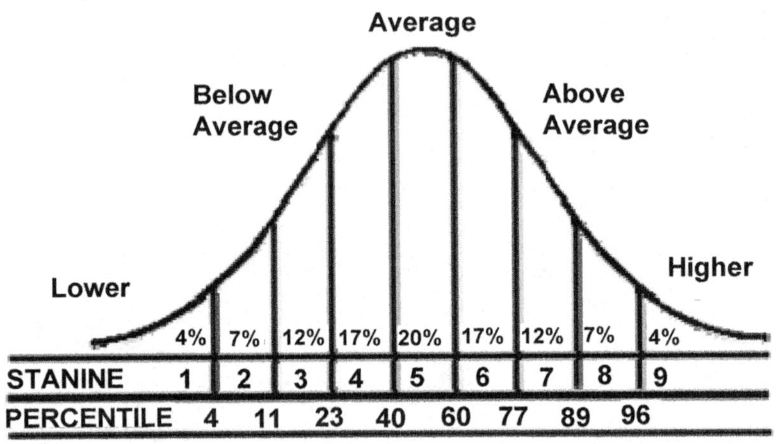

Quartiles divide the data into 4 parts. First find the median of the data set (Q2), then find the median of the upper (Q3) and lower (Q1) halves of the data set. If there are an odd number of values in the data set, include the median value in both halves when finding quartile values. For example, given the data set: {1, 4, 9, 16, 25, 36, 49, 64, 81} first find the median value, which is 25 this is the second quartile. Since there are an odd number of values in the data set (9), we include the median in both halves. To find the quartile values, we much find the medians of: {1, 4, 9, 16, 25} and {25, 36, 49, 64, 81}. Since each of these subsets had an odd number of elements (5), we use the middle value. Thus the first quartile value is 9 and the third quartile value is 49. If the data set had an even number of elements, average the middle two values. The quartile values are always either one of the data points, or exactly half way between two data points.

Sample problem:

1. Given the following set of data, find the percentile of the score 104.

   70, 72, 82, 83, 84, 87, 100, 104, 108, 109, 110, 115

Solution: Find the percentage of scores below 104.

7/12 of the scores are less than 104. This is 58.333%; therefore, the score of 104 is in the 58th percentile.

2. Find the first, second and third quartile for the data listed.

6, 7, 8, 9, 10, 12, 13, 14, 15, 16, 18, 23, 24, 25, 27, 29, 30, 33, 34, 37

Quartile 1: The 1st Quartile is the median of the lower half of the data set, which is 11.
Quartile 2: The median of the data set is the 2nd Quartile, which is 17.

Quartile 3: The 3rd Quartile is the median of the upper half of the data set, which is 28.

An **understanding of the definitions** is important in determining the validity and uses of statistical data. All definitions and applications in this section apply to ungrouped data.

Data item: each piece of data is represented by the letter X.

Mean: the average of all data represented by the symbol $\overline{X}$.

Range: difference between the highest and lowest value of data items.

Sum of the Squares: sum of the squares of the differences between each item and the mean. $Sx^2 = (X - \overline{X})^2$

Variance: the sum of the squares quantity divided by the number of items. (the lower case Greek letter sigma squared (σ^2)represents variance).

$$\frac{Sx^2}{N} = \sigma^2$$

The larger the value of the variance the larger the spread

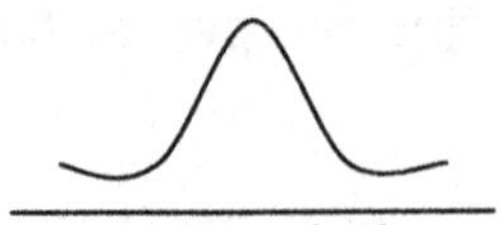

small variation

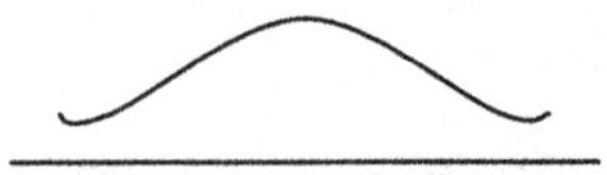

larger variation

Standard Deviation: the square root of the variance. The lower case Greek letter sigma (σ) is used to represent standard deviation.

$$\sigma = \sqrt{\sigma^2}$$

Most statistical calculators have standard deviation keys on them and should be used when asked to calculate statistical functions. It is important to become familiar with the calculator and the location of the keys needed.

Sample Problem:

Given the ungrouped data below, calculate the mean, range, standard deviation and the variance.

15	22	28	25	34	38
18	25	30	33	19	23

Mean ($\overline{X}$) = 25.8333333
Range: $38 - 15 = 23$
standard deviation $(\sigma) = 6.99137$
Variance $(\sigma^2) = 48.87879$

Statisticians use linear transformations of data sets to facilitate comparisons between data sets and equalize aberrant differences at very large and very small values. For example, to compare scores from different intelligence tests we transform raw test scores (i.e. number of correct answers) into scaled scores with uniform means and standard deviations. Linear transformations take the form

$$X' = a + bX_0.$$

X' is the transformed value, X_0 is the original value, *a* is the additive component and *b* is the multiplicative component.

The additive component of the transformation, *a*, shifts the original distribution to the right, if *a* is positive, or to the left, if *a* is negative. The additive component increases or decreases the mean of the distribution by *a* units, but does not affect the standard deviation. The following is an example of an additive transformation of a data set and a generic graphical representation.

a = 20, b = 1

	X_0	X'	
	12	32	
	14	34	
	21	41	
	23	43	
	27	47	
mean	19.4	39.4	→ difference of *a* units
standard deviation	6.27	6.27	

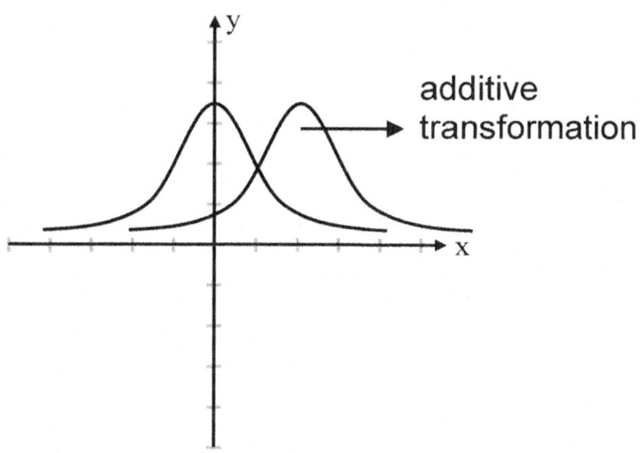

The multiplicative component of a linear transformation, *b*, horizontally stretches the distribution, if *b* is greater than one, and shrinks the distribution, if *b* is between zero and one. The multiplicative component increases or decreases the mean and standard deviation of a data set by a factor of *b*. The following is an example of a multiplicative transformation of a data set and a generic graphical representation.

$a = 0,\ b = 3$

	X_0	X'	
	12	36	
	14	42	
	21	63	
	23	69	
	27	81	
mean	19.4	58.2 →	increase by factor of *b* units
standard deviation	6.27	18.8 →	

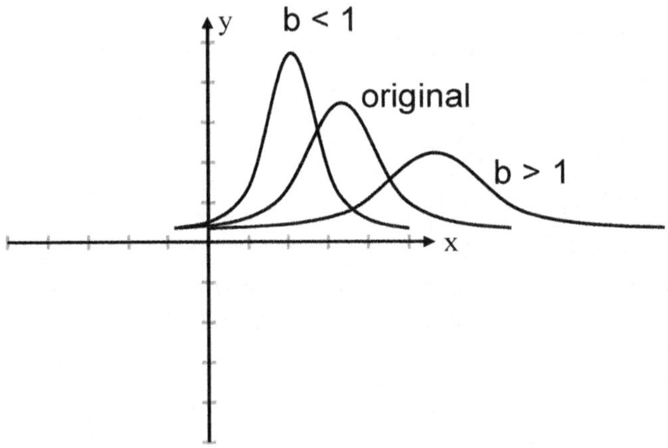

Different situations require different information. If we examine the circumstances under which an ice cream store owner may use statistics collected in the store, we find different uses for different information.

Over a 7-day period, the store owner collected data on the ice cream flavors sold. He found the mean number of scoops sold was 174 per day. The most frequently sold flavor was vanilla. This information was useful in determining how much ice cream to order in all and in what amounts for each flavor.

In the case of the ice cream store, the median and range had little business value for the owner.

Consider the set of test scores from a math class: 0, 16, 19, 65, 65, 65, 68, 69, 70, 72, 73, 73, 75, 78, 80, 85, 88, and 92. The mean is 64.06 and the median is 71. Since there are only three scores less than the mean out of the eighteen scores, the median (71) would be a more descriptive score.

Retail store owners may be most concerned with the most common dress size so they may order more of that size than any other.

Competency 16.0 The teacher understands concepts and applications of probability.

In probability, the **sample space** is a list of all possible outcomes of an experiment. For example, the sample space of tossing two coins is the set {HH, HT, TT, TH}, the sample space of rolling a six-sided die is the set {1, 2, 3, 4, 5, 6}, and the sample space of measuring the height of students in a class is the set of all real numbers {R}.

When conducting experiments with a large number of possible outcomes it is important to determine the size of the sample space. The size of the sample space can be determined by using the fundamental counting principle and the rules of combinations and permutations.

The **fundamental counting principle** states that if there are *m* possible outcomes for one task and *n* possible outcomes of another, there are (*m* x *n*) possible outcomes of the two tasks together.

A **permutation** is the number of possible arrangements of items, without repetition, where order of selection is important.

A **combination** is the number of possible arrangements, without repetition, where order of selection is not important.

Permutations and combinations are covered in detail in Competency 003.

Examples:

1. Find the size of the sample space of rolling two six-sided die and flipping two coins.

   Solution:

   List the possible outcomes of each event:
   each dice: {1, 2, 3, 4, 5, 6}
   each coin: {Heads, Tails}

   Apply the fundamental counting principle:
   size of sample space = 6 x 6 x 2 x 2 = 144

2. Find the size of the sample space of selecting three playing cards at random from a standard fifty-two card deck.

Solution:

Use the rule of combination –

$$_{52}C_3 = \frac{52!}{(52-3)!3!} = 22100$$

The **absolute probability** of some events cannot be determined. For instance, one cannot assume the probability of winning a tennis match is ½ because, in general, winning and losing are not equally likely. In such cases, past results of similar events can be used to help predict future outcomes. The **relative frequency** of an event is the number of times an event has occurred divided by the number of attempts.

$$\text{Relative frequency} = \frac{\text{number of successful trials}}{\text{total number of trials}}$$

For example, if a weighted coin flipped 50 times lands on heads 40 times and tails 10 times, the relative frequency of heads is 40/50 = 4/5. Thus, one can predict that if the coin is flipped 100 times, it will land on heads 80 times.

Example:

Two tennis players, John and David, have played each other 20 times.

John has won 15 of the previous matches and David has won 5.
(a) Estimate the probability that David will win the next match.
(b) Estimate the probability that John will win the next 3 matches.

Solution:

(a) David has won 5 out of 20 matches. Thus, the relative frequency of David winning is 5/20 or ¼. We can estimate that the probability of David winning the next match is ¼.

(b) John has won 15 out of 20 matches. The relative frequency of John winning is 15/20 or ¾. We can estimate that the probability of John winning a future match is ¾. Thus, the probability that John will win the next three matches is ¾ x ¾ x ¾ = 27/64.

Dependent events occur when the probability of the second event depends on the outcome of the first event. For example, consider the two events (A) it is sunny on Saturday and (B) you go to the beach. If you intend to go to the beach on Saturday, rain or shine, then A and B may be independent. If however, you plan to go to the beach only if it is sunny, then A and B may be dependent. In this situation, the probability of event B will change depending on the outcome of event A.

Suppose you have a pair of dice, one red and one green. If you roll a three on the red die and then roll a four on the green die, we can see that these events do not depend on the other. The total probability of the two independent events can be found by multiplying the separate probabilities.

$$\begin{aligned} P(A \text{ and } B) &= P(A) \times P(B) \\ &= 1/6 \times 1/6 \\ &= 1/36 \end{aligned}$$

Many times, however, events are not independent. Suppose a jar contains 12 red marbles and 8 blue marbles. If you randomly pick a red marble, replace it and then randomly pick again, the probability of picking a red marble the second time remains the same. However, if you pick a red marble, and then pick again without replacing the first red marble, the second pick becomes dependent upon the first pick.

$$P(\text{Red and Red}) \text{ with replacement} = P(\text{Red}) \times P(\text{Red})$$

$$\begin{aligned} &= 12/20 \times 12/20 \\ &= 9/25 \end{aligned}$$

$$P(\text{Red and Red}) \text{ without replacement} = P(\text{Red}) \times P(\text{Red})$$

$$\begin{aligned} &= 12/20 \times 11/19 \\ &= 33/95 \end{aligned}$$

Odds are defined as the ratio of the number of favorable outcomes to the number of unfavorable outcomes. The sum of the favorable outcomes and the unfavorable outcomes should always equal the total possible outcomes.

For example, given a bag of 12 red and 7 green marbles compute the odds of randomly selecting a red marble.

Odds of red = $\frac{12}{19}$

Odds of not getting red = $\frac{7}{19}$

In the case of flipping a coin, it is equally likely that a head or a tail will be tossed. The odds of tossing a head are 1:1. This is called even odds.

The Addition Principle of Counting states:

If A and B are events, $n(AorB) = n(A) + n(B) - n(A \cap B)$.

Example:

In how many ways can you select a black card or a Jack from an ordinary deck of playing cards?

Let B denote the set of black cards and let *J* denote the set of Jacks. Then, $n(B) = 26, n(J) = 4, n(B \cap J) = 2$ and

$$
\begin{aligned}
&n(BorJ) = n(B) + n(J) - n(B \cap A) \\
&= 26 + 4 - 2 \\
&= 28.
\end{aligned}
$$

The Addition Principle of Counting for Mutually Exclusive Events states:

If A and B are mutually exclusive events, $n(AorB) = n(A) + n(B)$.

Example:

A travel agency offers 40 possible trips: 14 to Asia, 16 to Europe and 10 to South America. In how many ways can you select a trip to Asia or Europe through this agency?

Let A denote trips to Asia and let E denote trips to Europe. Then, $A \cap E = \varnothing$ and

$$n(AorE) = 14 + 16 = 30.$$

Therefore, the number of ways you can select a trip to Asia or Europe is 30.

The Multiplication Principle of Counting for Dependent Events states:

Let A be a set of outcomes of Stage 1 and B a set of outcomes of Stage 2. Then the number of ways $n(AandB)$, that A and B can occur in a two-stage experiment is given by:

$$n(AandB) = n(A)n(B|A),$$

where $n(B|A)$ denotes the number of ways B can occur given that A has already occurred.

Example:

How many ways from an ordinary deck of 52 cards can two Jacks be drawn in succession if the first card is drawn but not replaced in the deck and then the second card is drawn?

This is a two-stage experiment for which we wish to compute $n(AandB)$, where A is the set of outcomes for which a Jack is obtained on the first draw and B is the set of outcomes for which a Jack is obtained on the second draw.

If the first card drawn is a Jack, then there are only three remaining Jacks left to choose from on the second draw. Thus, drawing two cards without replacement means the events A and B are dependent.

$$n(AandB) = n(A)n(B|A) = 4 \cdot 3 = 12$$

The Multiplication Principle of Counting for Independent Events states:

Let A be a set of outcomes of Stage 1 and B a set of outcomes of Stage 2. If A and B are independent events then the number of ways $n(AandB)$, that A and B can occur in a two-stage experiment is given by:

$$n(AandB) = n(A)n(B).$$

Example:

How many six-letter code "words" can be formed if repetition of letters is not allowed?

Since these are code words, a word does not have to look like a word; for example, abcdef could be a code word.
Since we must choose a first letter *and* a second letter *and* a third letter *and* a fourth letter *and* a fifth letter *and* a sixth letter, this experiment has six stages.

Since repetition is not allowed there are 26 choices for the first letter; 25 for the second; 24 for the third; 23 for the fourth; 22 for the fifth; and 21 for the sixth. Therefore, we have:

n(six-letter code words without repetition of letters)

$$= 26 \cdot 25 \cdot 24 \cdot 23 \cdot 22 \cdot 21$$

$$= 165,765,600$$

The **binomial distribution** is a sequence of probabilities with each probability corresponding to the likelihood of a particular event occurring. It is called a binomial distribution because each trial has precisely two possible outcomes. An **event** is defined as a sequence of Bernoulli trials that has within it a specific number of successes. The order of success is not important.

Note: There are two parameters to consider in a binomial distribution:

1. p = the probability of a success
2. n = the number of Bernoulli trials (i.e., the length of the sequence).

Example:

Toss a coin two times. Each toss is a Bernoulli trial as discussed above. Consider heads to be success. One event is one sequence of two coin tosses. Order does not matter.

There are two possibilities for each coin toss. Therefore, there are four (2·2) possible subevents: 00, 01, 10, 11 (where 0 = tail and 1 = head).

According to the multiplication rule, each subevent has a probability of $\frac{1}{4}\left(\frac{1}{2}\cdot\frac{1}{2}\right)$.

One subevent has zero heads, so the event of zero heads in two tosses is $p(h=0)=\frac{1}{4.}$

Two subevents have one head, so the event of one head in two tosses is $p(h=1)=\frac{2}{4}$.

One subevent has two heads, so the event of two heads in two tosses is $p(h=2)=\frac{1}{4}$.

So the binomial distribution for two tosses of a fair coin is:

$$p(h=0)=\frac{1}{4,}p(h=1)=\frac{2}{4,}p(h=2)=\frac{1}{4}.$$

A **normal distribution** is the distribution associated with most sets of real-world data. It is frequently called a **bell curve**. A normal distribution has a **random variable** X with mean μ and variance σ^2.

Example:

Albert's Bagel Shop's morning customer load follows a normal distribution, with **mean** (average) 50 and **standard deviation** 10. The standard deviation is the measure of the variation in the distribution. Determine the probability that the number of customers tomorrow will be less than 42.

First convert the raw score to a **z-score**. A z-score is a measure of the distance in standard deviations of a sample from the mean.

The z-score = $\frac{X_i=\bar{X}}{s}=\frac{42-50}{10}=\frac{-8}{10}=-.8$

Next, use a table to find the probability corresponding to the z-score. The table gives us .2881. Since our raw score is negative, we subtract the table value from .5.

$$.5-.2881=.2119$$

We can conclude that $P(x < 42) = .2119$. This means that there is about a 21% chance that there will be fewer than 42 customers tomorrow morning.

Example:

The scores on Mr. Rogers' statistics exam follow a normal distribution with mean 85 and standard deviation 5. A student is wondering what the probability is that she will score between a 90 and a 95 on her exam.

We wish to compute $P(90 < x < 95)$.

Compute the z-scores for each raw score.

$$\frac{90-85}{5} = \frac{5}{5} = 1 \quad \text{and} \quad \frac{95-85}{5} = \frac{10}{5} = 2.$$

Now we want $P(1 < z < 2)$.

Since we are looking for an occurrence between two values, we subtract:

$P(1 < z < 2) = P(z < 2) - P(z < 1)$.

We use a table to get:

$P(1 < z < 2) = .9772 - .8413 = .1359$. (Remember that since the z-scores are positive, we add .5 to each probability.)

We can then conclude that there is a 13.6% chance that the student will score between a 90 and a 95 on her exam.

Competency 17.0 The teacher understands the relationships among probability theory, sampling, and statistical inference, and how statistical inference is used in making and evaluating predictions.

Observation-inference is another mathematic process skill that used regularly in statistics. We can use the data gathered or observed from a sample of the population to make inferences about traits and qualities of the population as a whole. For example, if we observe that 40% of voters in our sample favors Candidate A, then we can infer that 40% of the entire voting population favors Candidate A. Successful use of observation-inference depends on accurate observation and representative sampling.

The **t-test** is the most commonly used method to evaluate the difference in means between two groups. The t-test assesses whether the means of two groups are statistically different from each other. The formula for the t-test is a ratio. The top part of the ratio is the difference between the two means or averages. The bottom part is a measure of the variability or dispersion of the scores.

$$\frac{\text{difference between group means}}{\text{variability of groups}} = \frac{\overline{X}_T - \overline{X}_C}{SE(\overline{X}_T - \overline{X}_C)} = \text{t-value}$$

In this example T refers to a treatment group and C refers to a control group.

To compute the top part of the formula, just find the difference between the means. The bottom part of the formula is called the **standard error of the difference**. To compute it, take the variance for each group and divide it by the number of people in that group. Add these two values and then take their square root.

$$SE(\overline{X}_T - \overline{X}_C) = \sqrt{\frac{\text{var}_T}{n_T} + \frac{\text{var}_C}{n_C}}$$

The final formula for the t-test is

$$t = \frac{\overline{X}_T - \overline{X}_C}{\sqrt{\frac{\text{var}_T}{n_T} + \frac{\text{var}_C}{n_C}}}$$

Once you compute the t-value, you have to look it up in a table of significance to test whether the ratio is large enough to say that the difference between the groups is not likely to have been a chance finding. To test the significance, you need to set a risk, or alpha level, which is usually .05. You also must determine the degrees of freedom (df) for the test. In the t-test, the df is the sum of the persons in both groups minus 2. Using the t-value, the alpha level, and the df, you can look up the t-value in a standard table of significance to determine whether the t-value is large enough to be significant. If it is, you can conclude that the difference between the means for the two groups is different.

The **chi-square test** is a method of determining the odds for or against a given deviation from expected statistical distribution.

Example:

We want to determine if the odds of flipping a coin heads-up is the same as tails-up; is the coin flipped fairly? We collect data by flipping the coin 200 times. The coin landed heads-up 92 times and tails-up 108 times.

To perform a chi-square test we first must establish a null hypothesis. In this example, the null hypothesis states that the coin should be equally likely to land head-up or tails-up, every time. The null hypothesis allows us to state expected frequencies. For 200 tosses we would expect 100 heads and 100 tails.

Next, prepare a table:

	Heads	Tails	Total
Observed	92	108	200
Expected	100	100	200
Total	192	208	400

The observed values are the data gathered. The expected values are the frequencies expected, based on the null hypothesis.

We calculate chi-squared:

$$\text{Chi-squared} = \frac{(\text{observed-expected})^2}{(\text{expected})}$$

We have two classes to consider in this example, heads and tails.

$$\text{Chi-squared} = \frac{(100-92)^2}{100} + \frac{(100-108)^2}{100}$$

$$= \frac{(8)^2}{100} + \frac{(-8)^2}{100}$$

$$= 0.64 + 0.64$$

$$= 1.28$$

We then consult a table of critical values of the chi-squared distribution. Here is a portion of such a table:

df/prob.	0.99	0.95	0.90	0.80	0.70	0.50	0.30	0.20	0.10	0.05
1	0.00013	0.0039	0.016	0.64	0.15	0.46	1.07	1.64	2.71	3.84
2	0.02	0.10	0.21	0.45	0.71	1.39	2.41	3.22	4.60	5.99
3	0.12	0.35	0.58	1.99	1.42	2.37	3.66	4.64	6.25	7.82
4	0.3	0.71	1.06	1.65	2.20	3.36	4.88	5.99	7.78	9.49
5	0.55	1.14	1.61	2.34	3.00	4.35	6.06	7.29	9.24	11.07

We determine the **degrees of freedom** (df) by subtracting one from the number of classes. In this example we have two classes (heads and tails), so df is 1. Our chi-squared value is 1.28. In the table our value lies between 1.07 (a probability of .30) and 1.64 (a probability of .20). Interpolation gives us an estimated probability of 0.26. This value means that there is a 74% chance that the coin is biased. Because the chi-squared value we obtained is greater than 0.05 (0.26 to be exact), we accept the null hypothesis as true and conclude that the coin is fair.

Random sampling is the process of studying an aspect of a population by selecting and gathering data from a segment of the population and making inferences and generalizations based on the results. Two main types of random sampling are simple and stratified. With simple random sampling, each member of the population has an equal chance of selection to the sample group. With stratified random sampling, each member of the population has a known but unequal chance of selection to the sample group, as the study selects a random sample from each population demographic. In general, stratified random sampling is more accurate because it provides a more representative sample group. Sample statistics are important generalizations about the entire sample such as mean, median, mode, range, and sampling error (standard deviation). Various factors affect the accuracy of sample statistics and the generalizations made from them about the larger population.

Sample size is one important factor in the accuracy and reliability of sample statistics. As sample size increases, sampling error (standard deviation) decreases. Sampling error is the main determinant of the size of the confidence interval. Confidence intervals decrease in size as sample size increases. A confidence interval gives an estimated range of values, which is likely to include a particular population parameter. The confidence level associated with a confidence interval is the probability that the interval contains the population parameter. For example, a poll reports 60% of a sample group prefers candidate A with a margin of error of $\pm$3% and a confidence level of 95%. In this poll, there is a 95% chance that the preference for candidate A in the whole population is between 57% and 63%.

The ultimate goal of sampling is to make generalizations about a population based on the characteristics of a random sample. Estimators are sample statistics used to make such generalizations. For example, the mean value of a sample is the estimator of the population mean. Unbiased estimators, on average, accurately predict the corresponding population characteristic. Biased estimators, on the other hand, do not exactly mirror the corresponding population characteristic. While most estimators contain some level of bias, limiting bias to achieve accurate projections is the goal of statisticians.

Correlation is a measure of association between two variables. It varies from -1 to 1, with 0 being a random relationship, 1 being a perfect positive linear relationship, and -1 being a perfect negative linear relationship.

The **correlation coefficient** (r) is used to describe the strength of the association between the variables and the direction of the association.

Example:

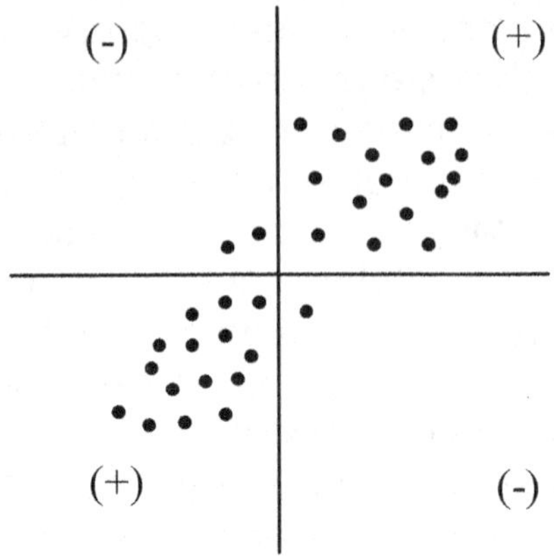

Horizontal and vertical lines are drawn through the point of averages which is the point on the averages of the x and y values. This divides the scatter plot into four quadrants.

If a point is in the lower left quadrant, the product of two negatives is positive; in the upper right, the product of two positives is positive. The positive quadrants are depicted with the positive sign (+). In the two remaining quadrants (upper left and lower right), the product of a negative and a positive is negative. The negative quadrants are depicted with the negative sign (-). If r is positive, then there are more points in the positive quadrants and if r is negative, then there are more points in the two negative quadrants.

Regression is a form of statistical analysis used to predict a dependent variable (y) from values of an independent variable (x). A regression equation is derived from a known set of data.

The simplest regression analysis models the relationship between two variables using the following equation: $y = a + bx$, where y is the dependent variable and x is the independent variable. This simple equation denotes a linear relationship between x and y. This form would be appropriate if, when you plotted a graph of x and y, you tended to see the points roughly form along a straight line.

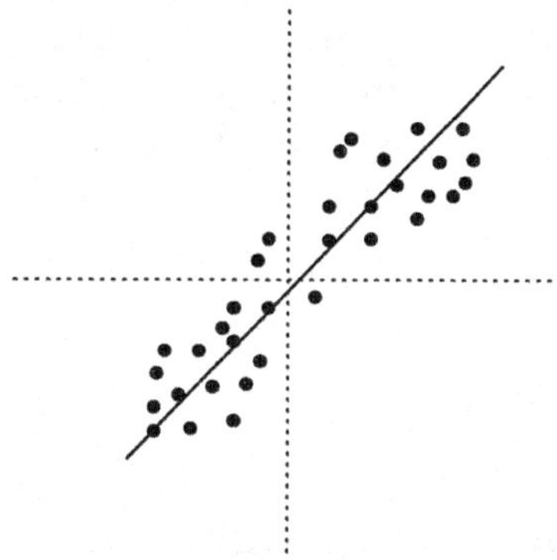

The line can then be used to make predictions.

If all of the data points fell on the line, there would be a perfect correlation ($r = 1.0$) between the x and y data points. These cases represent the best scenarios for prediction. A positive or negative r value represents how y varies with x. When r is positive, y increases as x increases. When r is negative y decreases as x increases.

A **linear regression** equation is of the for: $Y = a + bX$.

Example:

A teacher wanted to determine how a practice test influenced a student's performance on the actual test. The practice test grade and the subsequent actual test grade for each student are given in the table below:

Practice Test (x)	Actual Test (y)
94	98
95	94
92	95
87	89
82	85
80	78
75	73
65	67
50	45
20	40

We determine the equation for the linear regression line to be $y = 14.650 + 0.834x$.

A new student comes into the class and scores 78 on the practice test. Based on the equation obtained above, what would the teacher predict this student would get on the actual test?

$$y = 14.650 + 0.834(78)$$
$$y = 14.650 + 65.052$$
$$y = 80$$

A **logarithmic regression** equation is of the form: $y = a + b\ln x$.

Example:

The water, w, in an open container is evaporating. The number of ounces remaining after h hours is shown in the table below:

Hours (h)	2	5	10	15	19	30
Water (w)	13	11	9	8.5	7.5	6.5

We construct a scatter plot for this data and find a logarithmic regression equation to model the data.

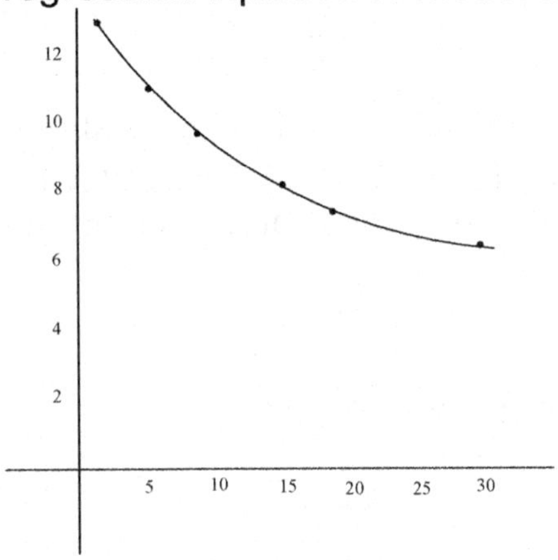

$$w = 14.71 - 2.41\ln h$$

Using this regression equation predict how many ounces of water are remaining in the container after 48 hours.

$$w = 14.71 - 2.41\ln(48)$$
$$w = 14.71 - 2.41(3.87)$$
$$w = 14.71 - 9.33$$
$$w = 5.4$$

An **exponential regression** equation is of the form: $y = a(b)^x$. An exponential function may be used to model either growth or decay, depending on the value of b. When $b > 1$ the function models exponential growth. When $0 < b < 1$ the function models exponential decay. You may fit an exponential curve to data and find the exponential function.

Example:
For the data in the table below, determine the exponential regression equation.

x	y
-5	10.22
1	7.98
2	6.63
5	5.51
8	4.57

$$y = 7.502(.94)^x$$

Using this equation predict what y will be when x = 12.

$$y = 7.502(.94)^{12}$$
$$y = 7.502(.476)$$
$$y = 3.57$$

Power regression is a functional form used for nonlinear regression. The formula is: $Y = ab^X$.

Example:
On a newly discovered planet the weight, w, of an object and the distance, d, this object is from the surface of the planet were recorded and are shown in the table below (Weight is in pounds and distance in miles).

d	w
2	116
3	58
4	36
5	24
6	18

We construct a scatter plot for the given data using distance as the independent variable (x) and find a power regression equation to model this data.

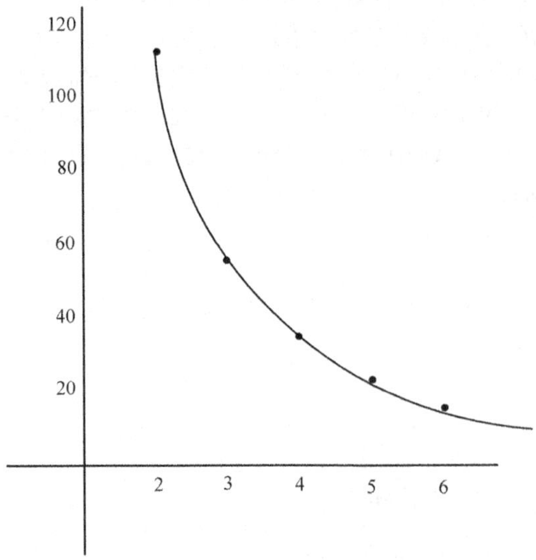

$$w = 377.336d^{-1.702}$$

Using the regression equation we find the predicted weight of the object to the nearest pound when it is 20 miles from the surface of the planet.

$$\begin{aligned} w &= 377.336(20)^{-1.702} \\ &= 377.336(.006) \\ &= 2 \end{aligned}$$

The **law of large numbers and the central limit theorem** are two fundamental concepts in statistics. The law of large numbers states that the larger the sample size, or the more times we measure a variable in a population, the closer the sample mean will be to the population mean. For example, the average weight of 40 apples out of a population of 100 will more closely approximate the population average weight than will a sample of 5 apples. The central limit theorem expands on the law of large numbers. The central limit theorem states that as the number of samples increases, the distribution of sample means (averages) approaches a normal distribution. This holds true regardless of the distribution of the population. Thus, as the number of samples taken increases the sample mean becomes closer to the population mean. This property of statistics allows us to analyze the properties of populations of unknown distribution.

In conclusion, the law of large numbers and central limit theorem show the importance of large sample size and large number of samples to the process of statistical inference. As sample size and the number of samples taken increase, the accuracy of conclusions about the population drawn from the sample data increases.

Basic statistical concepts can be applied without computations. For example, inferences can be drawn from a graph or statistical data. A bar graph could display which grade level collected the most money. Student test scores would enable the teacher to determine which units need to be remediated.

Conditional statements are frequently written in "**if-then**" form. The "if" clause of the conditional is known as the **hypothesis**, and the "then" clause is called the **conclusion**. In a proof, the hypothesis is the information that is assumed to be true, while the conclusion is what is to be proven true. A conditional is considered to be of the form:

If p, then q

p is the hypothesis. q is the conclusion.

Conditional statements can be diagrammed using a **Venn diagram**. A diagram can be drawn with one circle inside another circle. The inner circle represents the hypothesis. The outer circle represents the conclusion. If the hypothesis is taken to be true, then you are located inside the inner circle.

If you are located in the inner circle then you are also inside the outer circle, so that proves the conclusion is true.

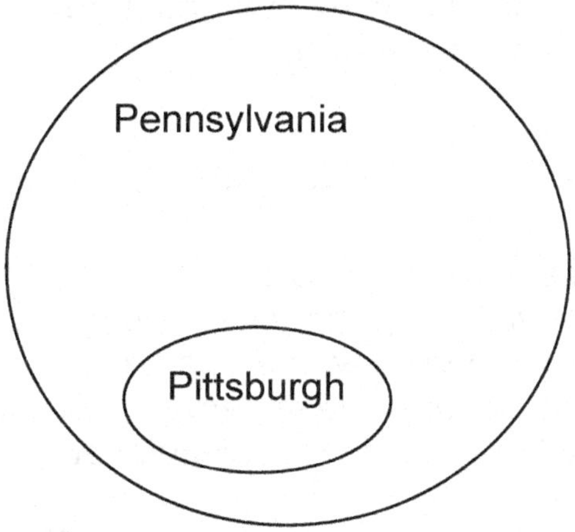

Example:
If an angle has a measure of 90 degrees, then it is a right angle.

In this statement "an angle has a measure of 90 degrees" is the hypothesis.
In this statement "it is a right angle" is the conclusion.

Example:
If you are in Pittsburgh, then you are in Pennsylvania.

In this statement "you are in Pittsburgh" is the hypothesis.
In this statement "you are in Pennsylvania" is the conclusion.

DOMAIN V. MATHEMATICAL PROCESSES AND PERSPECTIVES

Competency 18.0 The teacher understands mathematical reasoning and problem solving.

In **a 2 column proof,** the left side of the proof should be the given information, or statements that could be proved by deductive reasoning. The right column of the proof consists of the reasons used to determine that each statement to the left was verifiably true. The right side can identify given information, or state theorems, postulates, definitions or algebraic properties used to prove that particular line of the proof is true.

Assume the opposite of the conclusion. Keep your hypothesis and given information the same. Proceed to develop the steps of the proof, looking for a statement that contradicts your original assumption or some other known fact. This contradiction indicates that the assumption you made at the beginning of the proof was incorrect; therefore, the original conclusion has to be true.

Proofs by mathematical induction.

Proof by induction states that a statement is true for all numbers if the following two statements can be proven:

1. The statement is true for $n = 1$.
2. If the statement is true for $n = k$, then it is also true for $n = k+1$.

In other words, we must show that the statement is true for a particular value and then we can assume it is true for another, larger value (k). Then, if we can show that the number after the assumed value ($k+1$) also satisfies the statement, we can assume, by induction, that the statement is true for all numbers.

The four basic components of induction proofs are: (1) the statement to be proved, (2) the beginning step ("let $n = 1$"), (3) the assumption step ("let $n = k$ and assume the statement is true for k, and (4) the induction step ("let $n = k+1$").

Example:

Prove that the sum all numbers from 1 to *n* is equal to $\frac{(n)(n+1)}{2}$.

Let *n* = 1. — Beginning step.

Then the sum of 1 to 1 is 1.

And $\frac{(n)(n+1)}{2}$ = 1.

Thus, the statement is true for *n* = 1. — Statement is true in a particular instance.

Assumption:

Let *n* = *k* + 1

k = *n* - 1

Then [1 + 2 +...+ *k*] + (*k*+1) = $\frac{(k)(k+1)}{2}$ + (k+1) — Substitute the assumption.

$= \frac{(k)(k+1)}{2} + \frac{2(k+1)}{2}$ — Common denominator.

$= \frac{(k)(k+1)+2(k+1)}{2}$ — Add fractions.

$= \frac{(k+2)(k+1)}{2}$ — Simplify.

$= \frac{(k+1)+1)(k+1)}{2}$ — Write in terms of *k*+1.

For n = 4 , k = 3

$= \frac{(4+1)(4)}{2} = \frac{20}{2} = 10$

Conclude that the original statement is true for *n* = *k*+1 if it is assumed that the statement is true for *n* = *k*.

Proofs on a coordinate plane

Use proofs on the coordinate plane to prove properties of geometric figures. Coordinate proofs often utilize formulas such as the Distance Formula, Midpoint Formula, and the Slope Formula.

The most important step in coordinate proofs is the placement of the figure on the plane. Place the figure in such a way to make the mathematical calculations as simple as possible.

Example:

1. Prove that the square of the length of the hypotenuse of triangle ABC is equal to the sum of the squares of the lengths of the legs using coordinate geometry.

Draw and label the graph.

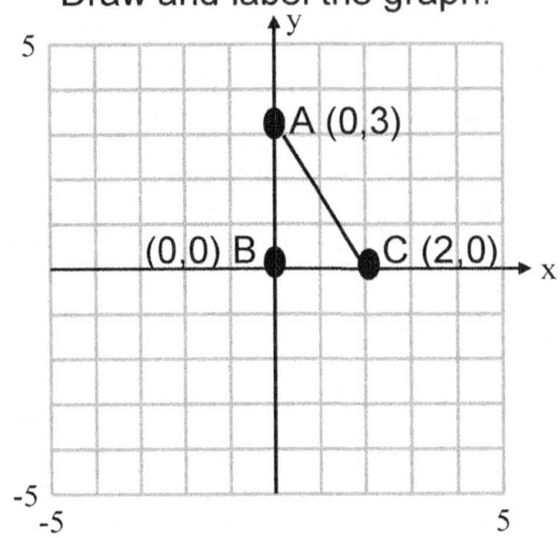

Use the distance formula to find the lengths of the sides of the triangle.

$$d = \sqrt{(x_2 - x_1)^2 + (y_2 - y_1)^2}$$

AB = $\sqrt{3^2} = 3$, BC = $\sqrt{2^2} = 2$, AC = $\sqrt{3^2 + 2^2} = \sqrt{13}$

Conclude

$(AB)^2 + (BC)^2 = 3^2 + 2^2 = 13$
$(AC)^2 = (\sqrt{13})^2 = 13$

Thus, $(AB)^2 + (BC)^2 = (AC)^2$

Conditional statements can be diagrammed using a Venn diagram. A diagram can be drawn with one figure inside another figure. The inner figure represents the hypothesis. The outer figure represents the conclusion. If the hypothesis is taken to be true, then you are located inside the inner figure. If you are located in the inner figure then you are also inside the outer figure, so that proves the conclusion is true. Sometimes that conclusion can then be used as the hypothesis for another conditional, which can result in a second conclusion.

Suppose that these statements were given to you, and you are asked to try to reach a conclusion. The statements are:

All swimmers are athletes.
All athletes are scholars.

In "if-then" form, these would be:
If you are a swimmer, then you are an athlete.
If you are an athlete, then you are a scholar.

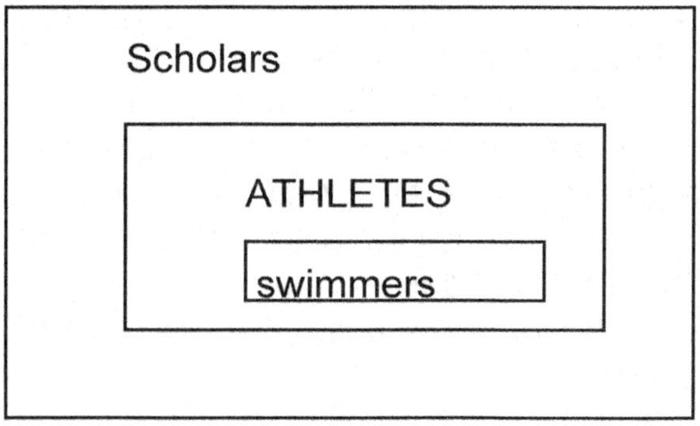

Clearly, if you are a swimmer, then you are also an athlete. This includes you in the group of scholars.

Suppose that these statements were given to you, and you are asked to try to reach a conclusion. The statements are:

All swimmers are athletes.
All wrestlers are athletes.

In "if-then" form, these would be:
If you are a swimmer, then you are an athlete.
If you are a wrestler, then you are an athlete.

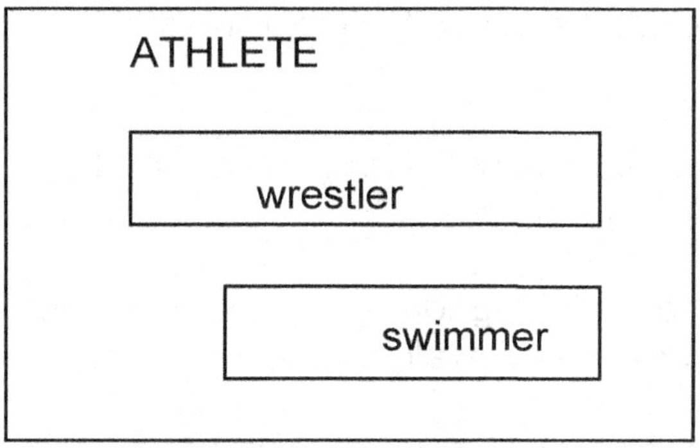

Clearly, if you are a swimmer or a wrestler, then you are also an athlete. This does NOT allow you to come to any other conclusions.

A swimmer may or may NOT also be a wrestler. Therefore, NO CONCLUSION IS POSSIBLE.

Suppose that these statements were given to you, and you are asked to try to reach a conclusion. The statements are:

All rectangles are parallelograms.
Quadrilateral ABCD is not a parallelogram.

In "if-then" form, the first statement would be:

If a figure is a rectangle, then it is also a parallelogram.

Note that the second statement is the negation of the conclusion of statement one. Remember also that the contrapositive is logically equivalent to a given conditional. That is, **"If $\sim$ q, then $\sim$ p"**. Since" ABCD is NOT a parallelogram " is like saying **"If $\sim$ q,"** then you can come to the conclusion **"then $\sim$ p".** Therefore, the conclusion is ABCD is not a rectangle. Looking at the Venn diagram below, if all rectangles are parallelograms, then rectangles are included as part of the parallelograms. Since quadrilateral ABCD is not a parallelogram, that it is excluded from anywhere inside the parallelogram box. This allows you to conclude that ABCD can not be a rectangle either.

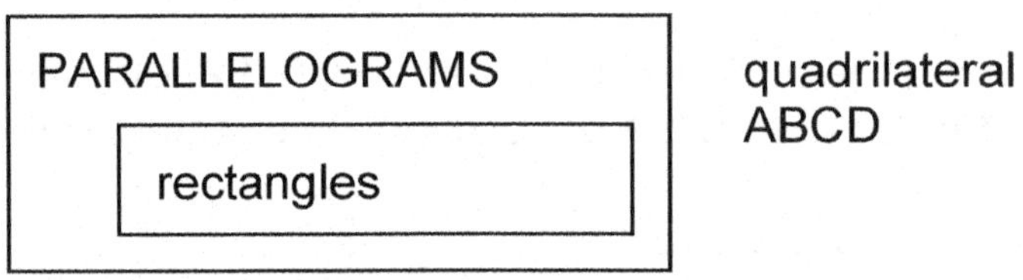

Try These:

What conclusion, if any, can be reached? Assume each statement is true, regardless of any personal beliefs.

1. If the Red Sox win the World Series, I will die.
   I died.

2. If an angle's measure is between 0° and 90°, then the angle is acute. Angle B is not acute.

3. Students who do well in geometry will succeed in college.
   Annie is doing extremely well in geometry.

4. Left-handed people are witty and charming.
   You are left-handed.

A **counterexample** is an exception to a proposed rule or conjecture that disproves the conjecture. For example, the existence of a single non-brown dog disproves the conjecture "all dogs are brown". Thus, any non-brown dog is a counterexample.

In searching for mathematic counterexamples, one should consider extreme cases near the ends of the domain of an experiment and special cases where an additional property is introduced. Examples of extreme cases are numbers near zero and obtuse triangles that are nearly flat. An example of a special case for a problem involving rectangles is a square because a square is a rectangle with the additional property of symmetry.

Example:

Identify a counterexample for the following conjectures.

1. If n is an even number, then $n + 1$ is divisible by 3.

   $n = 4$
   $n + 1 = 4 + 1 = 5$
   5 is not divisible by 3.

2. If n is divisible by 3, then $n^2 - 1$ is divisible by 4.

   $n = 6$
   $n^2 - 1 = 6^2 - 1 = 35$
   35 is not divisible by 4.

Successful math teachers introduce their students to multiple problem solving strategies and create a classroom environment where free thought and experimentation are encouraged. Teachers can promote problem solving by allowing multiple attempts at problems, giving credit for reworking test or homework problems, and encouraging the sharing of ideas through class discussion. There are several specific problem solving skills with which teachers should be familiar.

The **guess-and-check** strategy calls for students to make an initial guess at the solution, check the answer, and use the outcome of to guide the next guess. With each successive guess, the student should get closer to the correct answer. Constructing a table from the guesses can help organize the data.

Example:

There are 100 coins in a jar. 10 are dimes. The rest are pennies and nickels. There are twice as many pennies as nickels. How many pennies and nickels are in the jar?

There are 90 total nickels and pennies in the jar (100 coins – 10 dimes).

There are twice as many pennies as nickels. Make guesses that fulfill the criteria and adjust based on the answer found. Continue until we find the correct answer, 60 pennies and 30 nickels.

Number of Pennies	Number of Nickels	Total Number of Pennies and Nickels
40	20	60
80	40	120
70	35	105
60	30	90

When solving a problem where the final result and the steps to reach the result are given, students must **work backwards** to determine what the starting point must have been.

Example:

John subtracted seven from his age, and divided the result by 3. The final result was 4. What is John's age?

> Work backward by reversing the operations.
> 4 x 3 = 12;
> 12 + 7 = 19
> John is 19 years old.

Estimation and testing for **reasonableness** are related skills students should employ prior to and after solving a problem. These skills are particularly important when students use calculators to find answers.

Example:

Find the sum of 4387 + 7226 + 5893.

4300 + 7200 + 5800 = 17300 Estimation.
4387 + 7226 + 5893 = 17506 Actual sum.

By comparing the estimate to the actual sum, students can determine that their answer is reasonable.

Competency 19.0 **The teacher understands mathematical connections both within and outside of mathematics and how to communicate mathematical ideas and concepts.**

Students of mathematics must be able to recognize and interpret the different representations of arithmetic operations.

First, there are many different verbal descriptions for the operations of addition, subtraction, multiplication, and division. The table below identifies several words and/or phrases that are often used to denote the different arithmetic operations.

Operation	**Descriptive Words**
Addition	"plus", "combine", "sum", "total", "put together"
Subtraction	"minus", "less", "take away", "difference"
Multiplication	"product", "times", "groups of"
Division	"quotient", "into", "split into equal groups",

Second, diagrams of arithmetic operations can present mathematical data in visual form. For example, we can use the number line to add and subtract.

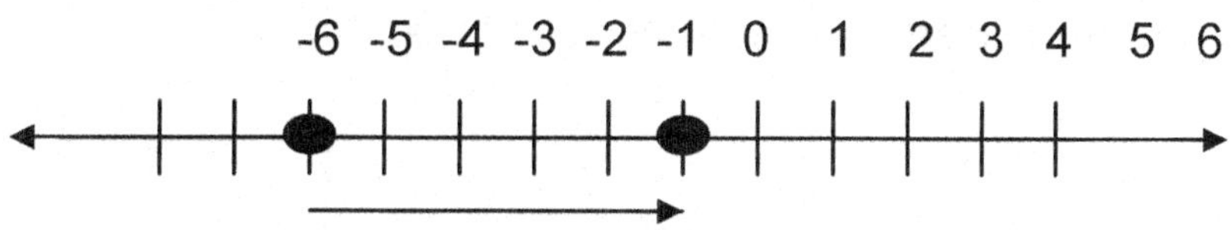

The addition of 5 to -4 on the number line; -4 + 5 = 1.

Finally, as shown in the examples below, we can use pictorial representations to explain all of the arithmetic processes.

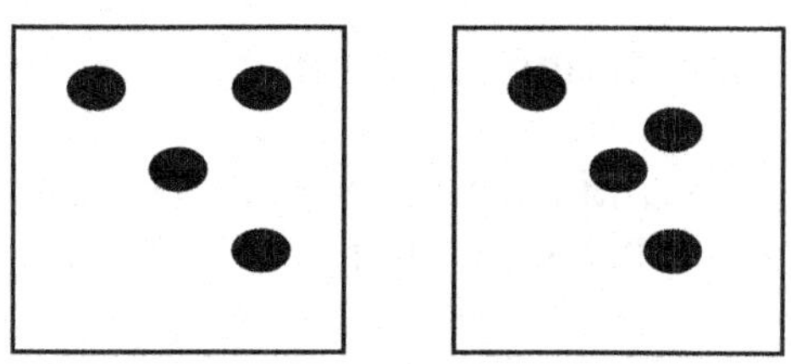

Two groups of four equals eight or 2 x 4 = 8 shown in picture form.

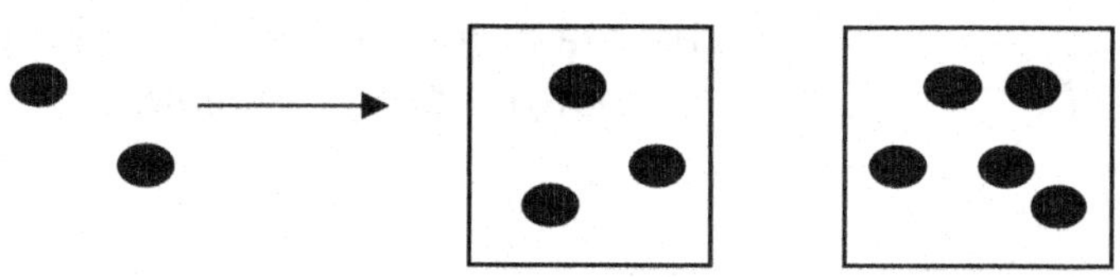

Adding three objects to two or 3 + 2 = 5 shown in picture form.

Examples, illustrations, and symbolic representations are useful tools in explaining and understanding mathematical concepts. The ability to create examples and alternative methods of expression allows students to solve real world problems and better communicate their thoughts.

Concrete examples are real world applications of mathematical concepts. For example, measuring the shadow produced by a tree or building is a real world application of trigonometric functions, acceleration or velocity of a car is an application of derivatives, and finding the volume or area of a swimming pool is a real world application of geometric principles.

Pictorial illustrations of mathematic concepts help clarify difficult ideas and simplify problem solving.

Examples:

1. Rectangle R represents the 300 students in School A. Circle P represents the 150 students that participated in band. Circle Q represents the 170 students that participated in a sport. 70 students participated in both band and a sport.

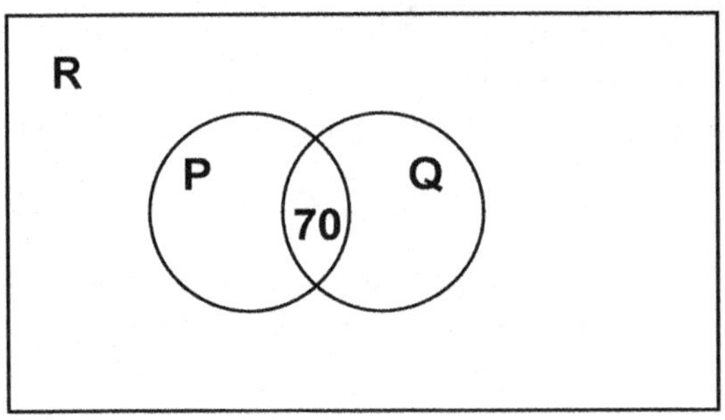

Pictorial representation of above situation.

2. A ball rolls up an incline and rolls back to its original position. Create a graph of the velocity of the ball.

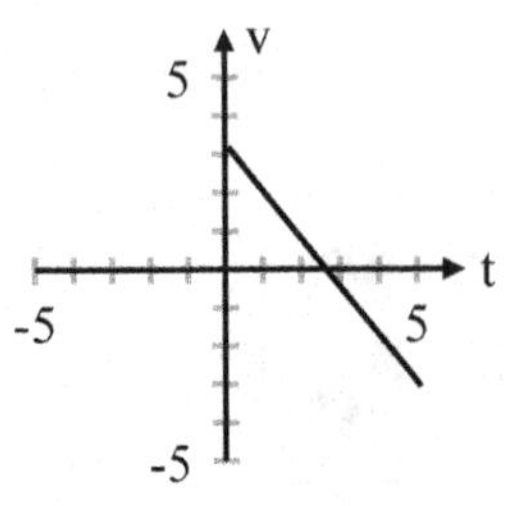

Velocity starts out at its maximum as the ball begins to roll, decreases to zero at the top of the incline, and returns to the maximum in the opposite direction at the bottom of the incline.

Symbolic representation is the basic language of mathematics. Converting data to symbols allows for easy manipulation and problem solving. Students should have the ability to recognize what the symbolic notation represents and convert information into symbolic form. For example, from the graph of a line, students should have the ability to determine the slope and intercepts and derive the line's equation from the observed data. Another possible application of symbolic representation is the formulation of algebraic expressions and relations from data presented in word problem form.

Interpolate information

To make a **bar graph** or a **pictograph**, determine the scale to be used for the graph. Then determine the length of each bar on the graph or determine the number of pictures needed to represent each item of information. Be sure to include an explanation of the scale in the legend.

Example: A class had the following grades:
4 A's, 9 B's, 8 C's, 1 D, 3 F's.
Graph these on a bar graph and a pictograph.

Pictograph

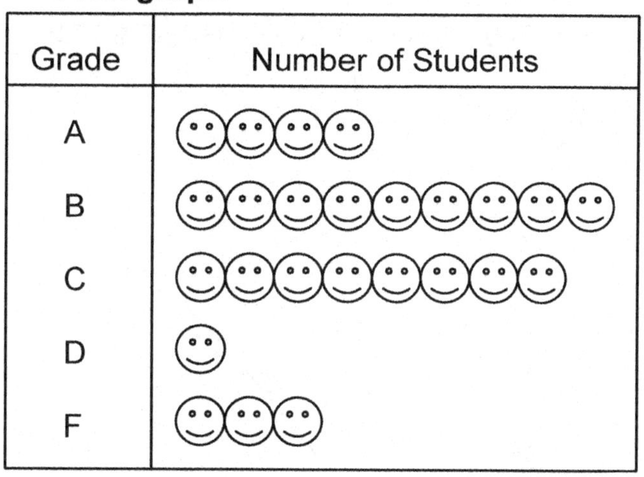

Bar graph

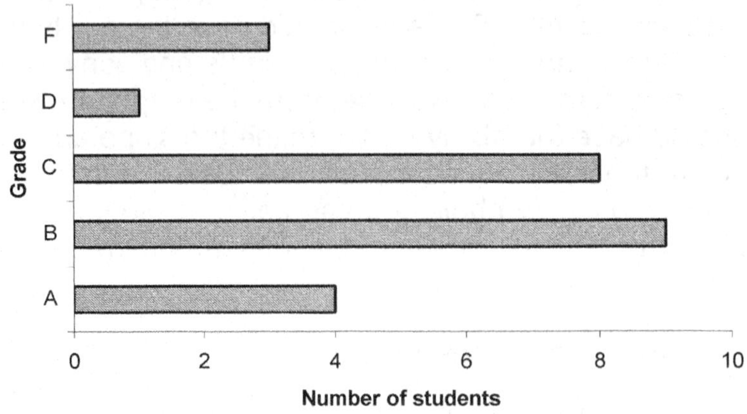

To make a **line graph**, determine appropriate scales for both the vertical and horizontal axes (based on the information to be graphed). Describe what each axis represents and mark the scale periodically on each axis. Graph the individual points of the graph and connect the points on the graph from left to right.

Example: Graph the following information using a line graph.

The number of National Merit finalists/school year

	90-'91	91-'92	92-'93	93-'94	94-'95	95-'96
Central	3	5	1	4	6	8
Wilson	4	2	3	2	3	2

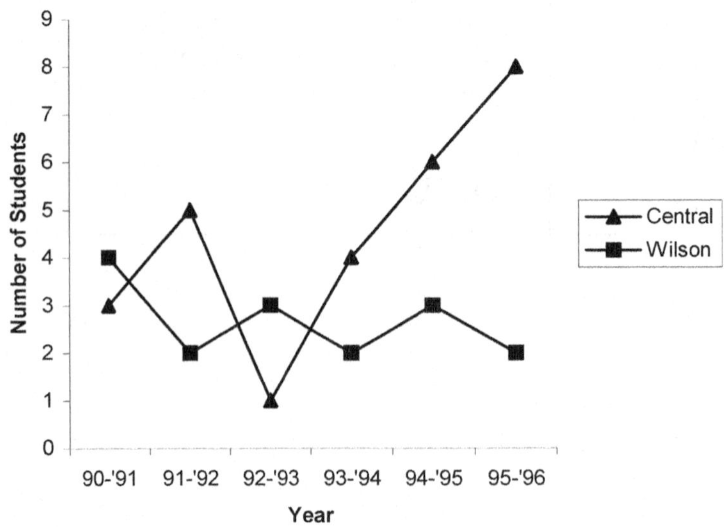

To make a **circle graph**, total all the information that is to be included on the graph. Determine the central angle to be used for each sector of the graph using the following formula:

$$\frac{\text{information}}{\text{total information}} \times 360^{\circ} = \text{degrees in central } \measuredangle$$

Lay out the central angles to these sizes, label each section and include its percent.

<u>Example:</u> Graph this information on a circle graph:

Monthly expenses:

Rent, \$400
Food, \$150
Utilities, \$75
Clothes, \$75
Church, \$100
Misc., \$200

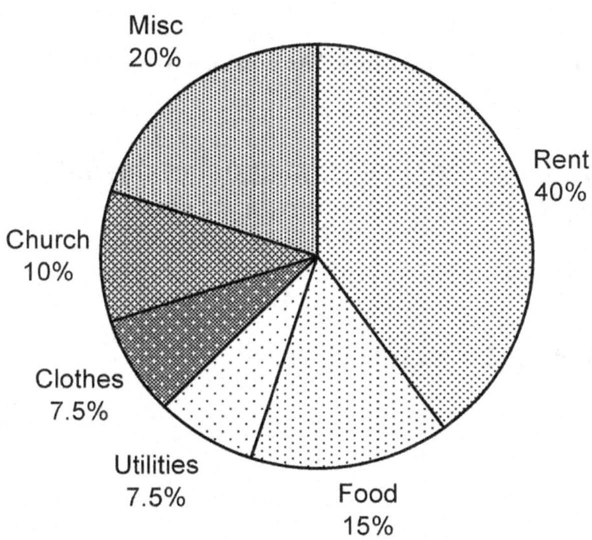

To read a bar graph or a pictograph, read the explanation of the scale that was used in the legend. Compare the length of each bar with the dimensions on the axes and calculate the value each bar represents. On a pictograph count the number of pictures used in the chart and calculate the value of all the pictures.

To read a circle graph, find the total of the amounts represented on the entire circle graph. To determine the actual amount that each sector of the graph represents, multiply the percent in a sector times the total amount number.

To read a chart read the row and column headings on the table. Use this information to evaluate the given information in the chart.

Because mathematics problems and concepts are often presented in written form students must have the ability to **interpret written presentations** and reproduce the concepts in symbolic form to facilitate manipulation and problem solving. Correct interpretation requires a sound understanding of the vocabulary of mathematics.

There are many types of written presentations of mathematics and the following are but two examples.

Examples:

1. The square of the hypotenuse of a right triangle is equivalent to the sum of the squares of the two legs.

$$a^2 + b^2 = c^2$$

2. Find the velocity of an object at time *t* given the objects position function is $f(t) = t^2 - 8t + 9$.

   The velocity at a given time (*t*) is equal to the value of the derivative of the position function at *t*. Thus...

   $$v(t) = f'(t) = 2t - 8$$

   The velocity after *t* seconds is $2t - 8$.

DOMAIN VI. MATHEMATICAL LEARNING, INSTRUCTION, AND ASSESSMENT

Competency 20.0 The teacher understands how children learn mathematics and plans, organizes, and implements instruction using knowledge of students, subject matter, and statewide curriculum (Texas Essential Knowledge and Skills [TEKS]).

The use of supplementary materials in the classroom can greatly enhance the learning experience by stimulating student interest and satisfying different learning styles. Manipulatives, models, and technology are examples of tools available to teachers.

Manipulatives are materials that students can physically handle and move. Manipulatives allow students to understand mathematic concepts by allowing them to see concrete examples of abstract processes. Manipulatives are attractive to students because they appeal to the students' visual and tactile senses. Available for all levels of math, manipulatives are useful tools for reinforcing operations and concepts. They are not, however, a substitute for the development of sound computational skills.

Models are another means of representing mathematical concepts by relating the concepts to real-world situations. Teachers must choose wisely when devising and selecting models because, to be effective, models must be applied properly. For example, a building with floors above and below ground is a good model for introducing the concept of negative numbers. It would be difficult, however, to use the building model in teaching subtraction of negative numbers.

Finally, there are many forms of **technology** available to math teachers. For example, students can test their understanding of math concepts by working on skill specific computer programs and websites. Graphing calculators can help students visualize the graphs of functions. Teachers can also enhance their lectures and classroom presentations by creating multimedia presentations.

Questioning Techniques

As the teacher's role in the classroom changes from lecturer to facilitator, the questions need to further stimulate students in various ways.

- Helping students work together
  What do you think about what John said?
  Do you agree? Disagree?
  Can anyone explain that differently?

- Helping students determine for themselves if an answer is correct
  What do you think that is true?
  How did you get that answer?
  Do you think that is reasonable? Why?

- Helping students learn to reason mathematically
  Will that method always work?
  Can you think of a case where it is not true?
  How can you prove that?
  Is that answer true in all cases?

- Helping student brainstorm and problem solve
  Is there a pattern?
  What else can you do?
  Can you predict the answer?
  What if...?

- Helping students connect mathematical ideas
  What did we learn before that is like this?
  Can you give an example?
  What math did you see on television last night? in the newspaper?

Competency 21.0 The teacher understands assessment and uses a variety of formal and informal assessment techniques to monitor and guide mathematics instruction and to evaluate student progress.

In addition to the traditional methods of performance assessment like multiple choice, true/false, and matching tests, there are many other methods of student assessment available to teachers. Alternative assessment is any type of assessment in which students create a response rather than choose an answer. It is sometimes know as formative assessment, due to the emphasis placed on feedback and the flow of communication between teacher and student. It is the opposite of summative assessment, which occurs periodically and consists of temporary interaction between teacher and student.

Short response and **essay** questions are alternative methods of performance assessment. In responding to such questions, students must utilize verbal, graphical, and mathematical skills to construct answers to problems. These multi-faceted responses allow the teacher to examine more closely a student's problem solving and reasoning skills.

Student **portfolios** are another method of alternative assessment. In creating a portfolio, students collect samples and drafts of their work, self-assessments, and teacher evaluations over a period of time. Such a collection allows students, parents, and teachers to evaluate student progress and achievements. In addition, portfolios provide insight into a student's thought process and learning style.

Projects, **demonstrations**, and **oral presentations** are means of alternative assessment that require students to use different skills than those used on traditional tests. Such assessments require higher order thinking, creativity, and the integration of reasoning and communication skills. The use of predetermined rubrics, with specific criteria for performance assessment, is the accepted method of evaluation for projects, demonstrations, and presentations.

ANSWER KEY TO PRACTICE PROBLEMS

COMPETENCY 4.0

page 21

Question #1 $S_5 = 75$

Question #2 $S_n = 28$

Question #3 $S_n = -\frac{-31122}{15625} \approx {}^{-}1.99$

COMPETENCY 5.0

page 24

Question #1 Domain = ${}^{-}\infty, \infty$ Range = ${}^{-}6, \infty$

Question #2 Domain = 1,4,7,6 Range = -2

Question #3 Domain = $x \neq 2, {}^{-}2$

Question #4 Domain = ${}^{-}\infty, \infty$ Range = -4, 4

Domain = ${}^{-}\infty, \infty$ Range = $2, \infty$

Question #5 Domain = ${}^{-}\infty, \infty$ Range = 5

Question #6 (3,9), (-4,16), (6,3), (1,9), (1,3)

COMPETENCY 6.0

page 35

Question #1 $y = \frac{3}{4}x + \frac{17}{4}$

Question #2 $x = 11$

Question #3 $y = \frac{3}{5}x + \frac{42}{5}$

Question #4 $y = 5$

page 36

Question #1 $\begin{pmatrix} x \\ y \end{pmatrix} = \begin{pmatrix} 3 \\ 1 \end{pmatrix}$

Question #2 $\begin{pmatrix} x \\ y \\ z \end{pmatrix} = \begin{pmatrix} 4 \\ 4 \\ 1 \end{pmatrix}$

page 40

Question #1 $x^2 - 10x + 25$

Question #2 $25x^2 - 10x - 48$

Question #3 $x^2 - 9x - 36$

COMPETENCY 7.0

page 47

Question #1

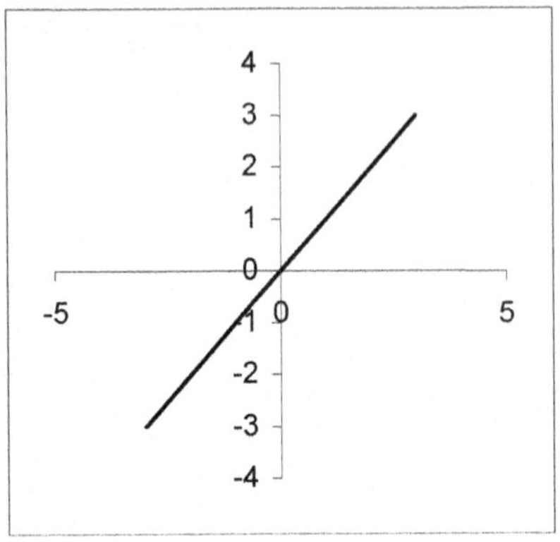

Question #2

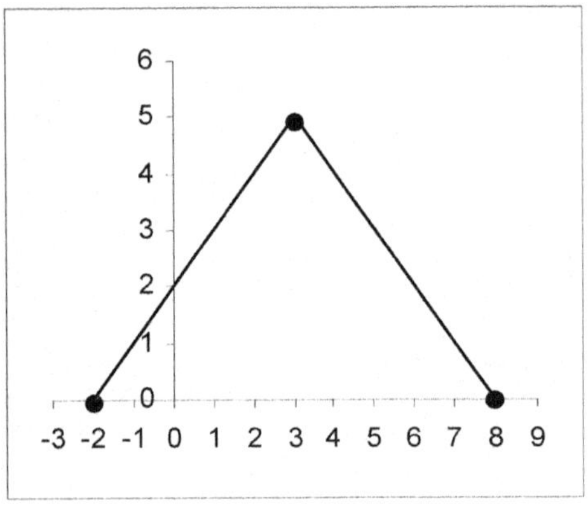

Question #3

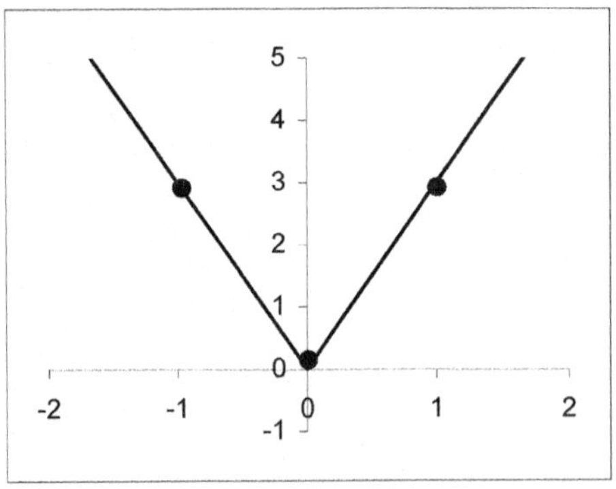

Questions #4

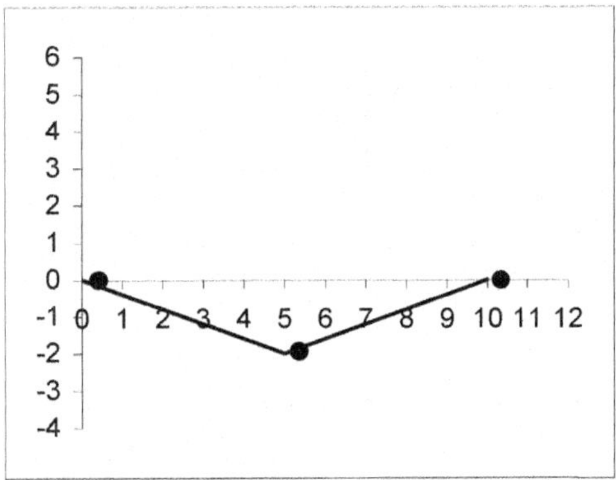

page 48

Question #1

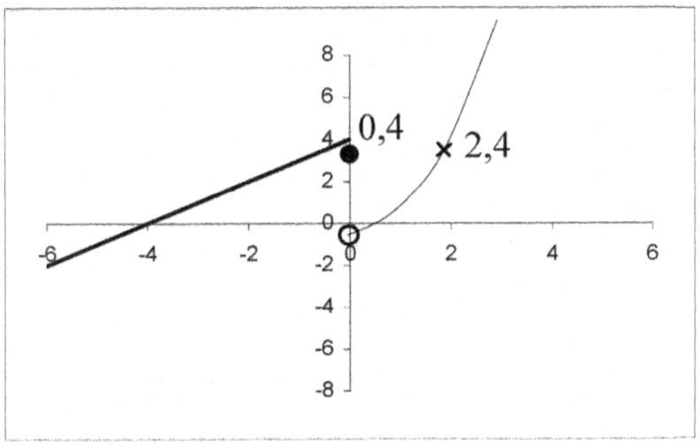

Question #2

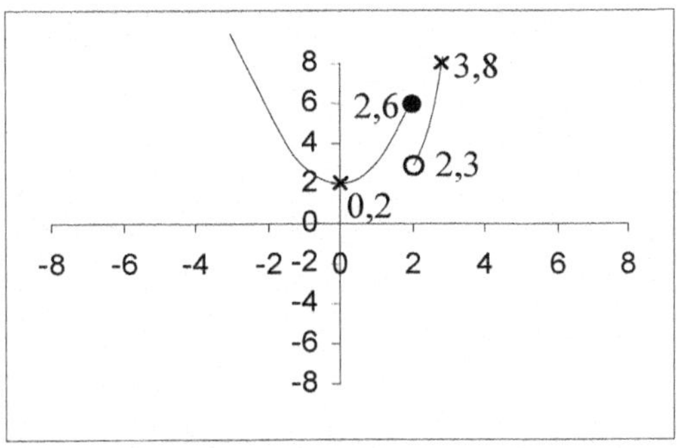

page 63

Question #1 It takes Curly 15 minutes to paint the elephant alone
Question #2 The original number is 5/15
Question #3 The car was traveling at 68mph and the truck was traveling at 62mph

COMPETENCY 8.0

page 69

Question #1 $x = 9$
Question #2 x = 3 or x = 6
Question #3 1

COMPETENCY 9.0

page 72

Question #1

$$\cot\theta = \frac{x}{y}$$

$$\frac{x}{y} = \frac{x}{r} \times \frac{r}{y} = \frac{x}{y} = \cot\theta$$

Question #2

$$1 + \cot^2\theta = \csc^2\theta$$

$$\frac{y^2}{y^2} + \frac{x^2}{y^2} = \frac{r^2}{y^2} = \csc^2\theta$$

COMPETENCY 10.0

page 80

Question #1 49.34
Question #2 1

page 82

Question #1 ∞
Question #2 -1

page 92

Question #1 $t(0) = -24$ m/sec
Question #2 $t(4) = 24$ m/sec

COMPETENCY 18.0

page 193

Question #1 The Red Sox won the World Series.
Question #2 Angle B is not between 0 and 90 degrees.
Question #3 Annie will do well in college.
Question #4 You are witty and charming.

Sample Test

1. **Change $.\overline{63}$ into a fraction in simplest form.**

A) 63/100
B) 7/11
C) 6 3/10
D) 2/3

2. **Which of the following sets is closed under division?**

I) {½, 1, 2, 4}
II) {-1, 1}
III) {-1, 0, 1}

A) I only
B) II only
C) III only
D) I and II

3. **Which of the following illustrates an inverse property?**

A) a + b = a - b
B) a + b = b + a
C) a + 0 = a
D) a + (-a) =0

4. $f(x) = 3x - 2;\ f^{-1}(x) =$

A) $3x + 2$
B) $x/6$
C) $2x - 3$
D) $(x + 2)/3$

5. **What would be the total cost of a suit for $295.99 and a pair of shoes for $69.95 including 6.5% sales tax?**

A) $389.73
B) $398.37
C) $237.86
D) $315.23

6. **A student had 60 days to appeal the results of an exam. If the results were received on March 23, what was the last day that the student could appeal?**

A) May 21
B) May 22
C) May 23
D) May 24

7. **Which of the following is always composite if *x* is odd, *y* is even, and both *x* and *y* are greater than or equal to 2?**

A) $x + y$
B) $3x + 2y$
C) $5xy$
D) $5x + 3y$

8. **Which of the following is incorrect?**

A) $(x^2y^3)^2 = x^4y^6$
B) $m^2(2n)^3 = 8m^2n^3$
C) $(m^3n^4)/(m^2n^2) = mn^2$
D) $(x + y^2)^2 = x^2 + y^4$

9. **Express .0000456 in scientific notation.**

A) $4.56x10^{-4}$
B) $45.6x10^{-6}$
C) $4.56x10^{-6}$
D) $4.56x10^{-5}$

10. **Compute the area of the shaded region, given a radius of 5 meters. 0 is the center.**

    A) 7.13 cm²
    B) 7.13 m²
    C) 78.5 m²
    D) 19.63 m²

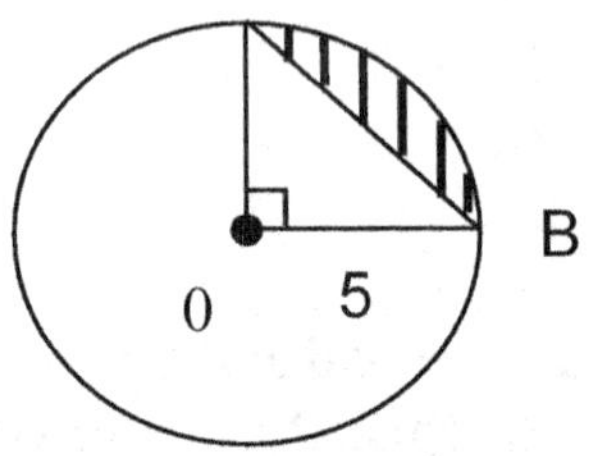

11. **If the area of the base of a cone is tripled, the volume will be**

    A) the same as the original
    B) 9 times the original
    C) 3 times the original
    D) 3 π times the original

12. **Find the area of the figure pictured below.**

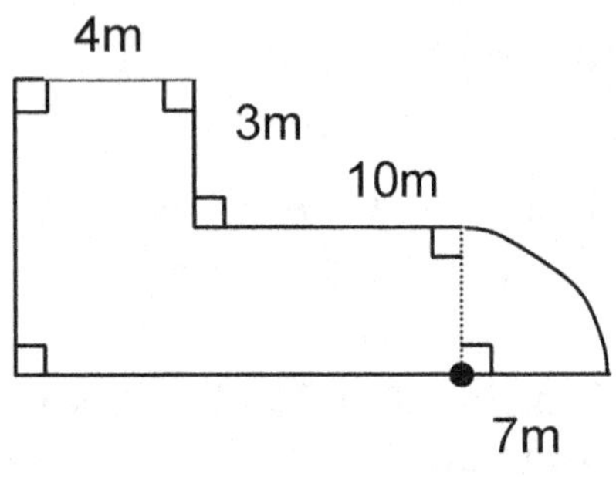

    A) 136.47 m²
    B) 148.48 m²
    C) 293.86 m²
    D) 178.47 m²

13. **The mass of a Chips Ahoy cookie would be to**

    A) 1 kilogram
    B) 1 gram
    C) 15 grams
    D) 15 milligrams

14. **Compute the median for the following data set:**

    {12, 19, 13, 16, 17, 14}

    A) 14.5
    B) 15.17
    C) 15
    D) 16

15. **Half the students in a class scored 80% on an exam, most of the rest scored 85% except for one student who scored 10%. Which would be the best measure of central tendency for the test scores?**

    A) mean
    B) median
    C) mode
    D) either the median or the mode because they are equal

16. What conclusion can be drawn from the graph below?

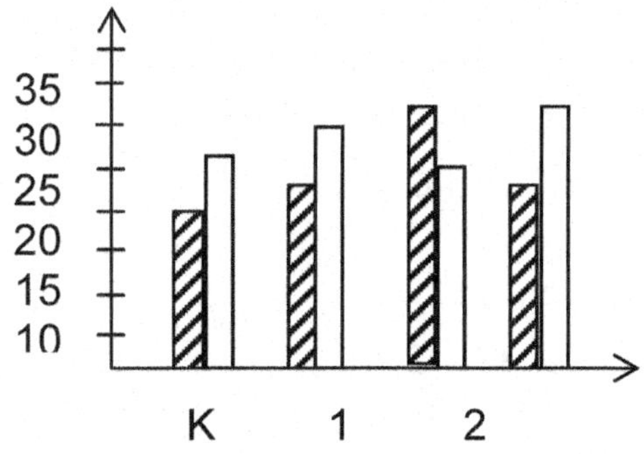

MLK Elementary Student Enrollment **Girls** **Boys**

A) The number of students in first grade exceeds the number in second grade.
B) There are more boys than girls in the entire school.
C) There are more girls than boys in the first grade.
D) Third grade has the largest number of students.

17) State the domain of the function $f(x)=\frac{3x-6}{x^2-25}$

A) $x \neq 2$
B) $x \neq 5, -5$
C) $x \neq 2, -2$
D) $x \neq 5$

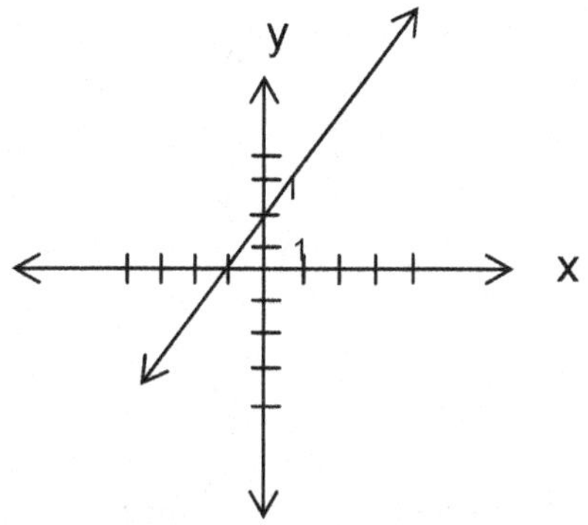

18. What is the equation of the above graph?

A) $2x+y=2$
B) $2x-y=-2$
C) $2x-y=2$
D) $2x+y=-2$

19. Solve for v_0 **:** $d=at(v_t-v_0)$

A) $v_0 = atd - v_t$
B) $v_0 = d - atv_t$
C) $v_0 = atv_t - d$
D) $v_0 = (atv_t - d)/at$

20. Which of the following is a factor of $6+48m^3$

A) (1 + 2m)
B) (1 - 8m)
C) (1 + m - 2m)
D) (1 - m + 2m)

21. Which graph represents the equation of $y = x^2 + 3x$?

A)

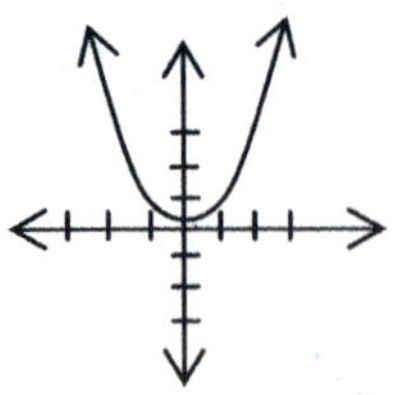

B)

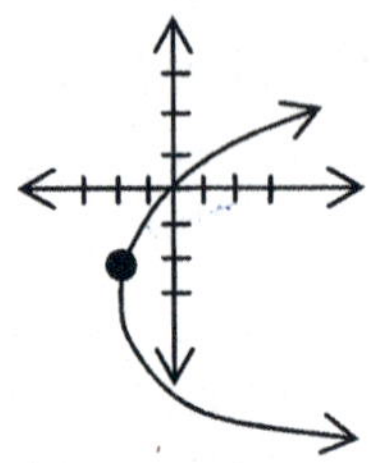

C)

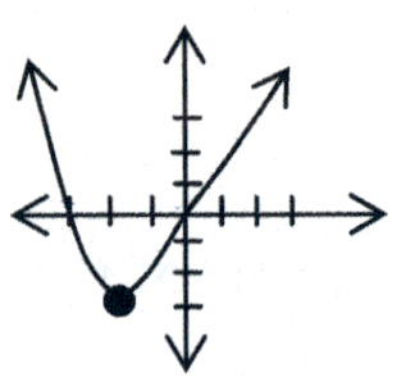

D)

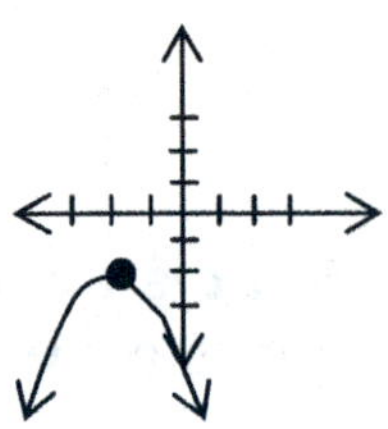

22. The volume of water flowing through a pipe varies directly with the square of the radius of the pipe. If the water flows at a rate of 80 liters per minute through a pipe with a radius of 4 cm, at what rate would water flow through a pipe with a radius of 3 cm?

A) 45 liters per minute
B) 6.67 liters per minute
C) 60 liters per minute
D) 4.5 liters per minute

23) Solve the system of equations for x, y **and** z.

$$3x + 2y - z = 0$$
$$2x + 5y = 8z$$
$$x + 3y + 2z = 7$$

A) $(-1,\ 2,\ 1)$
B) $(1,\ 2,\ -1)$
C) $(-3,\ 4,\ -1)$
D) $(0,\ 1,\ 2)$

24. Solve for x: $18 = 4 + |2x|$

A) $\{-11, 7\}$
B) $\{-7, 0, 7\}$
C) $\{-7, 7\}$
D) $\{-11, 11\}$

25. Which graph represents the solution set for $x^2 - 5x > -6$?

A)

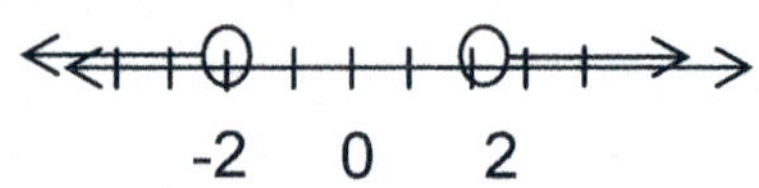

B)

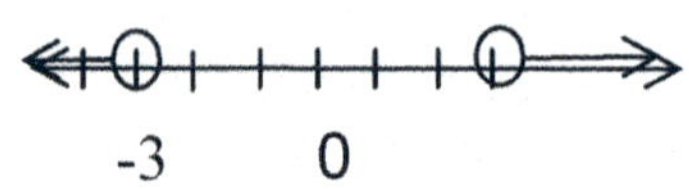

C)

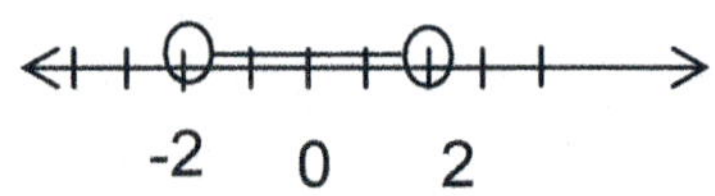

D)

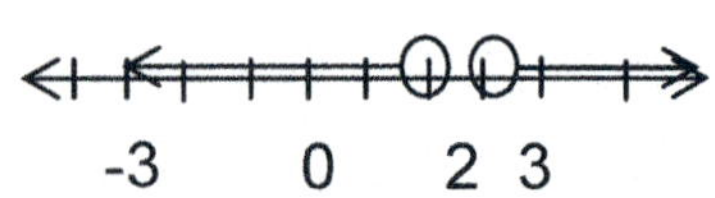

26. Find the zeroes of $f(x) = x^3 + x^2 - 14x - 24$

A) 4, 3, 2
B) 3, -8
C) 7, -2, -1
D) 4, -3, -2

27. Evaluate $3^{1/2}(9^{1/3})$

A) $27^{5/6}$
B) $9^{7/12}$
C) $3^{5/6}$
D) $3^{6/7}$

28. Simplify: $\sqrt{27} + \sqrt{75}$

A) $8\sqrt{3}$
B) 34
C) $34\sqrt{3}$
D) $15\sqrt{3}$

29. Simplify: $\dfrac{10}{1+3i}$

A) $-1.25(1-3i)$
B) $1.25(1+3i)$
C) $1+3i$
D) $1-3i$

30. Find the sum of the first one hundred terms in the progression. (-6, -2, 2 . . .)

A) 19,200
B) 19,400
C) -604
D) 604

31. How many ways are there to choose a potato and two green vegetables from a choice of three potatoes and seven green vegetables?

A) 126
B) 63
C) 21
D) 252

32. What would be the seventh term of the expanded binomial $(2a+b)^8$**?**

A) $2ab^7$
B) $41a^4b^4$
C) $112a^2b^6$
D) $16ab^7$

33. Which term most accurately describes two coplanar lines without any common points?

A) perpendicular
B) parallel
C) intersecting
D) skew

34. Determine the number of subsets of set *K*. *K* = {4, 5, 6, 7}

A) 15
B) 16
C) 17
D) 18

35. What is the degree measure of an interior angle of a regular 10 sided polygon?

A) 18°
B) 36°
C) 144°
D) 54°

36. If a ship sails due south 6 miles, then due west 8 miles, how far was it from the starting point?

A) 100 miles
B) 10 miles
C) 14 miles
D) 48 miles

37. What is the measure of minor arc AD, given measure of arc PS is 40° and $m<K=10^{\circ}$?

A) 50°
B) 20°
C) 30°
D) 25°

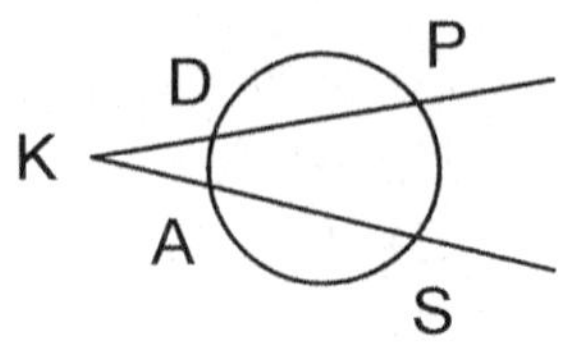

38. Choose the diagram which illustrates the construction of a perpendicular to the line at a given point on the line.

A)

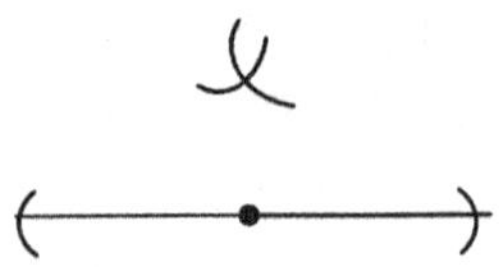

B)

C)

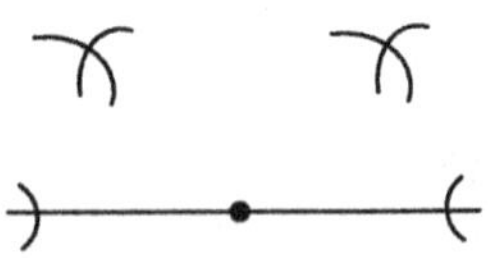

D)

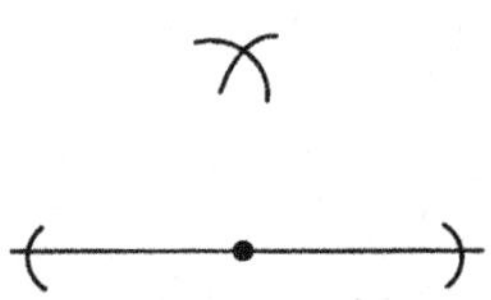

39. When you begin by assuming the conclusion of a theorem is false, then show that through a sequence of logically correct steps you contradict an accepted fact, this is known as

A) inductive reasoning
B) direct proof
C) indirect proof
D) exhaustive proof

40. Which theorem can be used to prove $\Delta BAK \cong \Delta MKA$?

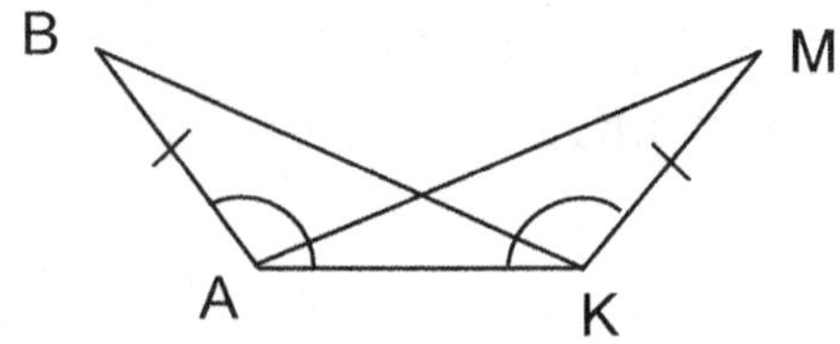

A) SSS
B) ASA
C) SAS
D) AAS

41. Given that QO⊥NP and QO=NP, quadrilateral NOPQ can most accurately be described as a

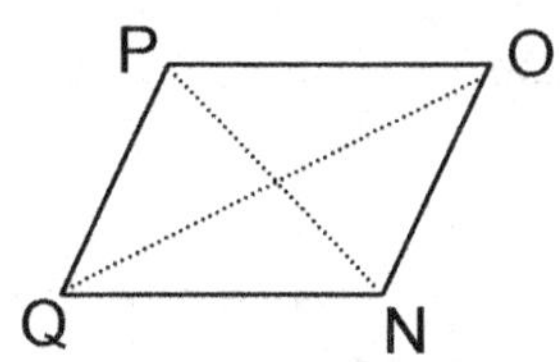

A) parallelogram
B) rectangle
C) square
D) rhombus

42. Choose the correct statement concerning the median and altitude in a triangle.

A) The median and altitude of a triangle may be the same segment.
B) The median and altitude of a triangle are always different segments.
C) The median and altitude of a right triangle are always the same segment.
D) The median and altitude of an isosceles triangle are always the same segment.

43. Which mathematician is best known for his work in developing non-Euclidean geometry?

A) Descartes
B) Riemann
C) Pascal
D) Pythagoras

44. Find the surface area of a box which is 3 feet wide, 5 feet tall, and 4 feet deep.

A) 47 sq. ft.
B) 60 sq. ft.
C) 94 sq. ft
D) 188 sq. ft.

45. Given a 30 meter x 60 meter garden with a circular fountain with a 5 meter radius, calculate the area of the portion of the garden not occupied by the fountain.

A) 1721 m²
B) 1879 m²
C) 2585 m²
D) 1015 m²

46. Determine the area of the shaded region of the trapezoid in terms of *x* and *y*.

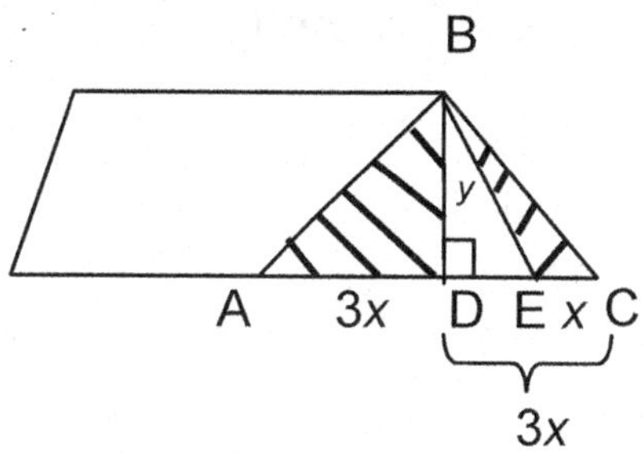

A) $4xy$
B) $2xy$
C) $3x^2y$
D) There is not enough information given.

47. Compute the standard deviation for the following set of temperatures. (37, 38, 35, 37, 38, 40, 36, 39)

A) 37.5
B) 1.5
C) 0.5
D) 2.5

48. Find the value of the determinant of the matrix.

$$\begin{vmatrix} 2 & 1 & -1 \\ 4 & -1 & 4 \\ 0 & -3 & 2 \end{vmatrix}$$

A) 0
B) 23
C) 24
D) 40

49. Which expression is equivalent to $1-\sin^2 x$?

A) $1-\cos^2 x$
B) $1+\cos^2 x$
C) $1/\sec x$
D) $1/\sec^2 x$

50. Determine the measures of angles A and B.

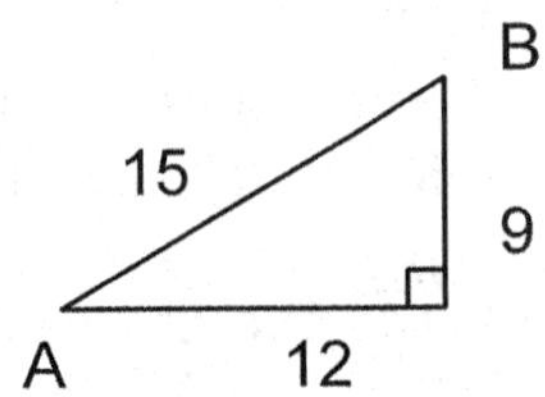

A) A = 30°, B = 60°
B) A = 60°, B = 30°
C) A = 53°, B = 37°
D) A = 37°, B = 53°

51. Given $f(x)=3x-2$ and $g(x)=x^2$, determine $g(f(x))$.

A) $3x^2-2$
B) $9x^2+4$
C) $9x^2-12x+4$
D) $3x^3-2$

52. Determine the rectangular coordinates of the point with polar coordinates (5, 60°).

A) (0.5, 0.87)
B) (-0.5, 0.87)
C) (2.5, 4.33)
D) (25, 150°)

53. Given a vector with horizontal component 5 and vertical component 6, determine the length of the vector.

A) 61
B) $\sqrt{61}$
C) 30
D) $\sqrt{30}$

54. Compute the distance from (-2,7) to the line *x* = 5.

A) -9
B) -7
C) 5
D) 7

55. Given $K(-4,y)$ and $M(2,-3)$ with midpoint $L(x,1)$, determine the values of x and y.

A) $x=-1,\ y=5$
B) $x=3,\ y=2$
C) $x=5,\ y=-1$
D) $x=-1,\ y=-1$

56. **Find the length of the major axis of $x^2 + 9y^2 = 36$.**

A) 4
B) 6
C) 12
D) 8

57. **Which equation represents a circle with a diameter whose endpoints are $(0,7)$ and $(0,3)$?**

A) $x^2 + y^2 + 21 = 0$
B) $x^2 + y^2 - 10y + 21 = 0$
C) $x^2 + y^2 - 10y + 9 = 0$
D) $x^2 - y^2 - 10y + 9 = 0$

58. **Which equation corresponds to the logarithmic statement: $\log_x k = m$?**

A) $x^m = k$
B) $k^m = x$
C) $x^k = m$
D) $m^x = k$

59. **Find the first derivative of the function: $f(x) = x^3 - 6x^2 + 5x + 4$**

A) $3x^3 - 12x^2 + 5x = f'(x)$
B) $3x^2 - 12x - 5 = f'(x)$
C) $3x^2 - 12x + 9 = f'(x)$
D) $3x^2 - 12x + 5 = f'(x)$

60. **Differentiate: $y = e^{3x+2}$**

A) $3e^{3x+2} = y'$
B) $3e^{3x} = y'$
C) $6e^3 = y'$
D) $(3x+2)e^{3x+1} = y'$

61. **Find the slope of the line tangent to $y = 3x(\cos x)$ at $(\pi/2, \pi/2)$.**

A) $-3\pi/2$
B) $3\pi/2$
C) $\pi/2$
D) $-\pi/2$

62. **Find the equation of the line tangent to $y = 3x^2 - 5x$ at $(1,-2)$.**

A) $y = x - 3$
B) $y = 1$
C) $y = x + 2$
D) $y = x$

63. **How does the function $y = x^3 + x^2 + 4$ behave from $x = 1$ to $x = 3$?**

A) increasing, then decreasing
B) increasing
C) decreasing
D) neither increasing nor decreasing

64. Find the absolute maximum obtained by the function $y = 2x^2 + 3x$ on the interval $x = 0$ to $x = 3$.

A) $-3/4$
B) $-4/3$
C) 0
D) 27

65. Find the antiderivative for $4x^3 - 2x + 6 = y$.

A) $x^4 - x^2 + 6x + C$
B) $x^4 - 2/3x^3 + 6x + C$
C) $12x^2 - 2 + C$
D) $\frac{4}{3}x^4 - x^2 + 6x + C$

66. Find the antiderivative for the function $y = e^{3x}$.

A) $3x(e^{3x}) + C$
B) $3(e^{3x}) + C$
C) $1/3(e^x) + C$
D) $1/3(e^{3x}) + C$

67. The acceleration of a particle is dv/dt = 6 m/s². Find the velocity at t=10 given an initial velocity of 15 m/s.

A) 60 m/s
B) 150 m/s
C) 75 m/s
D) 90 m/s

68. If the velocity of a body is given by v = 16 - t², find the distance traveled from t = 0 until the body comes to a complete stop.

A) 16
B) 43
C) 48
D) 64

69. Evaluate: $\int (x^3 + 4x - 5)dx$

A) $3x^2 + 4 + C$
B) $\frac{x^4}{4} + 2x^2 - 5x + C$
C) $x^{4/3} + 4x - 5x + C$
D) $x^3 + 4x^2 - 5x + C$

70. Evaluate $\int_0^2 (x^2 + x - 1)dx$

A) 11/3
B) 8/3
C) -8/3
D) -11/3

71. Find the area under the function $y = x^2 + 4$ from $x = 3$ to $x = 6$.

A) 75
B) 21
C) 96
D) 57

72. -3 + 7 = -4 6(-10) = - 60
-5(-15) = 75 -3+-8 = 11
8-12 = -4 7- -8 = 15

Which best describes the type of error observed above?

A) The student is incorrectly multiplying integers.
B) The student has incorrectly applied rules for adding integers to subtracting integers.
C) The student has incorrectly applied rules for multiplying integers to adding integers.
D) The student is incorrectly subtracting integers.

73. **About two weeks after introducing formal proofs, several students in your geometry class are having a difficult time remembering the names of the postulates. They cannot complete the reason column of the proof and as a result are not even attempting the proofs. What would be the best approach to help students understand the nature of geometric proofs?**

A) Give them more time; proofs require time and experience.
B) Allow students to write an explanation of the theorem in the reason column instead of the name.
C) Have the student copy each theorem in a notebook.
D) Allow the students to have open book tests.

74. **What would be the shortest method of solution for the system of equations below?**

$$3x + 2y = 38$$
$$4x + 8 = y$$

A) linear combination
B) additive inverse
C) substitution
D) graphing

75. **Identify the correct sequence of subskills required for solving and graphing inequalities involving absolute value in one variable, such as $|x+1| \leq 6$.**

A) understanding absolute value, graphing inequalities, solving systems of equations
B) graphing inequalities on a Cartesian plane, solving systems of equations, simplifying expressions with absolute value
C) plotting points, graphing equations, graphing inequalities
D) solving equations with absolute value, solving inequalities, graphing conjunctions and disjunctions

76. What would be the least appropriate use for handheld calculators in the classroom?

A) practice for standardized tests
B) integrating algebra and geometry with applications
C) justifying statements in geometric proofs
D) applying the law of sines to find dimensions

77. According to Piaget, what stage in a student's development would be appropriate for introducing abstract concepts in geometry?

A) concrete operational
B) formal operational
C) sensori-motor
D) pre-operational

78. A group of students working with trigonometric identities have concluded that $\cos 2x = 2\cos x$. How could you best lead them to discover their error?

A) Have the students plug in values on their calculators.
B) Direct the student to the appropriate chapter in the text.
C) Derive the correct identity on the board.
D) Provide each student with a table of trig identities.

79. Which of the following is the best example of the value of personal computers in advanced high school mathematics?

A) Students can independently drill and practice test questions.
B) Students can keep an organized list of theorems and postulates on a word processing program.
C) Students can graph and calculate complex functions to explore their nature and make conjectures.
D) Students are better prepared for business because of mathematics computer programs in high school.

80. Given the series of examples below, what is 5¢4?

4¢3=13	7¢2=47
3¢1=8	1¢5=-4

A) 20
B) 29
C) 1
D) 21

Answer Key

1)	B	17)	B	33)	B	49)	D	65)	A
2)	B	18)	B	34)	B	50)	D	66)	D
3)	D	19)	D	35)	C	51)	C	67)	C
4)	D	20)	A	36)	B	52)	C	68)	B
5)	A	21)	C	37)	B	53)	B	69)	B
6)	B	22)	A	38)	D	54)	D	70)	B
7)	C	23)	A	39)	C	55)	A	71)	A
8)	D	24)	C	40)	C	56)	C	72)	C
9)	D	25)	D	41)	C	57)	B	73)	B
10)	B	26)	D	42)	A	58)	A	74)	C
11)	C	27)	B	43)	B	59)	D	75)	D
12)	B	28)	A	44)	C	60)	A	76)	C
13)	C	29)	D	45)	A	61)	A	77)	B
14)	C	30)	A	46)	B	62)	A	78)	C
15)	B	31)	A	47)	B	63)	B	79)	C
16)	B	32)	C	48)	C	64)	D	80)	D

Rationales for Sample Questions

1. Let N = .636363.... Then multiplying both sides of the equation by 100 or 10^2 (because there are 2 repeated numbers), we get 100N = 63.636363... Then subtracting the two equations gives 99N = 63 or N = $\frac{63}{99}=\frac{7}{11}$. **Answer is B**

2. I is not closed because $\frac{4}{.5}=8$ and 8 is not in the set.

   III is not closed because $\frac{1}{0}$ is undefined.

   II is closed because $\frac{-1}{1}=-1, \frac{1}{-1}=-1, \frac{1}{1}=1, \frac{-1}{-1}=1$ and all the answers are in the set. **Answer is B**

3. **Answer is D** because a + (-a) = 0 is a statement of the Additive Inverse Property of Algebra.

4. To find the inverse, $f^{-1}(x)$, of the given function, reverse the variables in the given equation, y = 3x – 2, to get x = 3y – 2. Then solve for y as follows: x+2 = 3y, and y = $\frac{x+2}{3}$. **Answer is D.**

5. Before the tax, the total comes to \$365.94. Then .065(365.94) = 23.79. With the tax added on, the total bill is 365.94 + 23.79 = \$389.73. (Quicker way: 1.065(365.94) = 389.73.) **Answer is A**

6. Recall: 30 days in April and 31 in March. 8 days in March + 30 days in April + 22 days in May brings him to a total of 60 days on May 22. **Answer is B.**

7. A composite number is a number which is not prime. The prime number sequence begins 2,3,5,7,11,13,17,.... To determine which of the expressions is <u>always</u> composite, experiment with different values of x and y, such as x=3 and y=2, or x=5 and y=2. It turns out that 5xy will always be an even number, and therefore, composite, if y=2.
Answer is C.

8. Using FOIL to do the expansion, we get $(x + y^2)^2 = (x + y^2)(x + y^2) = x^2 + 2xy^2 + y^4$.
Answer is D.

9. In scientific notation, the decimal point belongs to the right of the 4, the first significant digit. To get from 4.56×10^{-5} back to 0.0000456, we would move the decimal point 5 places to the left.
Answer is D.

10. Area of triangle AOB is .5(5)(5) = 12.5 square meters. Since $\frac{90}{360} = .25$, the area of sector AOB (pie-shaped piece) is approximately $.25(\pi)5^2 = 19.63$. Subtracting the triangle area from the sector area to get the area of segment AB, we get approximately 19.63-12.5 = 7.13 square meters.
Answer is B.

11. The formula for the volume of a cone is $V = \frac{1}{3}Bh$, where B is the area of the circular base and h is the height. If the area of the base is tripled, the volume becomes
$V = \frac{1}{3}(3B)h = Bh$, or three times the original area. **Answer is C.**

12. Divide the figure into 2 rectangles and one quarter circle. The tall rectangle on the left will have dimensions 10 by 4 and area 40. The rectangle in the center will have dimensions 7 by 10 and area 70. The quarter circle will have area $.25(\pi)7^2 = 38.48$.
The total area is therefore approximately 148.48. **Answer is B.**

13. Since an ordinary cookie would not weigh as much as 1 kilogram, or as little as 1 gram or 15 milligrams, the only reasonable answer is 15 grams. **Answer is C.**

14. Arrange the data in ascending order: 12,13,14,16,17,19. The median is the middle value in a list with an odd number of entries. When there is an even number of entries, the median is the mean of the two center entries. Here the average of 14 and 16 is 15. **Answer is C.**

15. In this set of data, the median (see #14) would be the most representative measure of central tendency since the median is independent of extreme values. Because of the 10% outlier, the mean (average) would be disproportionately skewed. In this data set, it is true that the median and the mode (number which occurs most often) are the same, but the median remains the best choice because of its special properties. **Answer is B.**

16. In Kindergarten, first grade, and third grade, there are more boys than girls. The number of extra girls in grade two is more than made up for by the extra boys in all the other grades put together. **Answer is B.**

17. The values of 5 and –5 must be omitted from the domain of all real numbers because if x took on either of those values, the denominator of the fraction would have a value of 0, and therefore the fraction would be undefined. **Answer is B.**

18. By observation, we see that the graph has a y-intercept of 2 and a slope of 2/1 = 2. Therefore its equation is y = mx + b = 2x + 2. Rearranging the terms gives
2x – y = -2. **Answer is B.**

19. Using the Distributive Property and other properties of equality to isolate v_0 gives
$d = atv_t - atv_0$, $atv_0 = atv_t - d$, $v_0 = \frac{atv_t - d}{at}$. **Answer is D.**

20. Removing the common factor of 6 and then factoring the sum of two cubes gives
$6 + 48m^3 = 6(1 + 8m^3) = 6(1 + 2m)(1^2 - 2m + (2m)^2)$. **Answer is A.**

21. B is not the graph of a function. D is the graph of a parabola where the coefficient of x^2 is negative. A appears to be the graph of $y = x^2$. To find the x-intercepts of
$y = x^2 + 3x$, set y = 0 and solve for x: $0 = x^2 + 3x = x(x + 3)$ to get x = 0 or x = -3. Therefore, the graph of the function intersects the x-axis at x=0 and x=-3.
Answer is C.

22. Set up the direct variation: $\frac{V}{r^2} = \frac{V}{r^2}$. Substituting gives $\frac{80}{16} = \frac{V}{9}$. Solving for V gives 45 liters per minute. **Answer is A.**

23. Multiplying equation 1 by 2, and equation 2 by –3, and then adding together the two resulting equations gives -11y + 22z = 0. Solving for y gives y = 2z. In the meantime, multiplying equation 3 by –2 and adding it to equation 2 gives –y – 12z = -14. Then substituting 2z for y, yields the result z = 1. Subsequently, one can easily find that y = 2, and x = -1. **Answer is A.**

24. Using the definition of absolute value, two equations are possible: 18 = 4 + 2x or 18 = 4 – 2x. Solving for x gives x = 7 or x = -7. **Answer is C.**

25. Rewriting the inequality gives $x^2 - 5x + 6 > 0$. Factoring gives (x – 2)(x – 3) > 0.
The two cut-off points on the numberline are now at x = 2 and x = 3. Choosing a random number in each of the three parts of the numberline, we test them to see if they produce a true statement. If x = 0 or x = 4, (x-2)(x-3)>0 is true. If x = 2.5, (x-2)(x-3)>0 is false. Therefore the solution set is all numbers smaller than 2 or greater than 3.
Answer is D.

26. Possible rational roots of the equation $0 = x^3 + x^2 - 14x - 24$ are all the positive and negative factors of 24. By substituting into the equation, we find that –2 is a root, and therefore that x+2 is a factor. By performing the long division $(x^3 + x^2 - 14x - 24)/(x+2)$, we can find that another factor of the original equation is $x^2 - x - 12$ or (x-4)(x+3). Therefore the zeros of the original function are –2, -3, and 4. **Answer is D.**

27. Getting the bases the same gives us $3^{\frac{1}{2}}3^{\frac{2}{3}}$. Adding exponents gives $3^{\frac{7}{6}}$. Then some additional manipulation of exponents produces $3^{\frac{7}{6}} = 3^{\frac{14}{12}} = (3^2)^{\frac{7}{12}} = 9^{\frac{7}{12}}$. **Answer is B.**

28. Simplifying radicals gives $\sqrt{27} + \sqrt{75} = 3\sqrt{3} + 5\sqrt{3} = 8\sqrt{3}$. **Answer is A.**

29. Multiplying numerator and denominator by the conjugate gives $\frac{10}{1+3i} \times \frac{1-3i}{1-3i} = \frac{10(1-3i)}{1-9i^2} = \frac{10(1-3i)}{1-9(-1)} = \frac{10(1-3i)}{10} = 1-3i$. **Answer is D.**

30. To find the 100th term: $t_{100} = -6 + 99(4) = 390$. To find the sum of the first 100 terms: $S = \frac{100}{2}(-6+390) = 19200$. **Answer is A.**

31. There are 3 slots to fill. There are 3 choices for the first, 7 for the second, and 6 for the third. Therefore, the total number of choices is 3(7)(6) = 126. **Answer is A.**

32. The set-up for finding the seventh term is $\frac{8(7)(6)(5)(4)(3)}{6(5)(4)(3)(2)(1)}(2a)^{8-6}b^6$ which gives $28(4a^2b^6)$ or $112a^2b^6$. **Answer is C.**

33. By definition, parallel lines are coplanar lines without any common points. **Answer is B.**

34. A set of n objects has 2^n subsets. Therefore, here we have $2^4 = 16$ subsets. These subsets include four which each have 1 element only, six which each have 2 elements, four which each have 3 elements, plus the original set, and the empty set. **Answer is B.**

35. Formula for finding the measure of each interior angle of a regular polygon with n sides is $\frac{(n-2)180}{n}$. For n=10, we get $\frac{8(180)}{10} = 144$. **Answer is C.**

36. Draw a right triangle with legs of 6 and 8. Find the hypotenuse using the Pythagorean Theorem. $6^2 + 8^2 = c^2$. Therefore, c = 10 miles. **Answer is B.**

37. The formula relating the measure of angle K and the two arcs it intercepts is $m\angle K = \frac{1}{2}(mPS - mAD)$. Substituting the known values, we get $10 = \frac{1}{2}(40 - mAD)$. Solving for mAD gives an answer of 20 degrees. **Answer is B.**

38. Given a point on a line, place the compass point there and draw two arcs intersecting the line in two points, one on either side of the given point. Then using any radius larger than half the new segment produced, and with the pointer at each end of the new segment, draw arcs which intersect above the line. Connect this new point with the given point. **Answer is D**.

39. By definition this describes the procedure of an indirect proof. **Answer is C.**

40. Since side AK is common to both triangles, the triangles can be proved congruent by using the Side-Angle-Side Postulate. **Answer is C.**

41. In an ordinary parallelogram, the diagonals are not perpendicular or equal in length. In a rectangle, the diagonals are not necessarily perpendicular. In a rhombus, the diagonals are not equal in length. In a square, the diagonals are both perpendicular and congruent. **Answer is C.**

42. The most one can say with certainty is that the median (segment drawn to the midpoint of the opposite side) and the altitude (segment drawn perpendicular to the opposite side) of a triangle <u>may</u> coincide, but they more often do not. In an isosceles triangle, the median and the altitude to the <u>base</u> are the same segment. **Answer is A.**

43. In the mid-nineteenth century, Reimann and other mathematicians developed elliptic geometry. **Answer is B**.

44. Let's assume the base of the rectangular solid (box) is 3 by 4, and the height is 5. Then the surface area of the top and bottom together is 2(12) = 24. The sum of the areas of the front and back are 2(15) = 30, while the sum of the areas of the sides are 2(20)=40. The total surface area is therefore 94 square feet. **Answer is C**.

45. Find the area of the garden and then subtract the area of the fountain: $30(60)-\pi(5)^2$ or approximately 1721 square meters. **Answer is A.**

46. To find the area of the shaded region, find the area of triangle ABC and then subtract the area of triangle DBE. The area of triangle ABC is .5(6x)(y) = 3xy. The area of triangle DBE is .5(2x)(y) = xy. The difference is 2xy. **Answer is B.**

47. Find the mean: 300/8 = 37.5. Then, using the formula for standard deviation, we get

$$\sqrt{\frac{2(37.5-37)^2+2(37.5-38)^2+(37.5-35)^2+(37.5-40)^2+(37.5-36)^2+(37.5-39)^2}{8}}$$

which has a value of 1.5. **Answer is B.**

48. To find the determinant of a matrix without the use of a graphing calculator, repeat the first two columns as shown,

2	1	-1	2	1
4	-1	4	4	-1
0	-3	2	0	-3

Starting with the top left-most entry, 2, multiply the three numbers in the diagonal going down to the right: 2(-1)(2)=-4. Do the same starting with 1: 1(4)(0)=0. And starting with –1: -1(4)(-3) = 12. Adding these three numbers, we get 8. Repeat the same process starting with the top right-most entry, 1. That is, multiply the three numbers in the diagonal going down to the left: 1(4)(2) = 8. Do the same starting with 2: 2(4)(-3) = -24 and starting with –1: -1(-1)(0) = 0. Add these together to get
-16. To find the determinant, subtract the second result from the first: 8-(-16)=24.
Answer is C.

49. Using the Pythagorean Identity, we know $\sin^2 x + \cos^2 x = 1$. Thus $1 - \sin^2 x = \cos^2 x$, which by definition is equal to $1/\sec^2 x$. **Answer is D.**

50. Tan A = 9/12=.75 and $\tan^{-1}.75$ = 37 degrees. Since angle B is complementary to angle A, the measure of angle B is therefore 53 degrees. **Answer is D.**

51. The composite function $g(f(x)) = (3x-2)^2 = 9x^2 - 12x + 4$. **Answer is C.**

52. Given the polar point $(r,\theta) = (5, 60)$, we can find the rectangular coordinates this way: $(x,y) = (r\cos\theta, r\sin\theta) = (5\cos 60, 5\sin 60) = (2.5, 4.33)$. **Answer is C.**

53. Using the Pythagorean Theorem, we get $v = \sqrt{36+25} = \sqrt{61}$. **Answer is B.**

54. The line x = 5 is a vertical line passing through (5,0) on the Cartesian plane. By observation the distance along the horizontal line from the point (-2,7) to the line x=5 is 7 units. **Answer is D.**

55. The formula for finding the midpoint (a,b) of a segment passing through the points $(x_1,y_1) and (x_2,y_2) is (a,b) = (\frac{x_1+x_2}{2},\frac{y_1+y_2}{2})$. Setting up the corresponding equations from this information gives us x = $\frac{-4+2}{2}, and 1 = \frac{y-3}{2}$. Solving for x and y gives x = -1 and y = 5. **Answer is A.**

56. Dividing by 36, we get $\frac{x^2}{36}+\frac{y^2}{4}=1$, which tells us that the ellipse intersects the x-axis at 6 and –6, and therefore the length of the major axis is 12. (The ellipse intersects the y-axis at 2 and –2). **Answer is C.**

57. With a diameter going from (0,7) to (0,3), the diameter of the circle must be 4, the radius must be 2, and the center of the circle must be at (0,5). Using the standard form for the equation of a circle, we get $(x-0)^2 + (y-5)^2= 2^2$. Expanding, we get $x^2+y^2-10y+21=0$. **Answer is B.**

58. By definition of log form and exponential form, $\log_x k = m$ corresponds to $x^m = k$. **Answer is A.**

59. Use the Power Rule for polynomial differentiation: if y = ax^n, then y'=nax^{n-1}. **Answer is D.**

60. Use the Exponential Rule for derivatives of functions of e: if y = $ae^{f(x)}$, then y' = $f'(x)ae^{f(x)}$. **Answer is A.**

61. To find the slope of the tangent line, find the derivative, and then evaluate it at

x = $\frac{\pi}{2}$. y' = 3x(-sinx)+3cosx. At the given value of x,

y' = $3(\frac{\pi}{2})(-\sin\frac{\pi}{2})+3\cos\frac{\pi}{2}=\frac{-3\pi}{2}$. **Answer is A.**

62. To find the slope of the tangent line, find the derivative, and then evaluate it at x=1. y'=6x-5=6(1)-5=1. Then using point-slope form of the equation of a line, we get

y+2=1(x-1) or y = x-3. **Answer is A.**

63. To find critical points, take the derivative, set it equal to 0, and solve for x. f'(x) = $3x^2 + 2x$ = x(3x+2)=0. CP at x=0 and x=-2/3. Neither of these CP is on the interval from x=1 to x=3. Testing the endpoints: at x=1, y=6 and at x=3, y=38. Since the derivative is positive for all values of x from x=1 to x=3, the curve is increasing on the entire interval. **Answer is B.**

64. Find CP at x=-.75 as done in #63. Since the CP is not in the interval from x=0 to x=3, just find the values of the functions at the endpoints. When x=0, y=0, and when x=3, y = 27. Therefore 27 is the absolute maximum on the given interval.
Answer is D.

65. Use the rule for polynomial integration: given ax^n, the antiderivative is $\frac{ax^{n+1}}{n+1}$. **Answer is A.**

66. Use the rule for integration of functions of e: $\int e^x dx = e^x + C$. **Answer is D.**

67. Recall that the derivative of the velocity function is the acceleration function. In reverse, the integral of the acceleration function is the velocity function. Therefore, if a=6, then v=6t+C. Given that at t=0, v=15, we get v = 6t+15. At t=10, v=60+15=75m/s. **Answer is C.**

68. Recall that the derivative of the distance function is the velocity function. In reverse, the integral of the velocity function is the distance function. To find the time needed for the body to come to a stop when v=0, solve for t: $v = 16 - t^2 = 0$. Result: t = 4 seconds. The distance function is $s = 16t - \frac{t^3}{3}$. At t=4, s= 64 – 64/3 or approximately 43 units. **Answer is B.**

69. Integrate as described in #65. **Answer is B.**

70. Use the fundamental theorem of calculus to find the definite integral: given a continuous function f on an interval [a,b], then $\int_a^b f(x)dx = F(b) - F(a)$, where F is an antiderivative of f.

$\int_0^2 (x^2 + x - 1)dx = (\frac{x^3}{3} + \frac{x^2}{2} - x)$ Evaluate the expression at x=2, at x=0, and then subtract to get 8/3 + 4/2 – 2-0 = 8/3. **Answer is B.**

71. To find the area set up the definite integral: $\int_3^6 (x^2 + 4)dx = (\frac{x^3}{3} + 4x)$.
Evaluate the expression at x=6, at x=3, and then subtract to get (72+24)-(9+12)=75.
Answer is A.

72. The errors are in the following: -3+7=-4 and –3 + -8 = 11, where the student seems to be using the rules for signs when multiplying, instead of the rules for signs when adding. **Answer is C**.

73. **Answer is B**.

74. Since the second equation is already solved for y, it would be easiest to use the substitution method. **Answer is C.**

75. The steps listed in answer D would look like this for the given example: If $|x+1| \leq 6$, then $-6 \leq x+1 \leq 6$, which means $-7 \leq x \leq 5$. Then the inequality would be graphed on a number line and would show that the solution set is all real numbers between –7 and 5, including –7 and 5. **Answer is D**.

76. There is no need for calculators when justifying statements in a geometric proof.
Answer is C.

77. By observation the **Answer is B**.

78. **Answer is C.**

79. **Answer is C.**

80. By observation of the examples given, $a \not\subset b = a^2 - b$. Therefore, $5 \not\subset 4 = 25 - 4 = 21$. **Answer is D**.

DOMAIN VII. SCIENTIFIC INQUIRY AND PROCESSES

COMPETENCY 22.0 THE TEACHER UNDERSTANDS HOW TO SELECT AND MANAGE LEARNING ACTIVITIES TO ENSURE THE SAFETY OF ALL STUDENTS AND THE CORRECT USE AND CARE OF ORGANISMS, NATURAL RESOURCES, MATERIALS, EQUIPMENT, AND TECHNOLOGIES.

Skill 22.1 Uses current sources of information about laboratory safety, including safety regulations and guidelines for the use of science facilities.

Chemical purchase, use, and disposal

- Inventory all chemicals on hand at least annually. Keep the list up-to-date as chemicals are consumed and replacement chemicals are received.
- If possible, limit the purchase of chemicals to quantities that will be consumed within one year and that are packaged in small containers suitable for direct use in the lab without transfer to other containers.
- Label all chemicals to be stored with date of receipt or preparation and have labels initialed by the person responsible.
- Generally, bottles of chemicals should not remain:
    - Unused on shelves in the lab for more than one week. Move these chemicals to the storeroom or main stockroom.
    - Unused in the storeroom near the lab for more than one month. Move these chemicals to the main stockroom.
- Check shelf life of chemicals. Properly dispose of any out-dated chemicals.
- Ensure that the disposal procedures for waste chemicals conform to environmental protection requirements.
- Do not purchase or store large quantities of flammable liquids. Fire department officials can recommend the maximum quantities that may be kept on hand.
- Never open a chemical container until you understand the label and the relevant portions of the MSDS.

Chemical storage plan for laboratories

- Chemicals should be stored according to hazard class (ex. flammables, oxidizers, health hazards/toxins, corrosives, etc.).
- Store chemicals away from direct sunlight or localized heat.
- All chemical containers should be properly labeled, dated upon receipt, and dated upon opening.
- Store hazardous chemicals below shoulder height of the shortest person working in the lab.

- Shelves should be painted or covered with chemical-resistant paint or chemical-resistant coating.
- Shelves should be secure and strong enough to hold chemicals being stored on them. Do not overload shelves.
- Personnel should be aware of the hazards associated with all hazardous materials.
- Separate solids from liquids.

Below are examples of chemical groups that can be used to categorize storage. Use these groups as examples when separating chemicals for compatibility. Please note: reactive chemicals must be more closely analyzed since they have a greater potential for violent reactions. Contact Laboratory Safety if you have any questions concerning chemical storage.

Acids

- Make sure that all acids are stored by compatibility (ex. separate inorganics from organics).
- Store concentrated acids on lower shelves in chemical-resistant trays or in a corrosives cabinet. This will temporarily contain spills or leaks and protect shelving from residue.
- Separate acids from incompatible materials such as bases, active metals (ex. sodium, magnesium, potassium) and from chemicals which can generate toxic gases when combined (ex. sodium cyanide and iron sulfide).

Bases

- Store bases away from acids.
- Store concentrated bases on lower shelves in chemical-resistant trays or in a corrosives cabinet. This will temporarily contain spills or leaks and protect shelving from residue.

Flammables

- Approved flammable storage cabinets should be used for flammable liquid storage.
- You may store 20 gallons of flammable liquids per 100 sq.ft. in a properly fire separated lab. The maximum allowable quantity for flammable liquid storage in any size lab is not to exceed 120 gallons.
- You may store up to 10 gallons of flammable liquids outside of approved flammable storage cabinets.
- An additional 25 gallons may be stored outside of an approved storage cabinet if it is stored in approved safety cans not to exceed 2 gallons in size.

- Use only explosion-proof or intrinsically safe refrigerators and freezers for storing flammable liquids.

Peroxide-Forming Chemicals

- Peroxide-forming chemicals should be stored in airtight containers in a dark, cool, and dry place.
- Unstable chemicals such as peroxide-formers must always be labeled with date received, date opened, and disposal/expiration date.
- Peroxide-forming chemicals should be properly disposed of before the date of expected peroxide formation (typically 6-12 months after opening).
- Suspicion of peroxide contamination should be immediately investigated. Contact Laboratory Safety for procedures.

Water-Reactive Chemicals

- Water-reactive chemicals should be stored in a cool, dry place.
- Do not store water-reactive chemicals under sinks or near water baths.
- Class D fire extinguishers for the specific water-reactive chemical being stored should be made available.

Oxidizers

- Make sure that all oxidizers are stored by compatibility.
- Store oxidizers away from flammables, combustibles, and reducing agents.

Toxins

- Toxic compounds should be stored according to the nature of the chemical, with appropriate security employed when necessary.
- A "Poison Control Network" telephone number should be posted in the laboratory where toxins are stored. Color-coded labeling systems that may be found in your lab are shown below:

Hazard	Color Code
Flammables	Red
Health Hazards/Toxins	Blue
Reactives/Oxidizers	Yellow
Contact Hazards	White
General Storage	Gray, Green, Orange

Please Note: Chemicals with labels that are colored and striped may react with other chemicals in the same hazard class. See MSDS for more information.

Chemical containers which are not color-coded should have hazard information on the label. Read the label carefully and store accordingly.

Disposal of chemical waste

Schools are regulated by the Environmental Protection Agency, as well as state and local agencies, when it comes to disposing of chemical waste. Check with your state science supervisor, local college or university environmental health and safety specialists, and the Laboratory Safety Workshop for advice on the disposal of chemical waste. The American Chemical Society publishes an excellent guidebook, ***Laboratory Waste Management, A Guidebook*** (1994).

The following are merely guidelines for disposing of chemical waste.

You may dispose of hazardous waste as outlined below. It is the responsibility of the generator to ensure hazardous waste does not end up in ground water, soil, or the atmosphere through improper disposal.

1. **Sanitary Sewer** - Some chemicals (acids or bases) may be neutralized and disposed to the sanitary sewer. This disposal option must be approved by the local waste water treatment authority prior to disposal. This may not be an option for some small communities that do not have sufficient treatment capacity at the waste water treatment plant for these types of wastes. Hazardous waste may NOT be disposed of in this manner. This includes heavy metals.

2. **Household Hazardous Waste Facility** - Waste chemicals may be disposed through a county household hazardous waste facility (HHW) or through a county contracted household hazardous waste disposal company. Not all counties have a program to accept waste from schools. Verify with your county HHW facility that they can handle your waste prior to making arrangements.

3. **Disposal Through a Contractor** - A contractor may be used for disposal of waste chemicals. Remember that you must keep documentation of your hazardous waste disposal for at least three years. This information must include a waste manifest, reclamation agreement or any written record which describes the waste and how much was disposed, where it was disposed and when it was disposed. Waste analysis records must also be kept when it is necessary to make a determination of whether waste is hazardous. **Any unknown chemicals should be considered hazardous!**

Skill 22.2 Recognizes potential safety hazards in the laboratory and in the field and knows how to apply procedures, including basic first aid, for responding to accidents.

Safety is a learned behavior and must be incorporated into instructional plans. Measures of prevention and procedures for dealing with emergencies in hazardous situations have to be in place and readily available for reference. Copies of these must be given to all people concerned, such as administrators and students.

The single most important aspect of safety is planning and anticipating various possibilities and preparing for the eventuality. Any Physics teacher/educator planning on doing an experiment must try it before the students do it. In the event of an emergency, quick action can prevent many disasters. The teacher/educator must be willing to seek help at once without any hesitation because sometimes it may not be clear that the situation is hazardous and potentially dangerous.

There are a number of procedures to prevent and correct any hazardous situation. There are several safety aids available commercially such as posters, safety contracts, safety tests, safety citations, texts on safety in secondary classroom/laboratories, hand books on safety and a host of other equipment. Another important thing is to check the laboratory and classroom for safety and report it to the administrators before staring activities/experiments. It is important that teachers and educators follow these guidelines to protect the students and to avoid most of the hazards. They have a responsibility to protect themselves as well. **There should be not any compromises in issues of safety.**

All science labs should contain the following items of **safety equipment**.
-Fire blanket that is visible and accessible
-Ground Fault Circuit Interrupters (GCFI) within two feet of water supplies
-Signs designating room exits
-Emergency shower providing a continuous flow of water
-Emergency eye wash station that can be activated by the foot or forearm
-Eye protection for every student and a means of sanitizing equipment
-Emergency exhaust fans providing ventilation to the outside of the building
-Master cut-off switches for gas, electric and compressed air. Switches must have permanently attached handles. Cut-off switches must be clearly labeled.
-An ABC fire extinguisher
-Storage cabinets for flammable materials
-Chemical spill control kit
-Fume hood with a motor that is spark proof
-Protective laboratory aprons made of flame retardant material
-Signs that will alert potential hazardous conditions
-Labeled containers for broken glassware, flammables, corrosives, and waste.

Students should wear safety goggles when performing dissections, heating, or while using acids and bases. Hair should always be tied back and objects should never be placed in the mouth. Food should not be consumed while in the laboratory. Hands should always be washed before and after laboratory experiments. In case of an accident, eye washes and showers should be used for eye contamination or a chemical spill that covers the student's body. Small chemical spills should only be contained and cleaned by the teacher. Kitty litter or a chemical spill kit should be used to clean spill. For large spills, the school administration and the local fire department should be notified. Biological spills should only be handled by the teacher. Contamination with biological waste can be cleaned by using bleach when appropriate. Accidents and injuries should always be reported to the school administration and local health facilities. The severity of the accident or injury will determine the course of action to pursue.

Skill 22.3 Employs safe practices in planning and implementing all instructional activities and designs, and implements rules and procedures to maintain a safe learning environment.

The single most important factor in preparing materials and apparatus for use in the science laboratory is familiarity with the materials and equipment. Teachers must have theoretical as well as hands-on knowledge of how a particular chemical is to be used or how a piece of apparatus functions. Here are some things teachers can do in preparation that will make the laboratory experience safe and effective for students:

- Perform all experiments yourself before introducing them to the students.
- Set up equipment away from table edges and with enough space in between.
- Make sure all safety equipment (e.g. eye wash station) is in place. (See section VI.15 for a list of safety equipment).
- Provide clear written instructions on how to perform an experiment.
- For hazardous chemicals or delicate equipment post warning signs in bold and bright lettering.
- Post multiple copies of laboratory rules that spell out safe ways to do common tasks such as such as handling of bottle stoppers, pouring corrosive reagents, smelling substances.
- Monitor the conditions and stock levels of all materials regularly so that appropriate materials are always on hand.
- When you have a choice in what materials to use, make sure you select the least hazardous ones.

Skill 22.4 Understands procedures for selecting, maintaining, and safely using chemicals, tools, technologies, materials, specimens, and equipment, including procedures for the recycling, reuse, and conservation of laboratory resources and for the safe handling and ethical treatment of organisms.

All laboratory solutions should be prepared as directed in the lab manual. Care should be taken to avoid contamination. All glassware should be rinsed thoroughly with distilled water before using and cleaned well after use. All solutions should be made with distilled water as tap water contains dissolved particles that may affect the results of an experiment. Unused solutions should be disposed of according to local disposal procedures.

The "Right to Know Law" covers science teachers who work with potentially hazardous chemicals. Briefly, the law states that employees must be informed of potentially toxic chemicals. An inventory must be made available if requested. The inventory must contain information about the hazards and properties of the chemicals. This inventory is to be checked against the "Substance List". Training must be provided on the safe handling and interpretation of the **Material Safety Data Sheet (MSDS)**.

The following chemicals are potential carcinogens and not allowed in school facilities: Acrylonitriel, Arsenic compounds, Asbestos, Bensidine, Benzene, Cadmium compounds, Chloroform, Chromium compounds, Ethylene oxide, Ortho-toluidine, Nickel powder, and Mercury.

Chemicals should not be stored on bench tops or heat sources. They should be stored in groups based on their reactivity with one another and in protective storage cabinets. All containers within the lab must be labeled. Suspect and known carcinogens must be labeled as such and segregated within trays to contain leaks and spills.

Chemical waste should be disposed of in properly labeled containers. Waste should be separated based on their reactivity with other chemicals.

Biological material should never be stored near food or water used for human consumption. All biological material should be appropriately labeled. All blood and body fluids should be put in a well-contained container with a secure lid to prevent leaking. All biological waste should be disposed of in biological hazardous waste bags.

Material safety data sheets are available for every chemical and biological substance. These are available directly from the company of acquisition or the internet. The manuals for equipment used in the lab should be read and understood before using them.

Laboratory and field equipment used for scientific investigation must be handled with the greatest caution and care. The teacher must be completely familiar with the use and maintenance of a piece of equipment before it is introduced to the students. Maintenance procedures for equipment must be scheduled and recorded and each instrument must be calibrated and used strictly in accordance with the specified guidelines in the accompanying manual. Following are some safety precautions one can take in working with different types of equipment:

1. Electricity: Safety in this area starts with locating the main cut off switch. All the power points, switches, and electrical connections must be checked one by one. Batteries and live wires must be checked. All checking must be done with the power turned off. The last act of assembling is to insert the plug and the first act of disassembling is to take off the plug.

2. Motion and forces: All stationary devices must be secured by C-clamps. Protective goggles must be used. Care must be taken at all times while knives, glass rods and heavy weights are used. Viewing a solar eclipse must always be indirect. When using model rockets, NASA's safety code must be implemented.

3. Heat: The master gas valve must be off at all times except while in use. Goggles and insulated gloves are to be used whenever needed. Never use closed containers for heating. Burners and gas connections must be checked periodically. Gas jets must be closed soon after the experiment is over. Fire retardant pads and quality glassware such as Pyrex must be used.

4. Pressure: While using a pressure cooker, never allow pressure to exceed 20 lb/square inch. The pressure cooker must be cooled before it is opened. Care must be taken when using mercury since it is poisonous. A drop of oil on mercury will prevent the mercury vapors from escaping.

5. Light: Broken mirrors or those with jagged edges must be discarded immediately. Sharp-edged mirrors must be taped. Spectroscopic light voltage connections must be checked periodically. Care must be taken while using ultraviolet light sources. Some students may have psychological or physiological reactions to the effects of strobe like (e.g. epilepsy).

6. Lasers: Direct exposure to lasers must not be permitted. The laser target must be made of non-reflecting material. The movement of students must be restricted during experiments with lasers. A number of precautions while using lasers must be taken – use of low power lasers, use of approved laser goggles, maintaining the room's brightness so that the pupils of the eyes remain small. Appropriate beam stops must be set up to terminate the laser beam when needed. Prisms should be set up before class to avoid unexpected reflection.

7. Sound: Fastening of the safety disc while using the high speed siren disc is very important. Teacher must be aware of the fact that sounds higher than 110 decibels will cause damage to hearing.

8. Radiation: Proper shielding must be used while doing experiments with x-rays. All tubes that are used in a physics laboratory such as vacuum tubes, heat effect tubes, magnetic or deflection tubes must be checked and used for demonstrations by the teacher. Cathode rays must be enclosed in a frame and only the teacher should move them from their storage space. Students must watch the demonstration from at least eight feet away.

9. **Radioactivity:** The teacher must be knowledgeable and properly trained to handle the equipment and to demonstrate. Proper shielding of radioactive material and proper handling of material are absolutely critical. Disposal of any radioactive material must comply with the guidelines of NRC.

Use of equipment in the physics lab

Oscilloscope: An oscilloscope is a piece of electrical test equipment that allows signal voltages to be viewed as two-dimensional graphs of electrical potential differences plotted as a function of time.

The oscilloscope functions by measuring the deflection of a beam of electrons traveling through a vacuum in a cathode ray tube. The deflection of the beam can be caused by a magnetic field outside the tube or by electrostatic energy created by plates inside the tube. The unknown voltage or potential energy difference can be determined by comparing the electron deflection it causes to the electron deflection caused by a known voltage.

Oscilloscopes can also determine if an electrical circuit is oscillating and at what frequency. They are particularly useful for troubleshooting malfunctioning equipment. You can see the "moving parts" of the circuit and tell if the signal is being distorted. With the aid of an oscilloscope you can also calculate the "noise" within a signal and see if the "noise" changes over time.

Inputs of the electrical signal are usually entered into the oscilloscope via a coaxial cable or probes. A variety of transducers can be used with an oscilloscope that enable it to measure other stimuli including sound, pressure, heat, and light.

Voltmeter/Ohmmeter/Ammeter: A common electrical meter, typically known as a multimeter, is capable of measuring voltage, resistance, and current. Many of these devices can also measure capacitance (farads), frequency (hertz), duty cycle (a percentage), temperature (degrees), conductance (siemens), and inductance (henrys).

These meters function by utilizing the following familiar equations:

Across a resistor (Resistor R):

$$V_R = IR_R$$

Across a capacitor (Capacitor C):

$$V_C = IX_C$$

Across an inductor (Inductor L):

$$V_L = IX_L$$

Where V=voltage, I=current, R=resistance, X=reactance.

If any two factors in the equations are held constant or are known, the third factor can be determined and is displayed by the multimeter.

Signal Generator: A signal generator, also known as a test signal generator, function generator, tone generator, arbitrary waveform generator, or frequency generator, is a device that generates repeating electronic signals in either the analog or digital domains. They are generally used in designing, testing, troubleshooting, and repairing electronic devices.

A function generator produces simple repetitive waveforms by utilizing a circuit called an electronic oscillator or a digital signal processor to synthesize a waveform. Common waveforms are sine, sawtooth, step or pulse, square, and triangular. Arbitrary waveform generators are also available which allow a user to create waveforms of any type within the frequency, accuracy and output limits of the generator. Function generators are typically used in simple electronics repair and design where they are used to stimulate a circuit under test. A device such as an oscilloscope is then used to measure the circuit's output.

Spectrometer: A spectrometer is an optical instrument used to measure properties of light over a portion of the electromagnetic spectrum. Light intensity is the variable that is most commonly measured but wavelength and polarization state can also be determined. A spectrometer is used in spectroscopy for producing spectral lines and measuring their wavelengths and intensities. Spectrometers are capable of operating over a wide range of wavelengths, from short wave gamma and X-rays into the far infrared. In optics, a spectrograph separates incoming light according to its wavelength and records the resulting spectrum in some detector. In astronomy, spectrographs are widely used with telescopes.

Skill 22.5 Knows how to use appropriate equipment and technology (e.g., Internet, spreadsheet, calculator) for gathering, organizing, displaying, and communicating data in a variety of ways (e.g., charts, tables, graphs, diagrams, written reports, oral presentations).

Scientists use a variety of tools and technologies to perform tests, collect and display data, and analyze relationships. Data is commonly organized in table format and displayed in graphs. Examples of tools that aid in data gathering, organization and analysis include computer-linked probes, spreadsheets, and graphing calculators.

Scientists use **computer-linked probes** to measure various environmental factors including temperature, dissolved oxygen, pH, ionic concentration, and pressure. The advantage of computer-linked probes, as compared to more traditional observational tools, is that the probes automatically gather data and present it in an accessible format. This property of computer-linked probes eliminates the need for constant human observation and manipulation.

Spreadsheets are often used to organize, analyze, and display data. For example, conservation ecologists use spreadsheets to model population growth and development, apply sampling techniques, and create statistical distributions to analyze relationships. Spreadsheet use simplified data collection and manipulation and allows the presentation of data in a logical and understandable format.

Graphing calculators are another technology with many applications to science. For example, physicists use algebraic functions to analyze the time or space dependence of various processes. Graphing calculators can manipulate algebraic data and create graphs for analysis and observation. In addition, the matrix function of graphing calculators may be used to model certain problems. The use of graphing calculators simplifies the creation of displays such as histograms, scatter plots, and line graphs. Scientists can also transfer data and displays to computers for further analysis. Finally, scientists connect computer-linked probes, used to collect data, to graphing calculators to ease the collection, transmission, and analysis of data.

Linear regression is a common technique used by graphing software to translate tabular data into plots in which continuous lines are displayed even though data is available only at some discrete points. Regression is essentially a filling in of the gaps between data points by making a reasonable estimation of what the in-between values are using a standard mathematical process. In other words, it is finding the "best-fit" curve to represent experimental data. The graphs displayed below show how a straight line or a curve can be fitted to a set of discrete data points. There are many different regression algorithms that may be used and they differ from tool to tool. Some tools allow the user to decide what regression method will be used and what kind of curve (straight line, exponential etc.) will be used to fit the data.

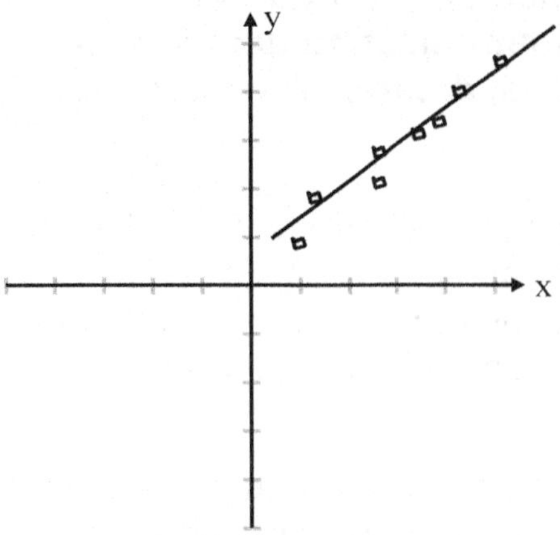

Linear Relationship

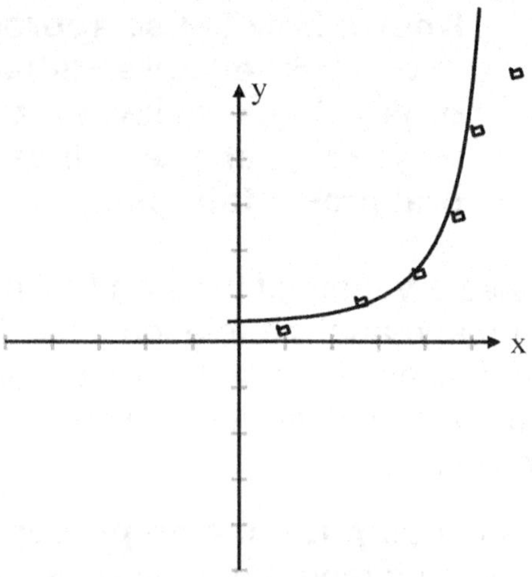

Non- Linear Relationship

Contrast the preceding graphs to the graph of a data set that shows no relationship between variables.

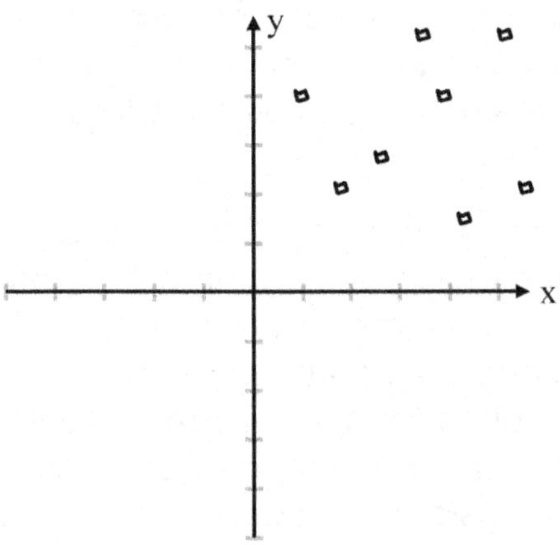

Extrapolation is the process of estimating data points outside a known set of data points. When extrapolating data of a linear relationship, we extend the line of best fit beyond the known values. The extension of the line represents the estimated data points. Extrapolating data is only appropriate if we are relatively certain that the relationship is indeed correctly represented by the best-fit curve.

Tables

Tables are excellent for organizing data as it is being recorded and for storing data that needs to be analyzed. In fact, almost all experimental data is initially organized into a table, such as in a lab notebook. Often, it is then entered into tables within spreadsheets for further processing. Tables can be used for presenting data to others if the data set is fairly small or has been summarized (for example, presenting average values). However, for larger data sets, tabular presentation may be overwhelming. Finally, tables are not particularly useful for recognizing trends in data or for making them apparent to others.

Charts and Graphs

Charts and graphs are the best way to demonstrate trends or differences between groups. They are also useful for summarizing data and presenting it. In most types of graphs, it is also simple to indicate uncertainty of experimental data using error bars. Many types of charts and graphs are available to meet different needs. Three of the most common are scatterplots, bar charts, and pie charts. An example of each is shown.

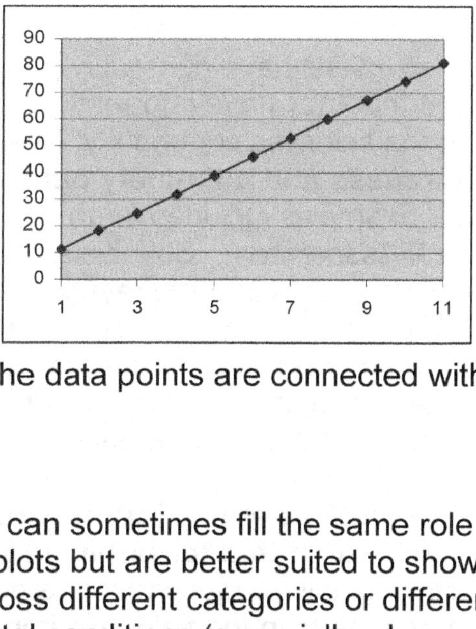

Scatterplots are typically shown on a Cartesian plane and are useful for demonstrating the relationship between two variables. A line chart (shown) is a special type of two-dimensional scatter plot in which the data points are connected with a line to make a trend more apparent.

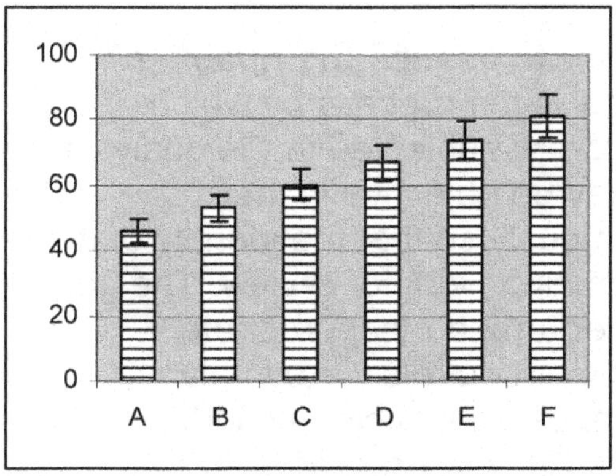

Bar charts can sometimes fill the same role as scatterplots but are better suited to show values across different categories or different experimental conditions (especially where those conditions are described qualitatively rather than quantified). Note the use of error bars in this example.

Finally, a pie chart is best used to present relative magnitudes or frequencies of several different conditions or events. They are most commonly used to show how various categories contribute to a whole.

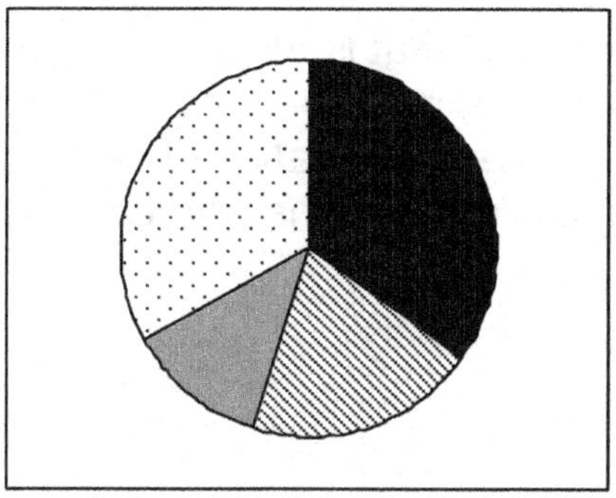

Diagrams
Diagrams are not typically used to present the specifics of data. However, they are very good for demonstrating phenomena qualitatively. Diagrams make it

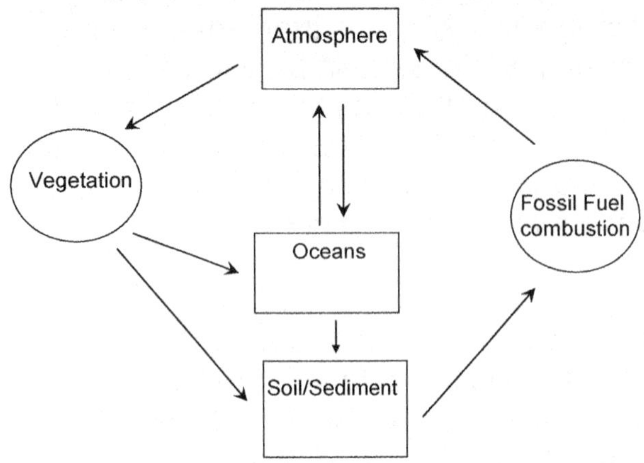

easy to visualize the connections and relationships between various elements. They may also be used to demonstrate temporal relationships. For example, diagrams can be used to illustrate the operation of an internal combustion engine or the complex biochemical pathways of an enzyme's action. The diagram shown is a simplified version of the carbon cycle.

Once scientists have rigorously performed a set of experiments, they may believe that they have information of significant value that should be shared within the scientific community. **New findings are often presented at scientific conferences and ultimately published in technical journals** (some well known examples include *Nature*, *Science*, and the *Journal of the American Medical Association*). Scientists prepare manuscripts detailing the conditions of their experiments and the results they obtained. They will typically also include their interpretation of those results and their impact on current theories in the field. These manuscripts are not wholly unlike lab reports, though they are considerably more polished, of course. Manuscripts are then submitted to appropriate technical journals.

All reputable scientific journals use peer review to assess the quality of research submitted for publication. Peer review is the process by which scientific results produced by one person or group are subjected to the analysis of other experts in the field. It is important that they be objective in their evaluations. Peer review is typically done anonymously so that the identity of the reviewer remains unknown by the scientists submitting work for review. The goal of peer review is to "weed out" any science not performed by appropriate standards. Reviewers will determine whether proper controls were in place, enough replicates were performed, and that the experiments clearly address the presented hypothesis. The reviewer will scrutinize the interpretations and how they fit into what is already known in the field. Often reviewers will suggest that additional experiments be done to further corroborate presented conclusions. If the reviewers are satisfied with the quality of the work, it will be published and made available to the entire scientific community.

Sometimes, there is significant opposition to new ideas, especially if they conflict with long-held ideas in a scientific field. However, with enough correct experimental support, scientific theories will in time be adjusted to reflect the newer findings. Thus science is an ongoing cycle: ideas are constantly being refined and modified to reflect new evidence and, ultimately, provide us with a more correct model of the world around us. There are many scientific theories that have experienced multiple revisions and expansions. Some well-known examples include the atomic theory, which was notable refined by Dalton, Thomson, Rutherford, and Bohr and the theory of natural selection, which was originally formulated by Darwin but has been enriched by discoveries in molecular biology.

There are many important reasons for scientists to report their detailed findings to the scientific community. The first and most important reason is to **ensure the correctness of scientific results and to advance our larger understanding of the natural world**. When new scientific findings are reported at conferences or peer reviewed for publication in technical journals, they are rigorously evaluated. This is to ensure that the experiments were performed with proper controls and that the results are repeatable. In controversial situations, experiments may even be fully repeated by other scientists to ensure that the same results are obtained. Once it is established that new findings are sound, the scientists can work together to determine how these results agree or disagree with previous experiments. They may also decide together what additional investigations would be useful to the field. **When open-minded scientific discourse is supported, opinions and information can be exchanged and theories can be refined.** Scientists will weigh the experimental evidence to determine what hypothesis has the most support. Given time, new theories may emerge and scientific knowledge can both broaden and deepen.

Skill 22.6 Understands how to use a variety of tools, techniques, and technology to gather, organize, and analyze data and how to apply appropriate methods of statistical measures and analysis.

To more easily manage large amounts of data, statistical measures are employed to characterize trends in the data. In many systems, the data fit a **normal distribution,** which has a concentration of data points in the center and two equally sized tails (the ends of the distributions).

This type of distribution looks like this:

A distribution is considered **skewed** if one tail is larger than the other. To further characterize distributions, a variety of statistical measures are used. The following are the most commonly used statistical measures:

Arithmetic Mean: The arithmetic mean is the same as the average of a distribution - the sum of all the data points divided by the number of data points. The arithmetic mean is a good measure of the central tendency of roughly normal distributions, but may be misleading in skewed distributions. In cases of skewed distributions, other statistics such as the median or geometric mean may be more informative.

Geometric Mean: The geometric mean is a better representation of the central tendency of a log-normal distribution or a distribution with a very wide range. The geometric mean is found by multiplying all the values together, then taking the nth root of the result, where n is the number of data points.

Median: The median is the middle of a distribution: half the scores are above the median and half are below it. Unlike the mean, the median is not highly sensitive to extreme data points. This makes the median a better measure than the mean for finding the central tendency of highly skewed distributions. The median is determined by organizing the data points from lowest to highest. When there is an odd number of numbers, the median is simply the middle number. For example, the median of 2, 4, and 7 is 4. When there is an even number of numbers, the median is the mean of the two middle numbers. Thus, the median of the numbers 2, 4, 7, and 12 is (4+7)/2 = 5.5.

Percentile: Percentiles are similar to a median, but may represent any point in the data set. For example, the 90th percentile represents that point at which 90% of the data points are below that value and 10% of the data points are above that value. Quartiles, representing the 25th, 50th, and 75th percentiles of a data set, are often used to describe a distribution.

Mode: The mode is the most frequently occurring data point in a distribution and is used as a measure of central tendency. The advantage of the mode as a measure of central tendency is that its meaning is obvious. However, the mode is greatly subject to sample fluctuations and so is not recommended for use as the only measure of central tendency. Additionally, many distributions have more than one mode. Note also that **in the case of a perfectly normal distribution, the mean, median, and mode are identical**.

Variance: The variance is used to give a measure of the variability in a distribution. It is computed as the average squared deviation of each number from its mean. For example, for the numbers 1, 2, and 3, the mean is 2 and the variance (σ^2) is:

$$\sigma^2= [(1-2)^2+(2-2)^2+(3-2)^2]/3=0.667$$

Standard deviation: Like variance, standard deviation is a measure of the spread of the distribution, but it is the more commonly used statistic. The standard deviation is simply the square root of the variance.

Note that the standard deviation can be used to compute the percentile rank associated with a given data point (if the mean and standard deviation of a normal distribution are known). In such a normal distribution, about 68% of the data points are within one standard deviation of the mean and about 95% of the data points are within two standard deviations of the mean.

Skill 22.7 Knows how to apply techniques to calibrate measuring devices and understands concepts of precision, accuracy, and error with regard to reading and recording numerical data from scientific instruments.

Scientific data can never be error-free. We can, however, gain useful information from our data by understanding what the sources of error are, how large they are and how they affect our results. Some errors are intrinsic to the measuring instrument, others are operator errors. Errors may be random (in any direction) or systematic (biasing the data in a particular way). In any measurement that is made, data must be quoted along with an estimate of the error in it.

Precision is a measure of how similar repeated measurements from a given device or technique are. Note that this is distinguished from **accuracy** which refers to how close to "correct" a measuring device or technique is. Thus, accuracy can be tested by measuring a known quantity (a standard) and determining how close the value provided by the measuring device is. To determine precision, however, we must make multiple measurements of the same sample. The precision of an instrument is typically given in terms of its standard error or standard deviation. Precision is typically divided into reproducibility and repeatability. These concepts are subtly different and are defined as follows:

Repeatability: Variation observed in measurements made over a short period of time while trying to keep all conditions the same (including using the same instrument, the same environmental conditions, and the same operator)

Reproducibility: Variation observed in measurements taken over a long time period in a variety of different settings (different places and environments, using different instruments and operators)

Both repeatability and reproducibility can be estimated by taking multiple measurements under the conditions specified above. Using the obtained values, standard deviation can be calculated using the formula:

$$\sigma = \sqrt{\frac{1}{N}\sum_{i=1}^{N}(x_i - \overline{x})^2}$$

where σ = standard deviation
N = the number of measurements
x_i = the individual measured values
$\overline{x}$ = the average value of the measured quantity

To obtain a reliable estimate of standard deviation, N, the number of samples, should be fairly large. We can use statistical methods to determine a confidence interval on our measurements. A typical confidence level for scientific investigations is 90% or 95%.

Often in scientific operations we want to determine a quantity that requires many steps to measure. Of course, each time we take a measurement there will be a certain associated error that is a function of the measuring device. Each of these errors contributes to an even greater one in the final value. This phenomenon is known as **propagation of error** or propagation of uncertainty.

A measured value is typically expressed in the form x±Δx, where Δx is the uncertainty or margin of error. What this means is that the value of the measured quantity lies somewhere between x-Δx and x+Δx, but our measurement techniques do not allow us any more precision. If several measurements are required to ultimately decide a value, we must use formulas to determine the total uncertainty that results from all the measurement errors. A few of these formulas for simple functions are listed below:

Formula	Uncertainty
$X = A \pm B$	$(\Delta X)^2 = (\Delta A)^2 + (\Delta B)^2$
$X = cA$	$\Delta X = c\Delta A$
$X = c(A \cdot B)$	$\left(\frac{\Delta X}{X}\right)^2 = \left(\frac{\Delta A}{A}\right)^2 + \left(\frac{\Delta B}{B}\right)^2$
$X = c\left(\frac{A}{B}\right)$	$\left(\frac{\Delta X}{X}\right)^2 = \left(\frac{\Delta A}{A}\right)^2 + \left(\frac{\Delta B}{B}\right)^2$

For example, if we wanted to determine the density of a small piece of metal we would have to measure its weight on a scale and then determine its volume by measuring the amount of water it displaces in a graduated cylinder. There will be error associated with measurements made by both the scale and the graduated cylinder. Let's suppose we took the following measurements:

Mass: 57± 0.5 grams
Volume: 23 ± 3 mm^3

Since density is simply mass divided by the volume, we can determine its value to be:

$$\rho = \frac{m}{V} = \frac{57g}{23mm^3} = 2.5\frac{g}{mm^3}$$

Now we must calculate the uncertainty on this measurement, using the formula above:

$$\left(\frac{\Delta x}{x}\right)^2 = \left(\frac{\Delta A}{A}\right)^2 + \left(\frac{\Delta B}{B}\right)^2$$

$$\Delta x = \left(\sqrt{\left(\frac{\Delta A}{A}\right)^2 + \left(\frac{\Delta B}{B}\right)^2}\right) x = \left(\sqrt{\left(\frac{0.5g}{57g}\right)^2 + \left(\frac{3mm^3}{23mm^3}\right)^2}\right) \times 2.5\frac{g}{mm^3} = 0.3\frac{g}{mm^3}$$

Thus, the final value for the density of this object is 2.5 ± 0.3 g/mm^3.

Skill 22.8 Uses the International System of Units (i.e., metric system) and performs unit conversions within and across measurement systems.

SI is an abbreviation of the French *Système International d'Unités* or the **International System of Units**. It is the most widely used system of units in the world and is the system used in science. The use of many SI units in the United States is increasing outside of science and technology. There are two types of SI units: **base units** and **derived units**. The base units are:

Quantity	Unit name	Symbol
Length	meter	m
Mass	kilogram	kg
Amount of substance	mole	mol
Time	second	s
Temperature	kelvin	K
Electric current	ampere	A
Luminous intensity	candela	cd

The name "kilogram" occurs for the SI base unit of mass for historical reasons. Derived units are formed from the kilogram, but appropriate decimal prefixes are attached to the word "gram." Derived units measure a quantity that may be **expressed in terms of other units**. Some derived units important for physics are:

Derived quantity	Unit name	Expression in terms of other units	Symbol
Area	square meter	m^2	
Volume	cubic meter liter	m^3 $dm^3 = 10^{-3}\ m^3$	 L or l
Mass	unified atomic mass unit	$(6.022 X 10^{23})^{-1}$ g	u or Da
Time	minute hour day	60 s 60 min=3600 s 24 h=86400 s	min h d
Speed	meter per second	m/s	
Acceleration	meter per second squared	m/s^2	
Temperature*	degree Celsius	K	°C
Mass density	gram per liter	$g/L = 1\ kg/m^3$	
Force	newton	$m \cdot kg/s^2$	N
Pressure	pascal standard atmosphere§	$N/m^2 = kg/(m \cdot s^2)$ 101325 Pa	Pa atm
Energy, Work, Heat	joule nutritional calorie§	$N \cdot m = m^3 \cdot Pa = m^2 \cdot kg/s^2$ 4184 J	J Cal
Heat (molar)	joule per mole	J/mol	
Heat capacity, entropy	joule per kelvin	J/K	
Heat capacity (molar), entropy (molar)	joule per mole kelvin	J/(mol•K)	
Specific heat	joule per kilogram kelvin	J/(kg•K)	
Power	watt	J/s	W
Electric charge	coulomb	s•A	C
Electric potential, electromotive force	volt	W/A	V
Viscosity	pascal second	Pa•s	
Surface tension	newton per meter	N/m	

*Temperature differences in Kelvin are the same as those differences in degrees Celsius. To obtain degrees Celsius from Kelvin, subtract 273.15. Differentiate *m* and meters (m) by context.

§These are commonly used non-SI units.

Decimal multiples of SI units are formed by attaching a **prefix** directly before the unit and a symbol prefix directly before the unit symbol. SI prefixes range from 10^{-24} to 10^{24}. Common prefixes you are likely to encounter in physics are shown below:

Factor	Prefix	Symbol	Factor	Prefix	Symbol
10^9	*giga—*	G	10^{-1}	*deci—*	d
10^6	*mega—*	M	10^{-2}	*centi—*	c
10^3	*kilo—*	k	10^{-3}	*milli—*	m
10^2	*hecto—*	h	10^{-6}	*micro—*	μ
10^1	*deca—*	da	10^{-9}	*nano—*	n
			10^{-12}	*pico—*	p

Example: 0.0000004355 meters is $4.355\text{X}10^{-7}$ m or $435.5\text{X}10^{-9}$ m. This length is also 435.5 nm or 435.5 nanometers.

Example: Find a unit to express the volume of a cubic crystal that is 0.2 mm on each side so that the number before the unit is between 1 and 1000.

Solution: Volume is length X width X height, so this volume is $(0.0002\text{ m})^3$ or $8\text{X}10^{-12}\text{ m}^3$. Conversions of volumes and areas using powers of units of length must take the power into account. Therefore:

$$1\text{ m}^3 = 10^3\text{ dm}^3 = 10^6\text{ cm}^3 = 10^9\text{ mm}^3 = 10^{18}\ \mu\text{m}^3,$$

The length 0.0002 m is $2\text{ X}10^2\ \mu$m, so the volume is also $8\text{X}10^6\ \mu\text{m}^3$. This volume could also be expressed as $8\text{X}10^{-3}\text{ mm}^3$. None of these numbers, however, is between 1 and 1000.

Expressing volume in liters is helpful in cases like these. There is no power on the unit of liters, therefore:

$$1\text{ L} = 10^3\text{ mL} = 10^6\ \mu\text{L} = 10^9\text{ nL}.$$

Converting cubic meters to liters gives

$$8 \times 10^{-12}\text{ m}^3 \times \frac{10^3\text{ L}}{1\text{ m}^3} = 8 \times 10^{-9}\text{ L}.$$

The crystal's volume is 8 nanoliters (8 nL).

Example: Determine the ideal gas constant, R, in L•atm/(mol•K) from its SI value of 8.3144 J/(mol•K).

Solution: One joule is identical to one m^3•Pa (see the table on the previous page).

$$8.3144\frac{\text{m}^3 \bullet \text{Pa}}{\text{mol} \bullet \text{K}} \times \frac{1000\text{ L}}{1\text{ m}^3} \times \frac{1\text{ atm}}{101325\text{ Pa}} = 0.082057\frac{\text{L} \bullet \text{atm}}{\text{mol} \bullet \text{K}}$$

The **order of magnitude** is a familiar concept in scientific estimation and comparison. It refers to a category of scale or size of an amount, where each category contains values of a fixed ratio to the categories before or after. The most common ratio is 10. Orders of magnitude are typically used to make estimations of a number. For example, if two numbers differ by one order of magnitude, one number is 10 times larger than the other. If they differ by two orders of magnitude the difference is 100 times larger or smaller, and so on. It follows that two numbers have the same order of magnitude if they differ by less than 10 times the size.

To estimate the order of magnitude of a physical quantity, you round the its value to the nearest power of 10. For example, in estimating the human population of the earth, you may not know if it is 5 billion or 12 billion, but a reasonable order of magnitude estimate is 10 billion. Similarly, you may know that Saturn is much larger than Earth and can estimate that it has approximately 100 times more mass, or that its mass is 2 orders of magnitude larger. The actual number is 95 times the mass of earth. Below are the dimensions of some familiar objects expressed in orders of magnitude.

Physical Item	Size	Order of Magnitude (meters)
Diameter of a hydrogen atom	100 picometers	10^{-10}
Size of a bacteria	1 micrometer	10^{-6}
Size of a raindrop	1 millimeter	10^{--3}
Width of a human finger	1 centimeter	10^{-2}
Height of Washington Monument	100 meters	10^{2}
Height of Mount Everest	10 kilometers	10^{4}
Diameter of Earth	10 million meters	10^{7}
One light year	1 light year	10^{16}

COMPETENCY 23.0 THE TEACHER UNDERSTANDS THE NATURE OF SCIENCE, THE PROCESS OF SCIENTIFIC INQUIRY, AND THE UNIFYING CONCEPTS THAT ARE COMMON TO ALL SCIENCES.

Skill 23.1 Understands the nature of science, the relationship between science and technology, the predictive power of science, and limitations to the scope of science (i.e., the types of questions that science can and cannot answer).

Modern science began around the late 16th century with a new way of thinking about the world. Few scientists will disagree with Carl Sagan's assertion that "science is a way of thinking much more than it is a body of knowledge" (Broca's Brain, 1979). Science is a process of inquiry and investigation. It is a way of thinking and acting, not just a body of knowledge to be acquired by memorizing facts and principles. This way of thinking, the scientific method, is based on the idea that scientists begin their investigations with observations. From these observations they develop a hypothesis, which is further developed into a prediction. The hypothesis is challenged through experimentation and further observations and is refined as necessary. Science has progressed in its understanding of nature through careful observation, a lively imagination, and increasingly sophisticated instrumentation. Science is distinguished from other fields of study in that it provides guidelines or methods for conducting research, and the research findings must be reproducible by other scientists for those findings to be validated.

It is important to recognize that scientific practice is not always this systematic. Discoveries have been made that are serendipitous and others have been predicted based on theory rather than observation of phenomena. Einstein's theory of relativity was developed not from the observation of data but with a kind of mathematical puzzle. Only later were experiments able to be conducted that validated his theory.

The scientific method is a logical set of steps that a scientist goes through to solve a problem. While an inquiry may start at any point in this method and may not involve all of the steps, the overall approach can be described as follows:

Making observations

Scientific questions frequently result from observation of events in nature or in the laboratory. An observation is not just a look at what happens. It also includes measurements and careful records of the event. Records could include photos, drawings, or written descriptions. The observations and data collection may provide answers, or they may lead to one or more questions. In chemistry, observations almost always deal with the behavior of matter.

Having arrived at a question, a scientist usually researches the scientific literature to see what is known about the question. Perhaps the question has already been answered, or another experimenter has found part of the solution. The scientist may want to test or reproduce the answer found in the literature. Or, the research might lead to a new question.

Sometimes the same observations are made over and over again and are always the same. For example, one can observe that daylight lasts longer in summer than in winter. This observation never varies. Such observations are called **laws of nature**. For example, one of the most important laws in chemistry was discovered in the late 1700s. Chemists observed that no mass was ever lost or gained in chemical reactions. This law became known as the law of conservation of mass. Explaining this law was a major topic of chemistry in the early 19th century.

Developing a hypothesis

If the question has not yet been answered, the scientist may prepare for an experiment by making a hypothesis. A hypothesis is a statement of a possible answer to the question. It is a tentative explanation for a set of facts and can be tested by experiments. Although hypotheses are usually based on observations, they may also be based on a sudden idea or intuition or a mathematical theory.

Conducting an experiment

An experiment tests the hypothesis to determine whether it may be a correct answer to the question or a solution to the problem. Some experiments may test the effect of one thing on another under controlled conditions. Such experiments have two variables. The experimenter controls one variable, called the independent variable. The other variable, the dependent variable, shows the result of changing the independent variable.

For example, suppose a researcher wanted to test the effect of Vitamin A on the ability of rats to see in dim light. The independent variable would be the dose of Vitamin A added to the rats' diet. The dependent variable would be the intensity of light to which the rats respond. All other factors, such as time, temperature, age, water and other nutrients given to the rats, are held constant.

Chemists sometimes do short experiments "just to see what happens" or to see what a certain reaction produces. Often, these are not formal experiments. Rather they are ways of making additional observations about the behavior of matter.

In most experiments scientists collect quantitative data, which are data that can be measured with instruments. They also collect qualitative data, descriptive information from observations other than measurements. Interpreting data and analyzing observations are an important part of the scientific method. If data are not organized in a logical manner, incorrect conclusions can be drawn. Also, other scientists may not be able to follow or reproduce the results.

Drawing conclusions

Finally, a scientist must draw conclusions from the experiment. A conclusion must address the hypothesis on which the experiment was based. The conclusions state whether or not the data support the hypothesis. If not, the conclusion should state what the experiment did show. If the hypothesis is not supported, the scientist uses the observations from the experiment to make a new or revised hypothesis and plan new experiments.

Developing a theory

When a hypothesis survives many experimental tests to determine its validity, the hypothesis may be developed into a **theory**. A theory explains a body of facts and laws that are based on the facts. A theory also reliably predicts the outcome of related events in nature. For example, the law of conservation of matter and many other experimental observations led to a theory proposed early in the 19th century. This theory explained the conservation law by proposing that all matter is made up of atoms which are never created or destroyed in chemical reactions, only rearranged. This atomic theory also successfully predicted the behavior of matter in chemical reactions that had not been studied at the time. As a result, the atomic theory has stood for 200 years with only minor modifications.

A theory also serves as a scientific **model**. A model can be a physical model made of wood or plastic, a computer program that simulates events in nature, or simply a mental picture of an idea. A model illustrates a theory and explains nature. For example, in your chemistry course, you will develop a mental (and possibly a physical) model of the atom and its behavior. Outside of science, the word theory is often used to describe an unproven notion. In science, theory means much more. It is a thoroughly tested explanation of things and events observed in nature.

A theory can never be proven true, but it can be proven untrue. All this requires is to demonstrate an exception to the theory.

The Scientific Method

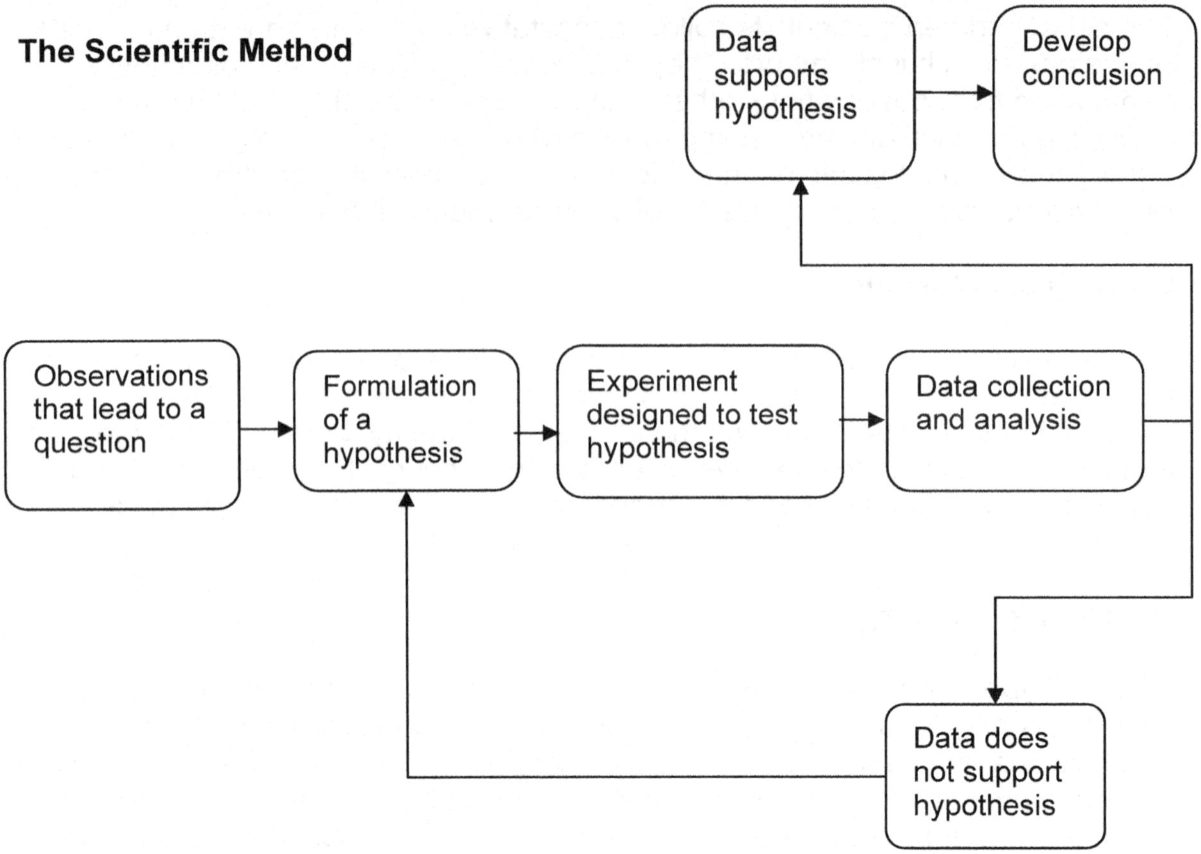

The limits of science

Throughout history, different ways of knowing have complimented each other and have been used by all civilizations. Different fields of study require different skills, from mathematics and science to arts and crafts, from medicine and law to religion and philosophy.

There are four basic ways of knowing:

- From personal experience
- From a trusted source
- From intuition or inspiration
- From reason or logical thinking.

Science is a combination of observational experience, reason, and logical thinking. Religion flows from a trusted source along with inspiration, while philosophy comes from logical thinking.

Science, including the natural sciences of physics, biology and chemistry along with social sciences like psychology and sociology, uses the scientific method to establish a standard of proof. There is a well-established series of steps that must be followed, beginning with research and experimentation and leading to peer review and publication. These standards of proof make it more difficult for science to be manipulated for any length of time. For example, March of 1989 brought the announcement of cold fusion, a long sought after phenomenon that set the scientific community ablaze. However, this fire was extinguished quickly amidst accusations of fraud and incompetence after many other research teams unsuccessfully attempted to replicate the results. The authors' research was discredited by rushed publication of incomplete results as well as errors in their data interpretation.

In other areas of knowing, information is much more difficult to evaluate since the evidence used for study is open to wide interpretation. For the most part, other areas of study rely on personal experience for knowing. Religious knowing relies on information from religious texts, personal inspiration, or from a deity. It can come from a variety of translations, or it can come through prayer or meditation or association with other believers and followers. Philosophy is a belief accepted by a school or group and relies on public discourse for validation. Assessments of the same piece of literature can vary widely from one person to another. Learning about history uses primary and secondary sources, and again relies on a great deal of elucidation.

While no method for learning is more right than another, each different way of knowing is necessary to fill our senses and explain our world. They all work together to provide a full picture of the world in which we live. In general, one type of investigation is not appropriate for evaluating other ways of knowing. For example, science is not an appropriate way of investigating religious truths, while religion or philosophy may not be adequate means of investigating scientific phenomena.

Skill 23.2 Knows the characteristics of various types of scientific investigations (e.g., descriptive studies, controlled experiments, comparative data analysis) and how and why scientists use different types of scientific investigations.

The design of chemical experiments must include every step needed to obtain the desired data. In other words, the design must be **complete** and it must include all required **controls**. These attributes are described further in Skill 2.3.

By definition, qualitative observations are descriptive in nature. For example, "the sample was yellow" is a descriptive statement and thus is an example of qualitative data. Quantitative data are measurements that are numerical in nature. "The sample had a mass of 1.15 grams" is a quantitative data point. Both types of observations have their place in scientific research.

Qualitative analysis (descriptive studies)

The objective of qualitative data analysis is a complete and detailed description, from which patterns or other useful information may be gained. For example, an ornithologist may observe individuals of a bird species over a period of time to identify the resources they need for food and shelter or to document mating behavior.

Qualitative analysis involves a continual interplay between theory and analysis. In analyzing qualitative data, we seek to discover patterns such as changes over time or possible causal links between variables. The main disadvantage of qualitative approaches to data analysis is that the results can not be extended to wider populations with the same degree of certainty that quantitative analyses can. This is because the findings of the research are not tested to discover whether they are statistically significant or due to chance.

Qualitative analysis is frequently used to study natural phenomena or activities that do not lend themselves as easily to quantitative analysis, such as the behavioral interactions of animals and people. In such cases, a qualitative analysis is often conducted prior to designing a more rigorous quantitative study.

In disciplines such as chemistry, which rely largely on quantitative approaches, qualitative observations are still very important and are nearly always recorded in laboratory notebooks alongside quantitative measurements. Qualitative information can be very important in developing hypotheses to explain unexpected results. For example, observations of color or texture changes during an experiment may identify impurities likely present in the reagents or overheating of the solution during a particular step.

Quantitative analysis

Analysis of quantitative data involves making detailed measurements, classifying features, counting them and constructing statistical models in an attempt to explain observations. These findings, then, can be generalized to a larger population or the ideal case, and direct comparisons can be made between two data sets so long as valid sampling and significance techniques have been used.

However, the picture of the data which emerges from quantitative analysis is less rich than that obtained from qualitative analysis. For statistical purposes, classifications must be strict. An item either belongs to class *x* or it doesn't. Quantitative analysis is therefore an *idealization* of the data in some cases. In addition, quantitative analysis tends to sideline rare occurrences. To ensure that certain statistical tests (such as chi-squared) provide reliable results, it is essential that minimum frequencies are obtained - meaning that categories may have to be collapsed into one another resulting in a loss of data richness. In addition, censoring of outliers may occur.

Basically, quantitative research is objective; qualitative is subjective. Quantitative research seeks explanatory laws; qualitative research aims at in-depth description. Quantitative research measures what it assumes to be a static reality in hopes of developing universal laws and is well suited to establishing cause-and-effect relationships. Qualitative research is an exploration of what is assumed to be a dynamic reality. It does not claim that what is discovered in the process is universal, and thus necessarily replicable.

Whether to choose a fundamentally quantitative or a qualitative design depends on the nature of the project, the type of information needed, the context of the study, and the availability of resources (time, money, and human). It is important to keep in mind that these are two different approaches, not necessarily polar opposites. In fact, elements of both designs can and should be used together in mixed-methods studies. In scientific disciplines such as chemistry, it is generally the norm to record at least some observations of both types.

Advantages of combining both types of data include:

1. Research development (one approach is used to inform the other, such as using qualitative research to develop an instrument to be used in quantitative research)
2. Increased validity (confirmation of results by means of different data types)
3. Complementarity (adding information, e.g., descriptions alongside measurements)
4. Providing additional resources for explaining unanticipated results or failed experiments.

Skill 23.3 Understands principles and procedures for designing and conducting a variety of scientific investigations, with emphasis on inquiry-based investigations, and how to communicate and defend scientific results.

Application of the scientific method requires familiarity with certain skills that are common to all disciplines. The tools used in each case will depend on the area of study and the specific subject of study. What is common is the mode and attitude with which each skill is applied. Needless to say, uncompromising honesty and reporting of observations with as much objectivity as possible is a fundamental requirement of the scientific process.

Observing: All scientific theories and laws ultimately rest on a strong foundation of experiment. Observation, whether by looking through a microscope or by measuring with a voltmeter, is the fundamental method by which a scientist interacts with the environment to gather the needed data. Scientific observations are not just casual scrutiny but are made in the context of a rigorously planned experiment that specifies precisely what is to be observed and how. Observations must be repeated and the conditions under which they are made clearly noted in order to ensure their validity.

Hypothesizing: Hypothesizing is proposing an answer to a scientific question in order to set the parameters for experiment and to decide what is to be observed and under what conditions. A hypothesis is not a random guess but an educated conjecture based on existing theories and related experiments and a process of rigorous logical reasoning from these basics.

Ordering: For experimental or calculated data to be useful and amenable to analysis, it must be organized appropriately. How data is ordered depends on the question under investigation and the observation process. Ordering may involve prioritizing or categorizing. A data set may be ordered in multiple ways with respect to different variables in order to perform different kinds of analysis on it.

Categorizing: Categorizing is part of the process of ordering or organizing data either into known groups or by identifying new groups through review of the data. The groups may be formed in multiple dimensions, i.e. with respect to more than one variable. For example, a group of objects may be categorized by color as well as size. For some scientific experiments, categorizing may be the goal of the investigation. Categorized data is typically presented in tabular form with the category names as headings.

Comparing: Comparing equivalent quantities is one of the fundamental processes of science. In some cases an observation may be compared with a known or standard number in order to ascertain whether it meets certain criteria. In other cases, data points may be compared with each other for the purpose of prioritizing, categorizing or graphing. Before comparing one must ensure that the numbers are expressed in the same units.

Inferring: Once data has been organized, graphed and analyzed, a scientist draws conclusions or inferences based on logical reasoning from what he/she sees. An inference is generally drawn in the context of the initial hypothesis. An inference addresses whether the data disproves or supports the hypothesis and to what extent. An inference may also be drawn about some aspect of the data that was not included in the hypothesis. It may lead to the formulation of new problems and new hypotheses or provide answers to questions other than those asked in the hypothesis.

Applying: Applying is the process of connecting a theory, law or thought process to a physical situation, experimental set up or data. Not all laws are applicable to all situations. Also, a theory may be applied to data, for instance, only when it is organized in a specific way. Science requires the ability to evaluate when and how ideas and theories are applicable in specific cases.

Communicating: Communicating, both orally and by writing, is a vital part of scientific activity. Science is never done in a vacuum and theories and experiments are validated only when other scientists can reproduce them and agree with the conclusions. Also, scientific theories can be put to practical use only when others are able to understand them clearly. Thus communication, particularly with peers, is critical to the success of science.

The scientific method is a logical set of steps that a scientist goes through to solve a problem. There are as many different scientific methods as there are scientists experimenting. However, there seems to be some pattern to their work. The scientific method is the process by which data is collected, interpreted and validated. While an inquiry may start at any point in this method and may not involve all of the steps here is the general pattern.

Formulating problems

Although many discoveries happen by chance, the standard thought process of a scientist begins with forming a question to research. The more limited and clearly defined the question, the easier it is to set up an experiment to answer it. Scientific questions result from observations of events in nature or events observed in the laboratory. An **observation** is not just a look at what happens. It also includes measurements and careful records of the event. Records could include photos, drawings, or written descriptions. The observations and data collection lead to a question. In physics, observations almost always deal with the behavior of matter. Having arrived at a question, a scientist usually researches the scientific literature to see what is known about the question. Maybe the question has already been answered. The scientist then may want to test the answer found in the literature. Or, maybe the research will lead to a new question.

Sometimes the same observations are made over and over again and are always the same. For example, you can observe that daylight lasts longer in summer than in winter. This observation never varies. Such observations are called **laws** of nature. One of the most important scientific laws was discovered in the late 1700s. Chemists observed that no mass was ever lost or gained in chemical reactions. This law became known as the law of conservation of mass. Explaining this law was a major topic of scientific research in the early 19th century.

Forming a hypothesis

Once the question is formulated, take an educated guess about the answer to the problem or question. This 'best guess' is your hypothesis. A **hypothesis is a statement of a possible answer to the question**. It is a tentative explanation for a set of facts and can be tested by experiments. Although hypotheses are usually based on observations, they may also be based on a sudden idea or intuition.

Experiment
An experiment tests the hypothesis to determine whether it may be a correct answer to the question or a solution to the problem. Some experiments may test the effect of one thing on another under controlled conditions. Such experiments have two variables. The experimenter controls one variable, called the **independent variable**. The other variable, the **dependent variable**, is the change caused by changing the independent variable. For example, suppose a researcher wanted to test the effect of vitamin A on the ability of rats to see in dim light. The independent variable would be the dose of Vitamin A added to the rats' diet. The dependent variable would be the intensity of light that causes the rats to react. All other factors, such as time, temperature, age, water given to the rats, the other nutrients given to the rats, and similar factors, are held constant. Scientists sometimes do short experiments "just to see what happens". Often, these are not formal experiments. Rather they are ways of making additional observations about the behavior of matter. A good test will try to manipulate as few variables as possible so as to see which variable is responsible for the result. This requires a second example of a **control**. A control is an extra setup in which all the conditions are the same except for the variable being tested.

In most experiments, scientists collect quantitative data, which is data that can be measured with instruments. They also collect qualitative data, descriptive information from observations other than measurements. Interpreting data and analyzing observations are important. If data is not organized in a logical manner, wrong conclusions can be drawn. Also, other scientists may not be able to follow your work or repeat your results.

Conclusion
Finally, a scientist must draw conclusions from the experiment. A conclusion must address the hypothesis on which the experiment was based. The conclusion states whether or not the data supports the hypothesis. If it does not, the conclusion should state what the experiment *did* show. If the hypothesis is not supported, the scientist uses the observations from the experiment to make a new or revised hypothesis. Then, new experiments are planned.

Theory
When a hypothesis survives many experimental tests to determine its validity, the hypothesis may evolve into a **theory**. A theory explains a body of facts and laws that are based on the facts. A theory also reliably predicts the outcome of related events in nature. For example, the law of conservation of matter and many other experimental observations led to a theory proposed early in the 19th century. This theory explained the conservation law by proposing that all matter is made up of atoms which are never created or destroyed in chemical reactions, only rearranged. This atomic theory also successfully predicted the behavior of matter in chemical reactions that had not been studied at the time. As a result, the atomic theory has stood for 200 years with only small modifications.

A theory also serves as a scientific **model**. A model can be a physical model made of wood or plastic, a computer program that simulates events in nature, or simply a mental picture of an idea. A model illustrates a theory and explains nature. For instance, in your science class you may develop a mental (and maybe a physical) model of the atom and its behavior. Outside of science, the word theory is often used to describe someone's unproven notion about something. In science, theory means much more. It is a thoroughly tested explanation of things and events observed in nature. A theory can never be proven true, but it can be proven untrue. All it takes to prove a theory untrue is to show an exception to the theory. The test of the hypothesis may be observations of phenomena or a model may be built to examine its behavior under certain circumstances.

Steps of a Scientific Method

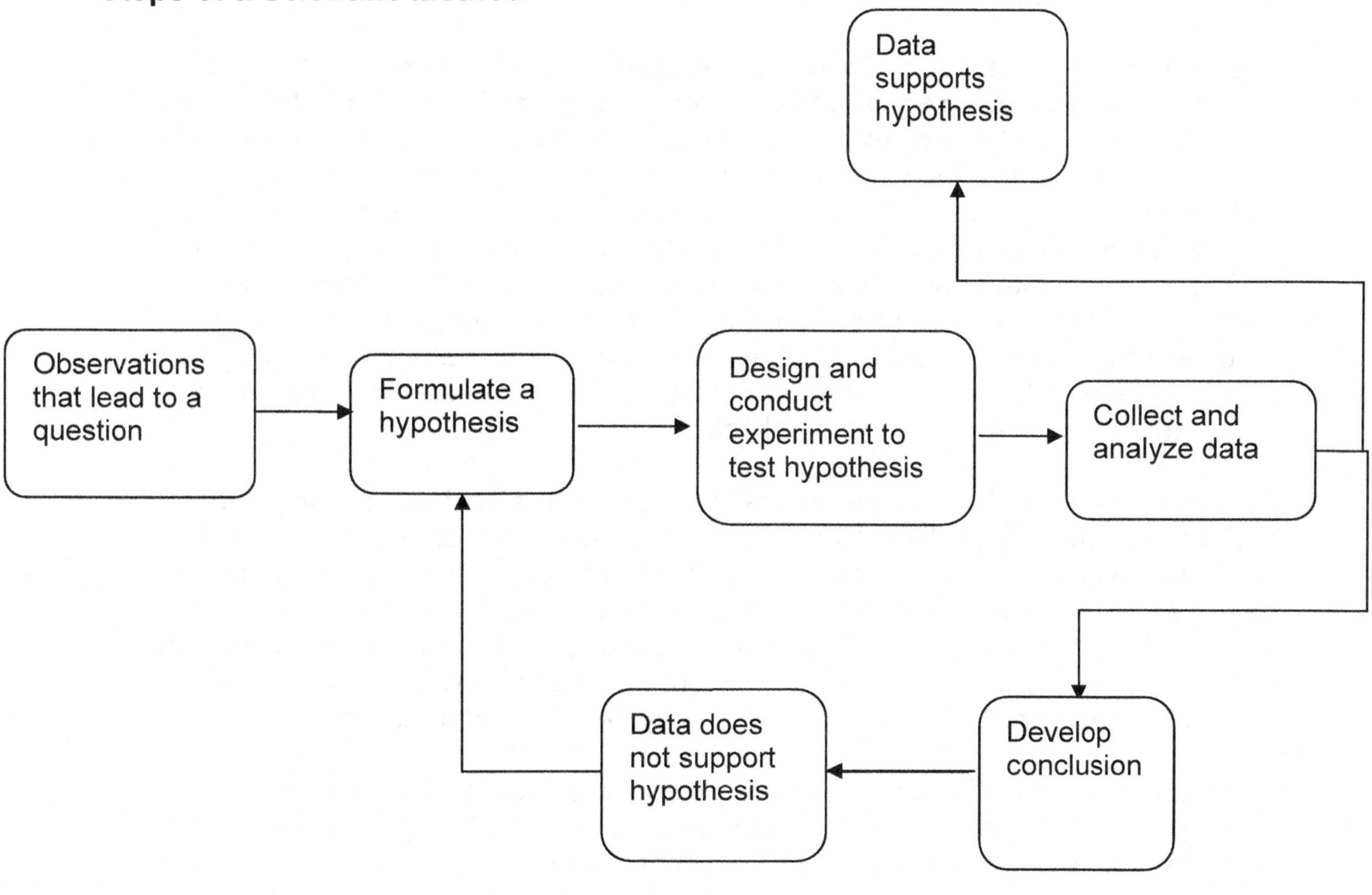

Skill 23.4 Understands how logical reasoning, verifiable observational and experimental evidence, and peer review are used in the process of generating and evaluating scientific knowledge.

Modern science began around the late 16th century with a new way of thinking about the world. **Science is a process of inquiry and investigation**. It is a way of thinking and acting, not just a body of knowledge to be acquired by memorizing facts and principles. This way of thinking, the scientific method, is based on the idea that scientists begin their investigations with observations.

Science is distinguished from other fields of study in that it **provides guidelines or methods for conducting research.** Of utmost importance is that the results of scientific research be **reproducible**, not just by the original investigators, but by any other person performing the identical experiment. Ideally the scientific community all works together to advance knowledge of the natural world. The process of peer review is central to this ideal.

Peer review is the process by which scientific results produced by one group are subjected to the analysis of other experts in the field. In practice it is most often used by scientific journals. Scientists author manuscripts detailing their experiments, results, and interpretations and these manuscripts are distributed by the journal editors to other researchers in the field for review prior to publication. The authors must address the comments and questions of the reviewers and make appropriate revisions for their work to be accepted for publication. Peer review is also the process by which applications for research funds are evaluated and awarded. Peer review may also be used informally by groups of researchers or graduate students wishing to get an evaluation of their research prior to writing it up for publication.

Reviewers of scientific work are typically experts in the field, but it is important that they be objective in their evaluations because it is possible that the results under review may contradict the ideas of the reviewers. Peer review is typically done **anonymously** so that the identities of the reviewers remain unknown by the scientists submitting work for review. However, less formal peer review may occur through lunch seminars, presentations at scientific conferences, and other venues where comments and responses may be provided in person.

The goal of peer review is to "weed out" science not performed to appropriate standards. This typically means the scientific method has been employed but also state of the art technical procedures have been followed and the conclusions that are drawn are fully supported by the results. Therefore, the reviewer will determine whether proper controls were in place, enough replicates were performed, and that the experiments clearly address the presented hypothesis. The reviewer will scrutinize the interpretations and how they fit into what is already known in the field. Often reviewers will suggest that additional experiments be done to further corroborate presented conclusions.

Occasionally, new scientific results may contradict long held ideas in a particular field. In these cases, in-depth and objective peer review is highly important. Scientists must work together to determine whether the new evidence is correct and how it might change current theories. Typically, there is resistance to the overthrow of scientific theories and many, many experiments must be arduously validated before new, contradictory hypotheses are accepted. Unfortunately, scientists are still people and so can be stubborn and slow to change their ideas. Therefore, **acceptance of new scientific results, even when experiments have been correctly performed, often takes some time.**

Skill 23.5 Understands how to identify potential sources of error in an investigation, evaluate the validity of scientific data, and develop and analyze different explanations for a given scientific result.

There are many sources of error in a scientific investigation. Being aware of the many types of error that can occur helps researchers design good studies, explain unexpected results, formulate and evaluate alternative hypotheses, and conduct peer review. Several broad categories of errors include the following:

Study design

Errors in study design may include a poorly developed problem statement or hypothesis, improper selection of controls, not controlling enough variables or allowing for too many independent variables at once, and improper selection of materials, equipment, or analytical tools. If you are sampling a population, care needs to be taken that your sampling protocol will result in a sample that is representative of the population you are sampling, to the degree possible. The sample size and/or replication also must be large enough to provide sufficient statistical power for analyzing the results.

Conduct of the study

Errors may occur if mistakes are made during the experiment, for example, miscalibration of an instrument, measuring incorrectly, setting the temperature too high, misreading an instrument, mislabeling a beaker, or mixing up reagents.

Random and systematic error

Use of measuring techniques may lead to either random or systematic errors, as discussed in Skill 1.7. It is important to differentiate random measurement error from natural variability in a population, and to be aware of which of the two you are observing. For example, water should always boil at the same temperature under the same environmental conditions, and if you observe variations in the measured temperature, it is most likely due to random error in the measurement.

On the other hand, students have natural variation in their heights, and if you measure such a variation you can be sure that natural variability in the population is what you are observing. Most likely there is also some random error in the measurement, but it is much smaller than the natural variability. If the population you are working with is expected to have natural variability, it is helpful to know the expected magnitude of this variability compared with the magnitude of random error in your measurements to determine whether the random error is likely to interfere with interpretation of the data.

Manipulation of the data

Errors may arise during manipulation of the data once the experiment is over. For example, a number may be incorrectly transcribed from a lab sheet to a spreadsheet. An equation may be set up incorrectly or an error may be made during a calculation. Special care should be used when manipulating data in spreadsheets or databases, as many software programs have inherent errors in their data manipulation routines. For example, the most widely used spreadsheet programs are known to have errors in rounding, treatment of significant digits, and generation of random numbers, as well as certain statistical manipulations. Databases may also automatically convert number fields to other types of data, such as text or dates. All calculations performed by a computer should be spot-checked by hand to ensure that calculations are being performed as intended and that errors (such as rounding errors) are not being propagated through the calculations.

Conclusions

Data analysis is one of the steps in which conceptual errors can be easily introduced if the researcher is not completely objective and open to alternative hypotheses. Even if the results seem to support the initial hypothesis, the researcher should ask whether alternative hypotheses or explanations are possible, and carefully study all quantitative and qualitative data collected to determine which hypothesis is best supported by all the information gathered. If some aspect of the results doesn't seem to fit, it should be emphasized in the discussion to allow broader review of the issue. Often another researcher may be able to shed light on the seeming discrepancy, but only if it is mentioned and discussed.

Presentation

One of the most important parts of conducting an experiment is presenting it to others. Care should be taken in the presentation to ensure that transcription errors are not introduced, tables and figures are labeled accurately and with complete units, and that the setup, conduct, and results of the study are accurately characterized.

Validation of results

In order to validate any scientific data, an experiment should be clearly documented and reproducible. Other researchers reviewing the study should be able to identify the steps taken to reduce all possible sources of error and should have access to the information needed to repeat the experiment. Original laboratory records and instrument print-outs should be archived by the researcher so that they can be made available if there are any questions that need to be investigated.

Skill 23.6 Knows the characteristics and general features of systems; how properties and patterns of systems can be described in terms of space, time, energy, and matter; and how system components and different systems interact.

The study of the properties and behavior of systems as a whole is known as **systems theory**. It is a highly interdisciplinary field ranging from physics to philosophy. **A system is composed of parts or activities that work together to form a whole**. The most basic definition of a system is a configuration of parts joined together by various relationships. Systems theory places emphasis on the recognition of the structure of systems and the dependence of its components on one another, even if in a time-delayed fashion. Typically, **the whole has unique properties not possessed by the parts alone.** As a result, systems theory prioritizes characterizing the behavior of the system and not the individual parts. This is occasionally at odds with the more traditional approach to science in which components are isolated as much as possible for study.

Systems theory supports the notion that these isolated components do not behave in the same manner if they are removed from their system. For instance, an individual cell in a Petri dish does not behave in the same manner as it would within an organ inside a person. **In some systems, it is not possible to explain the behavior of the whole in terms of the behavior of the parts**. If you consider the English alphabet, you can see the manner in which each letter is largely meaningless on its own but when used to together the letters form words which convey much information. As another example, it is very difficult to predict the properties of water based on the elemental properties of hydrogen and oxygen. Properties of the whole that go beyond what can be readily predicted by studying the parts are called **emergent properties**.

Systems theory has grown to encompass physics, chemistry, biology, engineering, economics, sociology, political science, management, psychotherapy, and many other disciplines. Therefore, systems-based models have been applied to a wide variety of instances in which multiple components interact. These systems can become quite complex. For instance, consider the human body. To understand how food is used to make energy, studying a single cell from the wall of the small intestine might give you some information about how free nutrients are absorbed. But you must also understand how all the organs in the digestive system work together in sequence to digest the food.

Next, you would study the equally complex process by which energy in sugar is converted to ATP (adenosine triphosphate). Again, simply examining the mechanisms of addition of a phosphate group to ADP (adenosine diphosphate) would not give you a full picture of what is happening. Only when relationships between components in the systems and the relationships between the systems are clear can the entire process be understood. This illustrates that components of systems may be separated by space and time and a single system may interact with other systems to form an even more complex system.

Skill 23.7 Knows how to apply and analyze the systems model (e.g., interacting parts, boundaries, input, output, feedback, subsystems) across the science disciplines.

One of the key aspects of systems is that the various parts *interact.* **A system is not simply a conglomeration of the various parts, but a description of the relationship between these parts**. Thus, when we model systems, it is important that both the parts and their interactions are elucidated. We can use the system theory approach to examine a man-made system: a savings account. A more traditional view of a savings account would simply show how interest is accrued. But a systems model might look more like this:

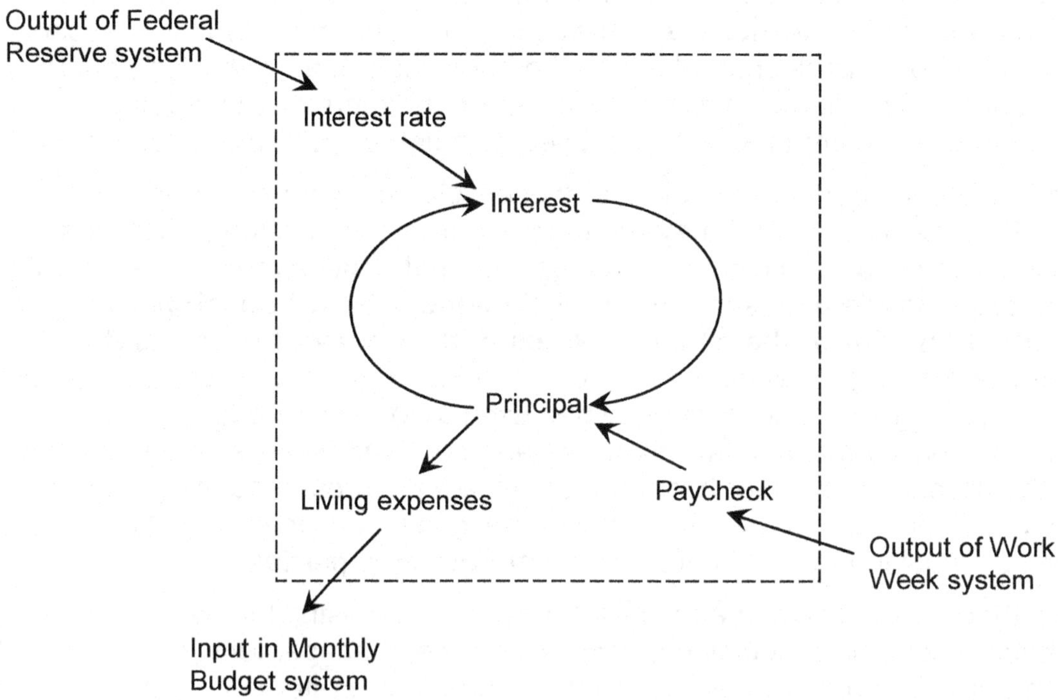

The dashed line encloses the savings account system and it is clear how the various components interact. The systems model makes it clear that principal doesn't simply grow forever; some principal is always lost to living expenses. We are also aware that each of these components belongs to other systems. **The system we have diagrammed provides outputs to and takes inputs from other systems**.

In the natural world, a similar level of complexity in interactions exists and ultimately requires an understanding of various scientific disciplines. For instance, if we wanted to fully explain an ecosystem, we would need to understand how climate (meteorology) and soil types (geology) influence the plant and animals (biology) that thrive and how they ultimately return nutrients to the environment (chemistry).

Our description of this ecosystem could become infinitely large as we include smaller and smaller subsystems within each system and feedback loops between these subsystems. At the same time, this ecosystem would be interacting with other ecosystems, and being acted on by large-scale effects such as weather, seasonal migrations of animals and birds, and human influences. When working with human or natural systems of such great complexity, it is important to define the boundaries of the system you are working with carefully.

Skill 23.8 Understands how shared themes and concepts (e.g., systems, order, and organization; evidence, models, and explanation; change, constancy, and measurements; evolution and equilibrium; and form and function) provide a unifying framework in science.

Math, science, and technology all have common themes in how they are applied and understood. Here are some of the fundamental concepts:

Systems

Because the natural world is so complex, the study of science involves the **organization** of items into smaller groups based on interaction or inter-dependence. These groups are known as **systems**. Systems consist of many separate parts interacting in specific ways to form a whole. It is these interactions that truly define the system. The complete system then has its own characteristics that go beyond the simple collection of its components (i.e., "the whole is more than the sum of the parts"). Natural phenomena and complex technologies can almost always be represented as systems. Examples of systems are the solar system, cardiovascular system, Newton's laws of force and motion, and the laws of conservation.

Models
Science and technology employ models to help simplify concepts. Models can be actual, small, physical mock-ups, mathematical equations, or diagrams that represent the fundamental relationships being studied. Models allow us to gain an understanding of these relationships and to make predictions. Similarly, diagrams, graphs, and charts are often employed to make these phenomena more readily understandable in a visual way.

Change and equilibrium
Another common theme among these three areas is the alternation between change and stability. These alternations occur in natural systems, which typically follow a pattern in which variation is introduced and then equilibrium is restored. Equilibrium is a state in which forces are balanced, resulting in stability. Static equilibrium is stability due to a lack of changes and dynamic equilibrium is stability due to a balance between opposite forces. Similarly, many technologies involve either creation or control of change. The process of change over a long period of time is known as **evolution**. While biological evolution is the most common example, one can also classify technological advancement, changes in the universe, and changes in the environment as evolution.

Scale
In science and technology, we must deal with quantities that have vastly different magnitudes. It is important to understand the relationships between such very different numbers. Specifically, it must be recognized that behavior, and even the laws of physics, may change with scale. Some relationships, for instance, the effect of friction on speed, are only valid over certain size scales. When developing new technology, such as nano-machines, we must keep in mind the importance of scale.

Form and function
Form and function are properties of systems that are closely related. The function of an object usually dictates its form and the form of an object usually facilitates its function. For example, the form of the heart (e.g. muscle, valves) allows it to perform its function of circulating blood through the body. The idea of function dictating the form is also used in architecture.

Skill 23.9 Understands how models are used to represent the natural world and how to evaluate the strengths and limitations of a variety of scientific models (e.g., physical, conceptual, mathematical).

A scientific model is a set of ideas that describes a natural process and are developed by empirical or theoretical methods. They help scientists focus on the basic fundamental processes. They may be physical representations, such as a space-filling model of a molecule or a map, or they may be mathematical algorithms.

Whatever form they take, scientific models are based on what is known about the systems or objects at the time that the models are constructed. Models usually evolve and are improved as scientific advances are made. Sometimes a model must be discarded because new findings show it to be misleading or incorrect.

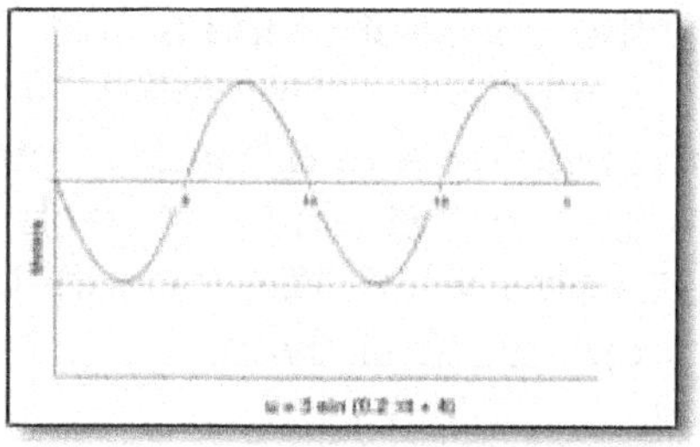

Models are developed in an effort to explain how things work in nature. Because models are not the "real thing", they can never correctly represent the system or object in all respects. The amount of detail that they contain depends upon how the model will be used as well as the sophistication and skill of the scientist doing the modeling. If a model has too many details left out, its usefulness may be limited. But too many details may make a model too complicated to be useful. So it is easy to see why models lack some features of the real system.

To overcome this difficulty, different models are often used to describe the same system or object. Scientists must then choose which model most closely fits the scientific investigation being carried out, which includes findings that are being described, and, in some cases, which one is compatible with the sophistication of the investigation itself. For example, there are many models of atoms. The solar system model described above is adequate for some purposes because electrons have properties of matter. They have mass and charge and they are found in motion in the space outside the nucleus. However, a highly mathematical model based on the field of quantum mechanics is necessary when describing the energy (or wave) properties of electrons in the atom.

Scientific models are based on physical observations that establish some facts about the system or object of interest. Scientists then combine these facts with appropriate laws or scientific principles and assumptions to produce a "picture" that mimics the behavior of the system or object to the greatest possible extent. It is on the basis of such models that science makes many of its most important advances because models provide a vehicle for making predictions about the behavior of a system or object. The predictions can then be tested as new measurements, technology or theories are applied to the subject. The new information may result in modification and refinement of the model, although certain issues may remain unresolved by the model for years. The goal, however, is to continue to develop the model in such a way as to move it ever closer to a true description of the natural phenomenon. In this way, models are vital to the scientific process.

In the West, the Greek philosophers Democritus and Leucippus first suggested the concept of the atom. They believed that all atoms were made of the same material but that varied sizes and shapes of atoms resulted in the varied properties of different materials. By the 19th century, John Dalton had advanced a theory stating that each element possesses atoms of a unique type. These atoms were also thought to be the smallest pieces of matter which could not be split or destroyed.

Atomic structure began to be better understood when, in 1897, JJ Thompson discovered the electron while working with cathode ray tubes. Thompson realized the negatively charged electrons were subatomic particles and formulated the "plum pudding model" of the atom to explain how the atom could still have a neutral charge overall. In this model, the negatively charged electrons were randomly present and free to move within a soup or cloud of positive charge. Thompson likened this to the dried fruit that is distributed within the English dessert plum pudding though the electrons were free to move in his model. Ernest Rutherford disproved this model with the discovery of the nucleus in 1909. Rutherford proposed a new "planetary" model of the atom in which electrons orbited around a positively charged nucleus like planets around the sun. Over the next 20 years, protons and neutrons (subnuclear particles) were discovered while additional experiments showed the inadequacy of the planetary model.

As quantum theory was developed and popularized (primarily by Max Planck and Albert Einstein), chemists and physicists began to consider how it might apply to atomic structure. Niels Bohr put forward a model of the atom in which electrons could only orbit the nucleus in circular orbitals with specific distances from the nucleus, energy levels, and angular momentums. In this model, electrons could only make instantaneous "quantum leaps" between the fixed energy levels of the various orbitals. The Bohr model of the atom was altered slightly by Arnold Sommerfeld in 1916 to reflect the fact that the orbitals were elliptical instead of round.

Though the Bohr model is still thought to be largely correct, it was discovered that electrons do not truly occupy neat, cleanly defined orbitals. Rather, they exist as more of an "electron cloud." The work of Louis de Broglie, Erwin Schrödinger, and Werner Heisenberg showed that an electron can actually be located at any distance from the nucleus. However, we can find the *probability* that the electrons exists at given energy levels (i.e., in particular orbitals) and those probabilities will show that the electrons are most frequently organized within the orbitals originally described in the Bohr model.

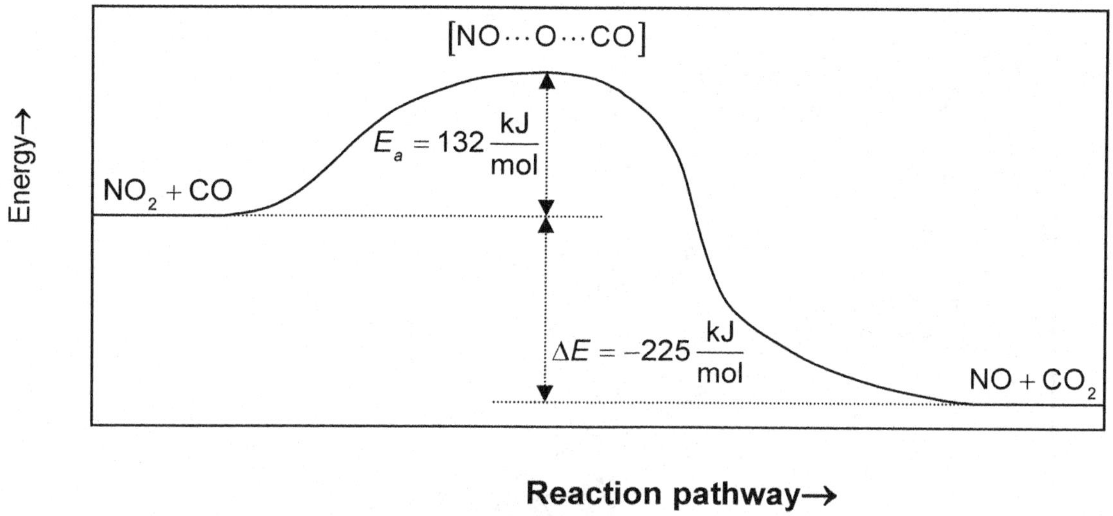

An energy diagram is a conceptual tool, so there is some variability in how its axes are labeled. The y-axis of the diagram is usually labeled energy (E), but it is sometimes labeled "enthalpy (H)" or (rarely) "free energy (G)." There is an even greater variability in how the x-axis is labeled. The terms "reaction pathway," "reaction coordinate," "course of reaction," or "reaction progress" may be used on the x-axis, or the x-axis may remain without a label.

The energy diagrams of an endothermic and exothermic reaction are compared below.

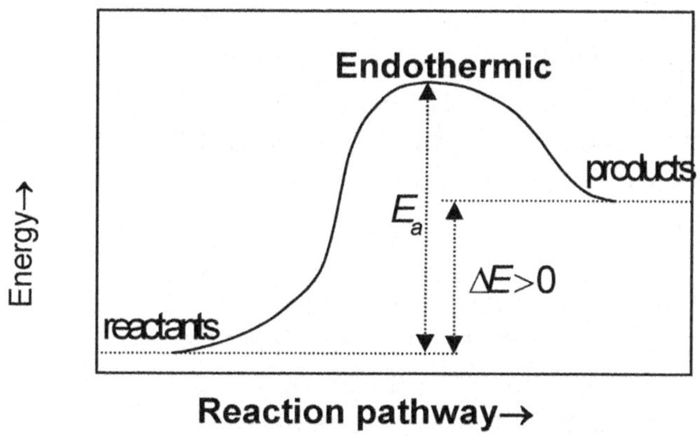

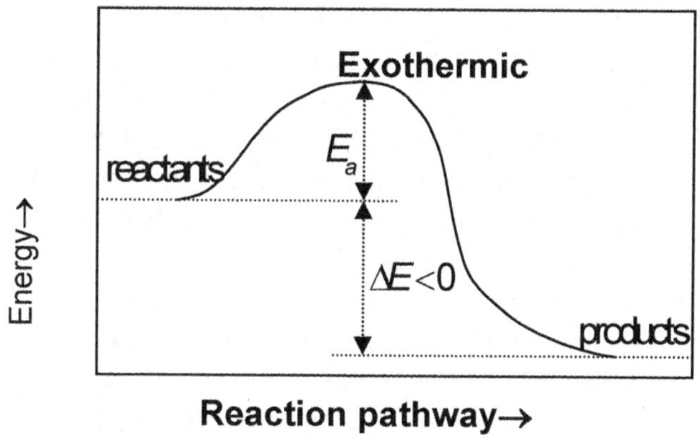

The rate of most simple reactions **increases with temperature** because a **greater fraction of molecules have the kinetic energy** required to overcome the reaction's activation energy. The chart below shows the effect of temperature on the distribution of kinetic energies in a sample of molecules. These curves are called **Maxwell-Boltzmann distributions**. The shaded areas represent the fraction of molecules containing sufficient kinetic energy for a reaction to occur. This area is larger at a higher temperature; so more molecules are above the activation energy and more molecules react per second.

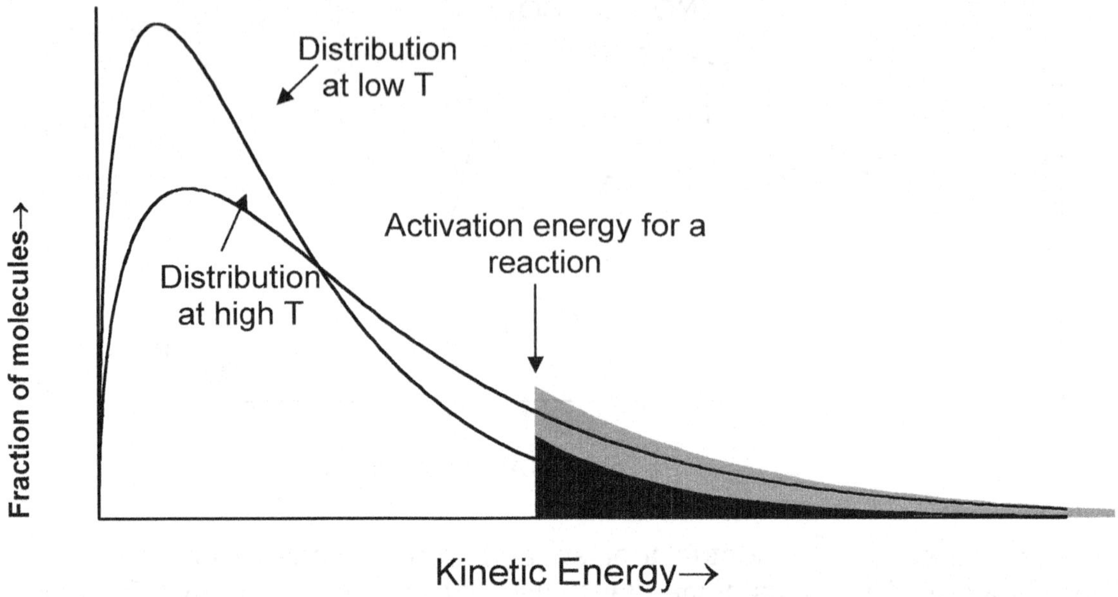

http://www.mhhe.com/physsci/chemistry/essentialchemistry/flash/activa2.swf provides an animated audio tutorial on energy diagrams.

Mathematics is a very broad field, encompassing various specific disciplines including calculus, trigonometry, algebra, geometry, complex analysis and other areas. **Many physical phenomena can be modeled mathematically using functions to relate a specific parameter or state of the system to one or more other parameters**. For example, the net force on a charged object is a function of the direction and magnitude of the forces acting upon it, such as electrical attraction or repulsion, tension from a string or spring, gravity and friction. When a phenomenon has been modeled mathematically as a set of expressions or equations, the general relationships of numbers can be applied to glean further information about the system for the purposes of greater understanding or prediction of future behavior.

The process of treating a system mathematically can involve use of empirical relationships (such as Ohm's law) or more *a priori* relationships (such as the Schrödinger equation). Given or empirically derived equations can be manipulated or combined, depending on the specific situation, to isolate a specific parameter as a function of other parameters (solution of the equation). This process can also lead to a different equation that presents a new or simplified relationship among specific parameters (derivation of an equation).

Both derivation and solution of equations can be pursued in either an **exact** or **approximate** manner. **In some cases, equations are intractable and certain assumptions must be made to facilitate finding a solution**. These assumptions are typically drawn from generalizations concerning the behavior of a particular system and will result in certain restrictions on the validity and applicability of a solution. An exact solution, presumably, has none of these limitations.

Another variation of approximate derivation and solution of an equation involves numerical techniques. Ideally, an analytical approach that employs no approximations is the best alternative; such an approach yields the broadest and most useful results. Nevertheless, the intractability of an equation can lead to the need for either an approximate or numerical solution. One particular example is from the field of electromagnetics, where determination of the scattering of radiation from a finite circular cylinder is either extremely difficult or impossible to perform analytically. Thus, either numerical or approximate approaches are required. The range of numerical techniques available depends largely on the type of problem involved. The approach for numerically solving simple algebraic equations, for example, is different from the approach for numerically solving complex integral equations.

Care must be taken with analytical (and numerical or approximate) approaches to physical situations **as it is possible to produce mathematically valid solutions that are physically unacceptable**. For instance, the equations that describe the wave patterns produced by a disturbance of the surface of water, upon solution, may yield an expression that includes both incoming and outgoing waves. Nevertheless, in some situations, there is no physical source of incoming waves; thus, the analytical solution must be tempered by the physical constraints of the problem. In this case, one of the solutions, although mathematically valid, must be rejected as unphysical.

COMPETENCY 24.0 THE TEACHER UNDERSTANDS THE HISTORY OF SCIENCE, HOW SCIENCE IMPACTS THE DAILY LIVES OF STUDENTS, AND HOW SCIENCE INTERACTS WITH AND INFLUENCES PERSONAL AND SOCIETAL DECISIONS.

Skill 24.1 Understands the historical development of science, key events in the history of science, and the contributions that diverse cultures and individuals of both genders have made to scientific knowledge.

Archimedes

Archimedes was a Greek mathematician, physicist, engineer, astronomer, and philosopher. He is credited with many inventions and discoveries some of which are still in use today such as the Archimedes screw. He designed the compound pulley, a system of pulleys used to lift heavy loads such as ships.

Although Archimedes did not invent the lever, he gave the first rigorous explanation of the principles involved which are the transmission of force through a fulcrum and moving the effort applied through a greater distance than the object to be moved. His Law of the Lever states that magnitudes are in equilibrium at distances reciprocally proportional to their weights.

He also laid down the laws of flotation and described Archimedes' principle which states that a body immersed in a fluid experiences a buoyant force equal to the weight of the displaced fluid.

Amedeo Avogadro

Avogadro was an Italian professor of physics born in the 18th century. He contributed to the understanding of the difference between atoms and molecules and the concept of molarity. The famous Avogadro's principle states that equal volumes of all gases at the same temperature and pressure contain an equal number of molecules.

Niels Bohr

Bohr was a Danish physicist who made fundamental contributions to understanding atomic structure and quantum mechanics. Bohr is widely considered one of the greatest physicists of the twentieth century.

Bohr's model of the atom was the first to place electrons in discrete quantized orbits around the nucleus.

Bohr also helped determine that the chemical properties of an element are largely determined by the number of electrons in the outer orbits of the atom. The idea that an electron could drop from a higher-energy orbit to a lower one emitting a photon of discrete energy originated with Bohr and became the basis for future quantum theory.

He also contributed significantly to the Copenhagen interpretation of quantum mechanics. He received the Nobel Prize for Physics for this work in 1922.

Robert Boyle
Robert Boyle was born in Ireland in 1627 and was one of the most prominent experimentalists of his time. He was the first scientist who kept accurate logs of his experiments and though an alchemist himself, gave birth to the science of chemistry as a separate rigorous discipline. He is well known for Boyle's law that describes the relationship between the pressure and volume of an ideal gas. It was one of the first mathematical expressions of a scientific principle.

Marie Curie
Curie was as a Polish-French physicist and chemist. She was a pioneer in radioactivity and the winner of two Nobel Prizes, one in Physics and the other in Chemistry. She was also the first woman to win the Nobel Prize.

Curie studied radioactive materials, particularly pitchblende, the ore from which uranium was extracted. The ore was more radioactive than the uranium extracted from it which led the Curies (Marie and her husband Pierre) to discover a substance far more radioactive then uranium. Over several years of laboratory work the Curies eventually isolated and identified two new radioactive chemical elements, polonium and radium. Curie refined the radium isolation process and continued intensive study of the nature of radioactivity.

Albert Einstein
Einstein was a German-born theoretical physicist who is widely considered one of the greatest physicists of all time. While best known for the theory of relativity, and specifically mass-energy equivalence, $E = mc^2$, he was awarded the 1921 Nobel Prize in Physics for his explanation of the photoelectric effect and "for his services to Theoretical Physics". In his paper on the photoelectric effect, Einstein extended Planck's hypothesis ($E = h\nu$) of discrete energy elements to his own hypothesis that electromagnetic energy is absorbed or emitted by matter in quanta and proposed a new law $E_{\max} = h\nu - P$ to account for the photoelectric effect.

He was known for many scientific investigations including the special theory of relativity which stemmed from an attempt to reconcile the laws of mechanics with the laws of the electromagnetic field. His general theory of relativity considered all observers to be equivalent, not only those moving at a uniform speed. In general relativity, gravity is no longer a force, as it is in Newton's law of gravity, but is a consequence of the curvature of space-time.

Other areas of physics in which Einstein made significant contributions, achievements or breakthroughs include relativistic cosmology, capillary action, critical opalescence, classical problems of statistical mechanics and problems in which they were merged with quantum theory (leading to an explanation of the Brownian movement of molecules), atomic transition probabilities, the quantum theory of a monatomic gas, the concept of the photon, the theory of radiation (including stimulated emission), and the geometrization of physics.

Einstein's research efforts after developing the theory of general relativity consisted primarily of attempts to generalize his theory of gravitation in order to unify and simplify the fundamental laws of physics, particularly gravitation and electromagnetism, which he referred to as the Unified Field Theory.

Michael Faraday

Faraday was an English chemist and physicist who contributed significantly to the fields of electromagnetism and electrochemistry. He established that magnetism could affect rays of light and that the two phenomena were linked. It was largely due to his efforts that electricity became viable for use in technology. The unit for capacitance, the farad, is named after him as is the Faraday constant, the charge on a mole of electrons (about 96,485 coulombs). Faraday's law of induction states that a magnetic field changing in time creates a proportional electromotive force.

Sir Isaac Newton

Newton was an English physicist, mathematician, astronomer, alchemist, and natural philosopher in the late 17th and early 18th centuries. He described universal gravitation and the three laws of motion laying the groundwork for classical mechanics. He was the first to show that the motion of objects on earth and in space is governed by the same set of mechanical laws. These laws became central to the scientific revolution that took place during this period of history. Newton's three laws of motion are:

> I. Every object in a state of uniform motion tends to remain in that state of motion unless an external force is applied to it.
> II. The relationship between an object's mass m, its acceleration a, and the applied force F is F = ma.
> III. For every action there is an equal and opposite reaction.

In mechanics, Newton developed the basic principles of conservation of momentum. In optics, he invented the reflecting telescope and discovered that the spectrum of colors seen when white light passes through a prism is inherent in the white light and not added by the prism as previous scientists had claimed. Newton notably argued that light is composed of particles. He also formulated an experimental law of cooling, studied the speed of sound, and proposed a theory of the origin of stars.

J. Robert Oppenheimer
Oppenheimer was an American physicist, best known for his role as the scientific director of the Manhattan Project, the effort to develop the first nuclear weapons. Sometimes called "the father of the atomic bomb", Oppenheimer later lamented the use of atomic weapons. He became a chief advisor to the United States Atomic Energy Commission and lobbied for international control of atomic energy. Oppenheimer was one of the founders of the American school of theoretical physics at the University of California, Berkeley. He did important research in theoretical astrophysics, nuclear physics, spectroscopy, and quantum field theory.

Wilhelm Ostwald
Wilhelm Ostwald, born in 1853 in Latvia, was one of the founders of classical physical chemistry which deals with the properties and reactions of atoms, molecules and ions. He developed the Ostwald process for the synthesis of nitric acid. In 1909 he won the Nobel prize for his work on catalysis, chemical equilibria and reaction velocities.

Linus Pauling
The American chemist Linus Pauling won the Nobel prize for chemistry in 1954 for his investigation of the nature of the chemical bond. He led the way in applying quantum mechanics to chemistry. Later in his career he focused on biochemical problems such as the structure of proteins and sickle cell anemia. He won the Nobel peace prize in 1962 for his contribution to nuclear disarmament.

Skill 24.2 Knows how to use examples from the history of science to demonstrate the changing nature of scientific theories and knowledge (i.e., that scientific theories and knowledge are always subject to revision in light of new evidence).

Scientific knowledge is based on a firm foundation of observation. Though mathematics and logic play a major role in defining and deducing scientific theories, ultimately even the most beautiful and intricate theory has to win the support of experiment. A single observation that contradicts an established theory can bring the whole edifice down if confirmed and reproduced. Thus scientific knowledge can never be totally certain and is always open to change based on some new evidence.

Sometimes advanced measuring devices and new equipment make it possible for scientists to detect phenomenon that noone had noted before. Nothing seemed more certain than classical Newtonian physics which explained everything from the motion of the planets to the behavior of earthly objects. At the end of the nineteenth century Lord Kelvin expressed the opinion that physics was complete except for the existence of "two small clouds"; the null result of the Michelson-Morley experiment and the failure of classical physics to predict the spectral distribution of blackbody radiation. The "two small clouds" turned out to be far more significant than Lord Kelvin could have imagined and led to the birth of relativity and quantum theory both of which totally changed the way we see the nature of reality.

If scientific knowledge is not inviolable, what keeps it from being vulnerable to challenge from anybody who thinks they have evidence to contradict a theory? Even though scientific knowledge is not sacred, the scientific process is. No observation is considered valid unless it can be reproduced by another scientist working independently under the same conditions. The peer-review process ensures that all results reported by a scientist undergo strict scrutiny by others working in the same field. Thus it is the integrity of the scientific process that keeps scientific knowledge, despite its openness to change, firmly grounded in objectivity and logic.

Skill 24.3 Knows that science is a human endeavor influenced by societal, cultural, and personal views of the world, and that decisions about the use and direction of science are based on factors such as ethical standards, economics, and personal and societal biases and needs.

Advances in science and technology create challenges and ethical dilemmas that national governments and society in general must attempt to solve. Local, state, national, and global governments and organizations must increasingly consider policy issues related to science and technology. For example, local and state governments must analyze the impact of proposed development and growth on the environment. Governments and communities must balance the demands of an expanding human population with the local ecology to ensure sustainable growth. Genetic research and manipulation, antibiotic resistance, stem cell research, and cloning are but a few of the issues facing national governments and global organizations today.

In all cases, policy makers must analyze all sides of an issue and attempt to find a solution that protects society while limiting scientific inquiry as little as possible. For example, policy makers must weigh the potential benefits of stem cell research, genetic engineering, and cloning (e.g. medical treatments) against the ethical and scientific concerns surrounding these practices. Many safety concerns have answered by strict government regulations. The FDA, USDA, EPA, and National Institutes of Health are just a few of the government agencies that regulate pharmaceutical, food, and environmental technology advancements.

Scientific and technological breakthroughs greatly influence other fields of study and the job market as well. Advances in information technology have made it possible for all academic disciplines to utilize computers and the internet to simplify research and information sharing. In addition, science and technology influence the types of available jobs and the desired work skills. For example, machines and computers continue to replace unskilled laborers and computer and technological literacy is now a requirement for many jobs and careers. Finally, science and technology continue to change the very nature of careers. Because of science and technology's great influence on all areas of the economy, and the continuing scientific and technological breakthroughs, careers are far less stable than in past eras. Workers can thus expect to change jobs and companies much more often than in the past.

Because people often attempt to use scientific evidence in support of political or personal agendas, the ability to evaluate the credibility of scientific claims is a necessary skill in today's society. The media and those with an agenda to advance often overemphasize the certainty and importance of experimental results. One should question any scientific claim that sounds fantastical or overly certain. Scientific, peer-reviewed journals are the most accepted source for information on scientific experiments and studies. Knowledge of experimental design and the scientific method is important in evaluating the credibility of studies. For example, one should look for the inclusion of control groups and the presence of data to support the given conclusions.

Skill 24.4 Understands the application of scientific ethics to the conducting, analyzing, and publishing of scientific investigations.

One form of scientific ethics is to be aware of the potential sources of error in an experiment, as discussed in Skill 3.2, and conduct research carefully to avoid such errors. This is important and a matter of ethics because all scientists build on the work of others. An error in one person's research may lead to incorrect conclusions that affect other scientists' research, and each scientist must accept a share of the responsibility for ensuring that the collective body of knowledge is accurate. For example, it was recently discovered that a computer model being used by one laboratory had transposed the arrangement of atoms in molecular structures to their mirror images. The published structures had already been relied upon by other researchers in the field for three years, and now many papers are having to be retracted and years of work redone. Had the original laboratory checked their computer model more carefully, this situation could have been avoided.

Scientists are expected to truthfully report the results of their experiments without fabricating data. Falsification of certain kinds of results has become easier with the advent of computer software that allows images to be manipulated. In the past decade, there has been an explosion of fraud in the scientific world. Some scientists feel great pressure to make important discoveries that will garner funding, further their career, or enable them to file a patent, and this clouds their judgment. However, scientists are forced to resign and forfeit their careers if they are found to have published false data. The consequences of fraud are expected to be severe, since other scientists worldwide will be utilizing their time and resources to build on the published work.

In most institutions, there are policies against scientists accepting gifts, honoraria, or payment from stakeholders in order to avoid a conflict of interest. Many scientific papers now include a disclosure paragraph stating whether any of the authors currently have any kind of relationship, monetary or otherwise, with a company or other entity that has an interest in the research presented. For example, if a paper is published with the latest experimental results on a new drug, the authors should be truly impartial scientists who have no relationship with the pharmaceutical company that invented the drug and do not stand to benefit financially from acceptance of the drug.

Skill 24.5 Applies scientific principles to analyze factors (e.g., diet, exercise, personal behavior) that influence personal and societal choices concerning fitness and health (e.g., physiological and psychological effects and risks associated with the use of substances and substance abuse).

Our knowledge of science is intimately connected with our understanding of personal and community health.

Nutrition

The science and technology of packaging helps keep foods fresh longer and alters the molecules in food. The thermochemistry of refrigeration helps food last longer. Other technologies such as pasteurization, drying, salting, and the addition of preservatives all prevent microbial contamination by altering the composition of food. Preservatives are substances added to food to prevent the growth of microorganisms and the spoilage they cause. For example, potassium and sodium nitrites and nitrates are often used as preservatives for vegetables, fruits, and processed meats.

Food may also be sterilized and preserved by food irradiation. Gamma rays from a sealed source of ^{60}Co or ^{137}Cs are used to kill microorganisms in over 40 countries. This process is less expensive than refrigeration, canning, or use of additives, and it does not make food radioactive.

Opponents of irradiation fear the risks involved in the transport and use of nuclear materials to build the facilities and to maintain them. A potential health risk in the food itself is the possibility that the radiation required to kill organisms may alter a biological molecule to produce a harmful by-product, but no evidence has been found of such a toxin. Another concern is that irradiation will lead to a permissive attitude about safe food-handling procedures that can lead to other types of contamination. Food irradiation is still under study in the United States to conclusively prove its safety, particularly for meats.

Eating too much processed food over time has a long history of causing harm because of the substances it lacks or contains. Whole, fresh food usually has a better nutritional value. For example, in the late 1800s many infants in the US developed scurvy (vitamin C deficiency) from drinking heat-treated milk that controlled bacterial infections but destroyed vitamin C. Local production of food with minimal time-to-market and proper preparation is a healthier approach to food safety than chemical modification and long-distance transport of the food, but processed food will remain popular for the foreseeable future because it is usually cheaper to buy, more profitable to sell, and more convenient to obtain, store, prepare and use.

Environment

Most scientists believe the emission of greenhouse gases has already led to global warming due to an increase in the greenhouse effect. The greenhouse effect occurs when these gases in the atmosphere warm the planet by absorbing heat to prevent it from escaping into space. This is similar—but not identical—to what occurs in greenhouse buildings. Greenhouse buildings warm an interior space by preventing mixing with colder air outside. Most greenhouse gases such as water vapor occur naturally and are important for life to exist on Earth, but are being added to the atmosphere is much larger quantities by human activities.

Human production of carbon dioxide from combustion of fossil fuels has increased the concentration of this important greenhouse gas to its highest value since millions of years ago. The precise impact of these changes in the atmosphere is difficult to predict with certainty, but is likely to include a rise in sea level, an increase in extreme weather events, reduction of glaciers and snowpack, increasing drought, and changes in habitat and species distributions.

Rain with a pH less than 5.6 is known as acid rain. Acid rain is caused by burning fossil fuels (especially coal) and by fertilizers used in intensive agriculture. These activities emit sulfur and nitrogen in gaseous compounds that are converted to sulfur oxides and nitrogen oxides. These in turn create sulfuric acid and nitric acid in rain. Acid rain may also be created from gases emitted by volcanoes and other natural sources. Acid rain harms fish and trees and triggers the release metal ions from minerals into water that can harm people. It can also have destructive effects on building surfaces, statues, and sculptures.

The problem of acid rain in the United States has been addressed to some extent in recent decades by the use of scrubbers in coal burning power plants and catalytic converters in vehicles.

The ozone layer is a region of the stratosphere that contains higher concentrations of ozone (O_3) than other parts of the atmosphere. The ozone layer is important for human health because it blocks ultraviolet radiation from the sun, helping to protect us from skin cancer.

Research in the 1970s revealed that several gases used for refrigeration and other purposes were depleting the ozone layer. Many of these ozone-destroying molecules are short alkyl halides known as chlorofluorocarbons or CFCs. CCl_3F is one example of these compounds. The widespread use of ozone-destroying compounds was banned by an international agreement in the early 1990s. Other substances are used in their place such as CF_3CH_2F, a hydrofluorocarbon. Since that time the concentration of ozone-depleting gases in the atmosphere has been declining and the rate of ozone destruction has been decreasing. Many see this improvement as one of the most important positive examples of international cooperation to solve a global environmental issue.

Substance abuse

Science can help us more fully understand substance abuse. Abuse and addiction can involve legal or illegal substances including alcohol, nicotine, prescription pain relievers, and a variety of illegal drugs. Addiction is characterized by the continued pathological use of these substances in conjunction with adverse social effects. Research indicates that substance abuse diseases, such as alcoholism, are heritable. However, a combination of genetic and environmental factors are typically responsible for leading to a substance abuse problem. Nonetheless, people with a history of addiction in their families are typically advised to exercise caution with substances.

Advances in medicine have been helpful in the treatment of substance abuse. Prescription medication can mitigate the effects of withdrawal. Others, such as Antabuse, can even serve as a deterrent to further use of the substance. Progress in psychiatric and psychological treatment has also been helpful in longer term treatment of addicts. For instance, talk therapy and the use of psychoactive medication may assist patients in dealing with some underlying causes of their addiction. A variety of rehabilitation and support programs currently exist to treat the immediate physical effects of substance abuse and aid in the recovery process.

Skill 24.6 Applies scientific principles, the theory of probability, and risk/benefit analysis to analyze the advantages of, disadvantages of, or alternatives to a given decision or course of action.

Individuals, on a daily basis, make judgments about risks and benefits as part of many decisions. For instance, a person may decide to spend one dollar on a lottery ticket because the potential benefit (millions of dollars in prize money) is great even though the chance of winning is small. Another person may decide that the potentially large pay-off is too unlikely to make it worth spending a dollar. **The science of risk-benefit analysis is simply a more formalized system to help make these decisions**. Many corporations employ risk/benefit analysis in determining future business strategies. Government also evaluates risks and benefits in many of its activities, such as deciding to approve a new drug or to carry out a manned space-flight mission.

Risk combines the probability of an event occurring and the results of that event occurring. For the risk assessment to be accurate, quantitative data are needed on both the probability and the magnitude of the effect. In the example of the lottery ticket above, both the amount of the possible loss ($1) and the probability of losing are known, based on information printed on the ticket. Similarly, the amount of the possible payoff is known, as well as the odds of winning. From this information, each person decides whether or not to take the risk based on the perceptions he or she has of the risk and reward involved. Persons or other entities seeking to avoid risk are said to be risk-averse, while those more comfortable with risk may be considered risk-tolerant or risk-seeking.

When companies are developing new products, this type of risk analysis is performed to determine what potential problems must be addressed; both the probability of an event and its impact must be considered. Therefore, if a certain malfunction is extremely unlikely but could cause consumer death, it must be remedied. Companies typically have an established **tolerance for risk**, usually based on financial considerations, but also taking into account reputation and other non-monetary values.

A simplified standard equation for calculating risk might look like this:

Risk = probability of accident x average cost of accident

We can use a similar technique to determine whether we should undertake a course of action that might have either a positive or a negative outcome. For instance, to determine whether an investment is worthwhile:

Risk/benefit = (probability of positive outcome x pay-off of positive outcome) – (probability of negative outcome x loss associated with negative outcome)

These calculations cannot actually tell someone what he or she should do; they only clarify the risk inherent in doing so. If a person were strictly risk-neutral, he or she would take the risk if the quantity above was positive and would not take the risk if the quantity above was negative. However, most of us are not strictly risk-neutral. We may be more or less risk-averse depending on the amount of money or danger involved and the seriousness of the outcome.

Risk assessment professionals may develop extremely complex models to determine the risks associated with scientific or engineering questions, for example, the risk associated with a chemical exposure or the risk of a dam failure. To perform these calculations, we **must have good data about probabilities**. These data come from properly performed scientific studies. There are some areas of science where the probabilities are not well-known, for example, in epidemiology. In these cases, safety factors are often added to risk assessment equations to account for the uncertainties involved, especially when the adverse effect being evaluated is severe, such as cancer or mortality.

Skill 24.7 Understands the role science can play in helping resolve personal, societal, and global issues (e.g., population growth, disease prevention, resource use).

Science and technology have the center-stage in our daily lives. More and more, it is becoming impossible for people in developed societies to exist without the necessities (e.g. cell phone, home appliances) and conveniences (e.g. satellite TV) afforded by technology. In fact, every day things that used to be conveniences are becoming necessities. Apart from the things that science and technology provide for us, they also represent a mind set and way of thinking such as the application of objectivity or rational thinking in evaluating events and options in our lives. Here are some of the ways in which science and technology are applied in our daily lives:

Health care: In this area, we can see many of the fruits of science and technology in nutrition, genetics, and the development of therapeutic agents. We can see an example of the adaptation of organisms in the development of resistant strains of microbes in response to use of antibiotics. Organic chemistry and biochemistry have been exploited to identify therapeutic targets and to screen and develop new medicines. Advances in molecular biology and our understanding of inheritance have led to the development of genetic screening and allowed us to sequence the human genome.

Environment: There are two broad happenings in environmental science and technology. First, there are many studies being conducted to determine the effects of changing environmental conditions and pollutions. New instruments and monitoring systems have increased the accuracy of these results. Second, advances are being made to mitigate the effects of pollution, develop sustainable methods of agriculture and energy production, and improve waste management.

Agriculture: Development of new technology in agriculture is particularly important as we strive to feed more people with less arable land. Again we see the importance of genetics in developing hybrids that have desirable characteristics. New strains of plants and farming techniques may allow the production of more nutrient rich food and/or allow crops to be grown successfully in harsh conditions. However, it is also important to consider the environmental impact of transgenic species and the use of pesticides and fertilizers. Scientific reasoning and experimentation can assist us in ascertaining the real effect of modern agricultural practices and ways to minimize their impact.

Information technology: The internet has become a new space in our lives. It is the global commons. It is where we conduct business, meet friends, obtain information and find entertainment. It affects all areas of human endeavor by allowing people worldwide to communicate easily and share ideas. It has also spawned a variety of new businesses.

With advances in technology come those in society who oppose it. Ethical questions come into play when discussing issues such as stem cell research or animal research for example. Does it need to be done? What are the effects on humans and animals? The answers to these questions are not always clear and often dependent on circumstance. Is the scientific process of organizing and weighing evidence applicable to these questions or do they lie beyond the domain of science and technology? These are the difficult issues that we have to face as technology moves forward.

DOMAIN VIII. **PHYSICS**

COMPETENCY 25.0 THE TEACHER UNDERSTANDS THE DESCRIPTION OF MOTION IN ONE AND TWO DIMENSIONS.

Skill 25.1 Analyzes and interprets graphs describing the motion of a particle.

The relationship between time, position or distance, velocity and acceleration can be understood conceptually by looking at a graphical representation of each as a function of time. Simply, the velocity is the slope of the position vs. time graph and the acceleration is the slope of the velocity vs. time graph. If you are familiar with calculus then you know that this relationship can be generalized: velocity is the first derivative and acceleration the second derivative of position.

Here are three examples:

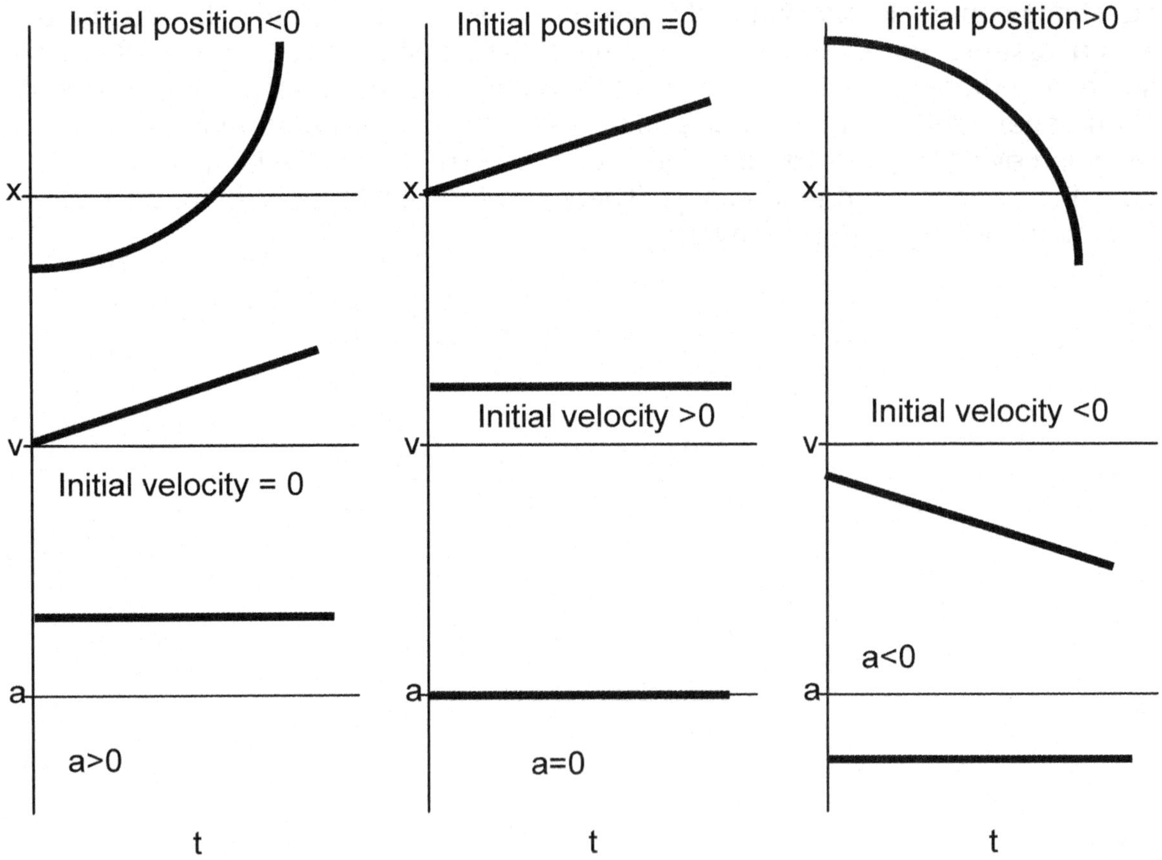

There are three things to notice:

1)In each case acceleration is constant. This isn't always the case, but a simplification for this illustration.
2)A non-zero acceleration produces a position curve that is a parabola.
3)In each case the initial velocity and position are specified separately. The acceleration curve gives the shape of the velocity curve, but not the initial value and the velocity curve gives the shape of the position curve but not the initial position.

Skill 25.2 Applies vector concepts to displacement, velocity, and acceleration in order to analyze and describe the motion of a particle.

Kinematics is the part of mechanics that seeks to understand the motion of objects, particularly the relationship between position, velocity, acceleration and time.

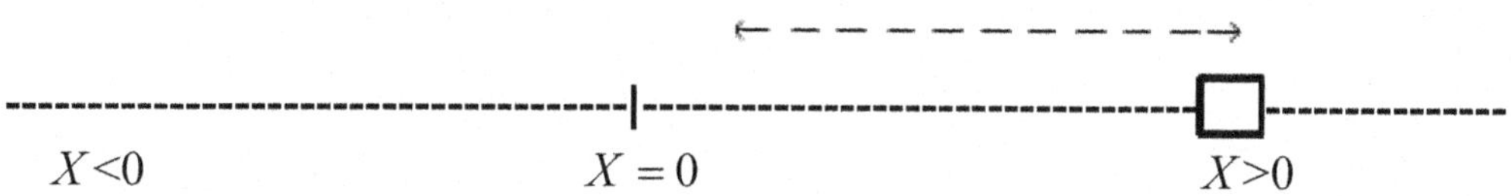

The above figure represents an object and its displacement along one linear dimension.

First we will define the relevant terms:

1. Position or Distance is usually represented by the variable *x*. It is measured relative to some fixed point or datum called the origin in linear units, meters, for example.

2. Displacement is defined as the change in position or distance which an object has moved and is represented by the variables D, d or Δx. Displacement is a vector with a magnitude and a direction.

3. Velocity is a vector quantity usually denoted with a V or v and defined as the rate of change of position. Typically units are distance/time, m/s for example. Since velocity is a vector, if an object changes the direction in which it is moving it changes its velocity even if the speed (the scalar quantity that is the magnitude of the velocity vector) remains unchanged.

i) Average velocity: $\vec{v} \equiv \frac{\Delta d}{\Delta t} = d_1 - d_0 / t_1 - t_0$

The ratio $\Delta d / \Delta t$ is called the average velocity. Average here denotes that this quantity is defined over a period Δt.

ii) Instantaneous velocity is the velocity of an object at a particular moment in time. Conceptually, this can be imagined as the extreme case when Δt is infinitely small.

5. Acceleration represented by *a* is defined as the rate of change of velocity and the units are m/s^2. Both an average and an instantaneous acceleration can be defined similarly to velocity.

From these definitions we develop the kinematic equations. In the following, subscript i denotes initial and subscript f denotes final values for a time period. Acceleration is assumed to be constant with time.

$$v_f = v_i + at \quad (1)$$

$$d = v_i t + \frac{1}{2}at^2 \quad (2)$$

$$v_f^2 = v_i^2 + 2ad \quad (3)$$

$$d = \left(\frac{v_i + v_f}{2}\right)t \quad (4)$$

Skill 25.3 Solves problems involving uniform and accelerated motion using scalar and vector quantities.

Simple problems involving distance, displacement, speed, velocity, and constant acceleration can be solved by applying the kinematics equations from the proceeding section. The following steps should be employed to simplify a problem and apply the proper equations:

1. Create a simple diagram of the physical situation.
2. Ascribe a variable to each piece of information given.
3. List the unknown information in variable form.
4. Write down the relationships between variables in equation form.
5. Substitute known values into the equations and use algebra to solve for the unknowns.
6. Check your answer to ensure that it is reasonable.

Example:
A man in a truck is stopped at a traffic light. When the light turns green, he accelerates at a constant rate of 10 m/s^2. **a)** How fast is he going when he has gone 100 m? **b)** How fast is he going after 4 seconds? **c)** How far does he travel in 20 seconds?

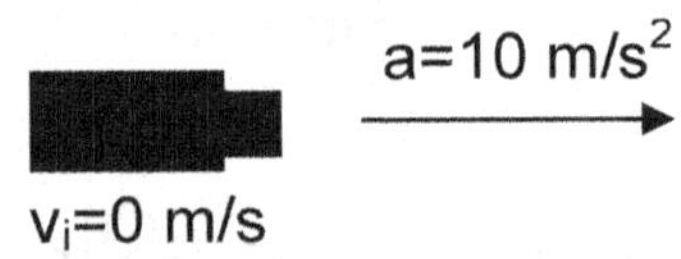

Solution:
We first construct a diagram of the situation. In this example, the diagram is very simple, only showing the truck accelerating at the given rate. Next we define variables for the known quantities (these are noted in the diagram):

$$a=10\ m/s^2;\quad v_i=0\ m/s$$

Now we will analyze each part of the problem, continuing with the process outlined above.

For part **a)**, we have one additional known variable: d=100 m

The unknowns are: v_f (the velocity after the truck has traveled 100m)

Equation (3) will allow us to solve for v_f, using the known variables:

$$v_f^{\,2} = v_i^{\,2} + 2ad$$

$$v_f^{\,2} = (0m/s)^2 + 2(10m/s^2)(100m) = 2000\frac{m^2}{s^2}$$

$$v_f = 45\frac{m}{s}$$

We use this same process to solve part **b)**. We have one additional known variable: t=4 s

The unknowns are: v_f (the velocity after the truck has traveled for 4 seconds)

Thus, we can use equation (1) to solve for v_f:

$$v_f = v_i + at$$

$$v_f = 0m/s + (10m/s^2)(4s) = 40m/s$$

For part **c)**, we have one additional known variable: t= 20 s

The unknowns are: d (the distance after the truck has traveled for 20 seconds)

Equation (2) will allow us to solve this problem:

$$d = v_i t + \frac{1}{2}at^2$$

$$d = (0m/s)(20s) + \frac{1}{2}(10m/s^2)(20s)^2 = 2000m$$

Finally, we consider whether these solutions seem physically reasonable. In this simple problem, we can easily say that they do.

Skill 25.4 Analyzes and solves problems involving projectile motion.

The most common example of an object moving in two dimensions is a projectile. A projectile is an object upon which the only force acting is gravity. Some examples:

i) An object dropped from rest.
ii) An object thrown vertically upwards at an angle
iii) A canon ball.

Once a projectile has been put in motion (say, by a canon or hand) the only force acting it is gravity, which near the surface of the earth implies it experiences $a=g=9.8m/s^2$.

This is most easily considered with an example such as the case of a bullet shot horizontally from a standard height at the same moment that a bullet is dropped from exactly the same height. Which will hit the ground first? If we assume wind resistance is negligible, then the acceleration due to gravity is our only acceleration on either bullet and we must conclude that they will hit the ground at the same time. The horizontal motion of the bullet is not affected by the downward acceleration.

Example:
I shoot a projectile at 1000 m/s from a perfectly horizontal barrel exactly 1 m above the ground. How far does it travel before hitting the ground?

Solution:
First figure out how long it takes to hit the ground by analyzing the motion in the vertical direction. In the vertical direction, the initial velocity is zero so we can rearrange kinematic equation 2 from the previous section to give:

$t = \sqrt{\frac{2d}{a}}$. Since our displacement is 1 m and $a=g=9.8m/s^2$, t=0.45 s.

Now use the time to hitting the ground from the previous calculation to calculate how far it will travel horizontally. Here the velocity is 1000m/s and there is no acceleration. So we simple multiply velocity with time to get the distance of 450m.

Skill 25.5 Analyzes and solves problems involving uniform circular and rotary motion.

Motion on an arc can also be considered from the view point of the kinematic equations. As pointed out earlier, displacement, velocity and acceleration are all vector quantities, i.e. they have magnitude (the speed is the magnitude of the velocity vector) and direction. This means that if one drives in a circle at constant speed one still experiences an acceleration that changes the direction. We can define a couple of parameters for objects moving on circular paths and see how they relate to the kinematic equations.

1. Tangential speed: The tangent to a circle or arc is a line that intersects the arc at exactly one point. If you were driving in a circle and instantaneously moved the steering wheel back to straight, the line you would follow would be the tangent to the circle at the point where you moved the wheel. The tangential speed then is the instantaneous magnitude of the velocity vector as one moves around the circle.

2. Tangential acceleration: The tangential acceleration is the component of acceleration that would change the tangential speed and this can be treated as a linear acceleration if one imagines that the circular path is unrolled and made linear.

3. Centripetal acceleration: Centripetal acceleration corresponds to the constant change in the direction of the velocity vector necessary to maintain a circular path. Always acting toward the center of the circle, centripetal acceleration has a magnitude proportional to the tangential speed squared divided by the radius of the path.

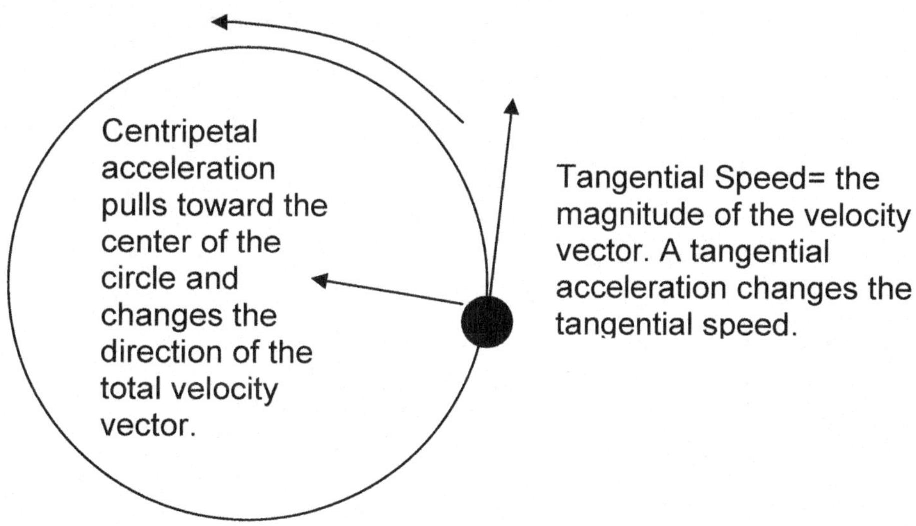

Uniform circular motion describes the motion of an object as it moves in a circular path at constant speed. There are many everyday examples of this behavior though we may not recognize them if the object does not complete a full circle. For example, a car rounding a curve (that is an arc of a circle) often exhibits uniform circular motion.

The following diagram and variable definitions will help us to analyze uniform circular motion.

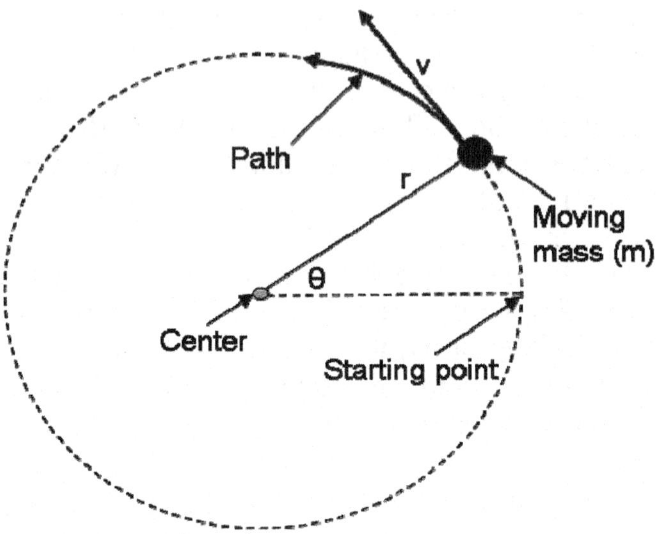

Above we see that the mass is traveling a path with constant radius (r) from some center point (x_0, y_0). By defining a variable (θ) that is a function of time (t) and is the angle between the mass's present position and original position on the circular path, we can write the following equations for the mass's position in a Cartesian plane.

$$x = r\cos(\theta) + x_0$$
$$y = r\sin(\theta) + y_0$$

Next observe that, because we are discussing uniform circular motion, the *magnitude* of the mass's velocity (v) is constant. However, the velocity's direction is always tangent to the circle and so always changing. We know that a changing velocity means that the mass must have a positive acceleration. This acceleration is directed toward the center of the circular path and is always perpendicular to the velocity, as shown below:

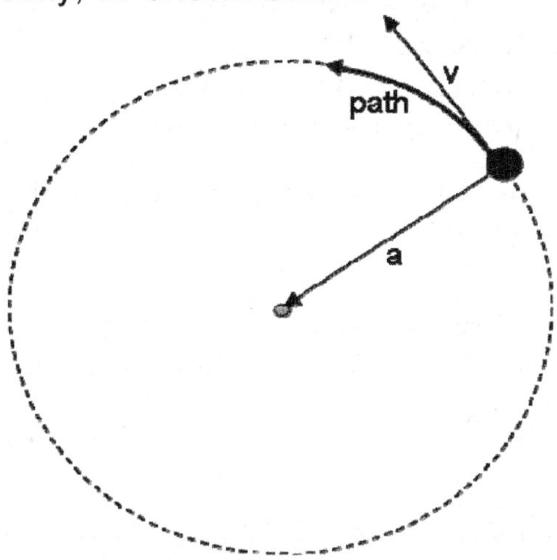

This is known as centripetal acceleration and is mathematically expressed as:

$$a = \frac{v^2}{r} = \frac{4\pi^2 r}{t^2}$$

where t is the period of the motion or the time taken for the mass to travel once around the circle. The force (F) experienced by the mass (m) is known as centripetal force and is always directed towards the center of the circular path. It has constant magnitude given by the following equation:

$$F = ma = m\frac{v^2}{r}$$

Skill 25.6 Understands motion of fluids.

The weight of a column of fluid creates hydrostatic pressure. Common situations in which we might analyze hydrostatic pressure include tanks of fluid, a swimming pool, or the ocean. Also, atmospheric pressure is an example of hydrostatic pressure.

Because hydrostatic pressure results from the force of gravity interacting with the mass of the fluid or gas, for an incompressible fluid it is governed by the following equation:

$$P = \rho g h$$

where P=hydrostatic pressure
ρ=density of the fluid
g=acceleration of gravity
h=height of the fluid column

Example: How much pressure is exerted by the water at the bottom of a 5 meter swimming pool filled with water?

Solution: We simply use the equation from above, recalling that the acceleration due to gravity is 9.8m/s^2 and the density of water is 1000 kg/m^3.

$$P = \rho g h = 1000\frac{kg}{m^3} \times 9.8\frac{m}{s^2} \times 5m = 49{,}000Pa = 49kPa$$

According to **Pascal's principle**, when pressure is applied to an enclosed fluid, it is transmitted undiminished to all parts of the fluid. For instance, if an additional pressure P_0 is applied to the top surface of a column of liquid of height *h* as described above, the pressure at the bottom of the liquid will increase by P_0 and will be given by $P = P_0 + \rho g h$. This principle is used in devices such as a **hydraulic lift** (shown below) which consists of two fluid-filled cylinders, one narrow and one wide, connected at the bottom. Pressure P (force = P X A1) applied on the surface of the fluid in the narrow cylinder is transmitted undiminished to the wider cylinder resulting in a larger net force (P X A2) transmitted through its surface. Thus a relatively small force is used to lift a heavy object. This does not violate the conservation of energy since the small force has to be applied through a large distance to move the heavy object a small distance.

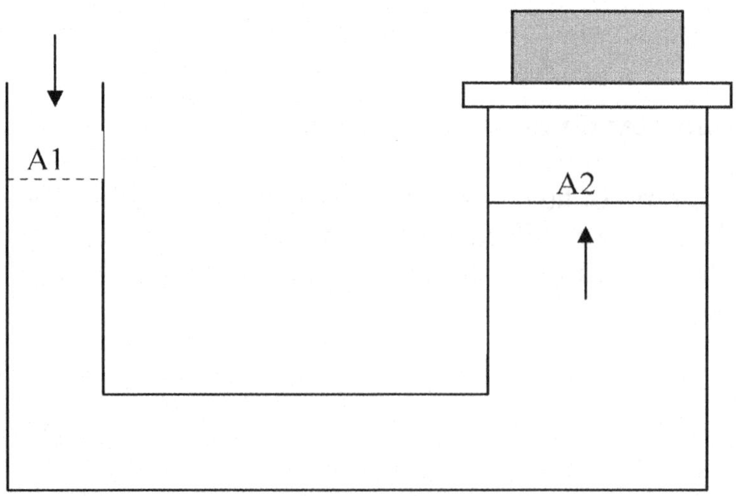

Archimedes' Principle states that, for an object in a fluid, "the upthrust is equal to the weight of the displaced fluid" and the weight of displaced fluid is directly proportional to the volume of displaced fluid. The second part of his discovery is useful when we want to, for instance, determine the volume of an oddly shaped object. We determine its volume by immersing it in a graduated cylinder and measuring how much fluid is displaced. We explore his first observation in more depth below.

Today, we call Archimedes' "upthrust" **buoyancy**. Buoyancy is the force produced by a fluid on a fully or partially immersed object. The buoyant force ($F_{buoyant}$) is found using the following equation:

$$F_{bouyant} = \rho V g$$

where ρ=density of the fluid
V=volume of fluid displaced by the submerged object
g=the acceleration of gravity

Notice that the buoyant force opposes the force of gravity. For an immersed object, a gravitational force (equal to the object's mass times the acceleration of gravity) pulls it downward, while a buoyant force (equal to the weight of the displaced fluid) pushes it upward.

Also note that, from this principle, we can predict *whether* an object will sink or float in a given liquid. We can simply compare the density of the material from which the object is made to that of the liquid. If the material has a lower density, it will float; if it has a higher density it will sink. Finally, if an object has a density equal to that of the liquid, it will neither sink nor float.

Example: Will gold (ρ=19.3 g/cm^3) float in water?

Solution: We must compare the density of gold with that of water, which is 1 g/cm^3.

$$\rho_{gold} > \rho_{water}$$

So, gold will sink in water.

Example: Imagine a 1 m^3 cube of oak (530 kg/m^3) floating in water. What is the buoyant force on the cube and how far up the sides of the cube will the water be?

Solution: Since the cube is floating, it has displaced enough water so that the buoyant force is equal to the force of gravity. Thus the buoyant force on the cube is equal to its weight 1X530X9.8 N = 5194 N.

To determine where the cube sits in the water, we simply the find the ratio of the wood's density to that of the water:

$$\frac{\rho_{oak}}{\rho_{water}} = \frac{530\,{}^{kg}/_{m^3}}{1000\,{}^{kg}/_{m^3}} = 0.53$$

Thus, 53% of the cube will be submerged. Since the edges of the cube must be 1m each, the top 0.47m of the cube will appear above the water.

The study of moving fluids is contained within fluid mechanics which is itself a component of continuum mechanics. Some of the most important applications of fluid mechanics involve liquids and gases moving in tubes and pipes.

Fluid flow may be laminar or turbulent. One cannot predict the exact path a fluid particle will follow in turbulent or erratic flow. Laminar flow, however, is smooth and each fluid particle follows a continuous path. Lines, know as **streamlines,** are drawn to show the path of a laminar fluid. Streamlines never cross one another and higher fluid velocity is depicted by drawing streamlines closer together.

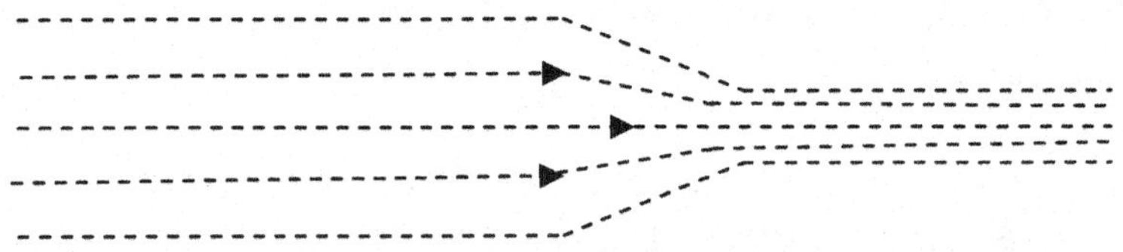

To understand the movement of laminar fluids, one of the first quantities we must define is volumetric flow rate which may have units of gallons per min (gpm), liters/s, cubic feet per min (cfm), gpf, or m^3/s:

$$Q = Av\cos\theta$$

Where Q=volumetric flow rate
A=cross sectional area of the pipe
v=fluid velocity
θ=the angle between the direction of the fluid flow and a vector normal to A

Note that in situations in which the fluid velocity is perpendicular to the cross sectional area, this equation is simply:

$$Q = Av$$

It is also convenient to sometimes discuss mass flow rate ($\dot{m}$), which we can easily find using the density (ρ) of the fluid:

$$\dot{m} = \rho v A = \rho Q$$

Usually, we make an assumption that the fluid is incompressible, that is, the density is constant. Like many commonly used simplifications, this assumption is largely and typically correct though real fluids are, of course, compressible to varying extents. When we do assume that density is constant, we can use conservation of mass to determine that when a pipe is expanded or restricted, the mass flow rate will remain the same. Let's see how this pertains to an example:

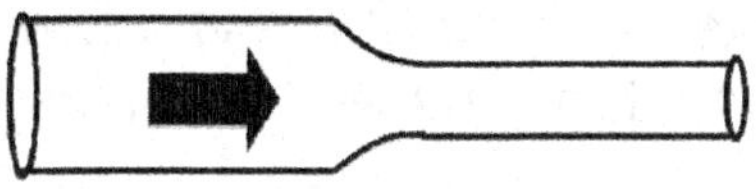

Given conservation of mass, it must be true that:

$$v_1 A_1 = v_2 A_2$$

This is known as the **equation of continuity**. Note that this means the fluid will flow faster in the narrower portions of the pipe and more slowly in the wider regions. An everyday example of this principle is seen when one holds their thumb over the nozzle of a garden hose; the cross sectional area is reduced and so the water flows more quickly.

Much of what we know about fluid flow today was originally discovered by Daniel Bernoulli. His most famous discovery is known as Bernoulli's Principle which states that, if no work is performed on a fluid or gas, an increase in velocity will be accompanied by a decrease in pressure. The mathematical statement of the Bernoulli's Principle for incompressible flow is:

$$\frac{v^2}{2} + gh + \frac{p}{\rho} = \text{constant}$$

where v= fluid velocity
g=acceleration due to gravity
h=height
p=pressure
ρ=fluid density

Though some physicists argue that it leads to the compromising of certain assumptions (i.e., incompressibility, no flow motivation, and a closed fluid loop), most agree it is correct to explain "lift" using Bernoulli's principle. This is because Bernoulli's principle can also be thought of as predicting that the pressure in moving fluid is less than the pressure in fluid at rest. Thus, there are many examples of physical phenomenon that can be explained by Bernoulli's Principle:

- The lift on airplane wings occurs because the top surface is curved while the bottom surface is straight. Air must therefore move at a higher velocity on the top of the wing and the resulting lower pressure on top accounts for lift.
- The tendency of windows to explode rather than implode in hurricanes is caused by the pressure drop that results from the high speed winds blowing across the outer surface of the window. The higher pressure on the inside of the window then pushes the glass outward, causing an explosion.
- The ballooning and fluttering of a tarp on the top of a semi-truck moving down the highway is caused by the flow of air across the top of the truck. The decrease in pressure causes the tarp to "puff up."
- A perfume atomizer pushes a stream of air across a pool of liquid. The drop in pressure caused by the moving air lifts a bit of the perfume and allows it to be dispensed.

Skill 25.7 Understands motion in terms of frames of reference and relativity concepts.

When we analyze a situation using the laws of physics, we must first consider the perspective from which it is viewed. This is known as the frame of reference. The principles which describe the relationships between different frames of reference are known as relativities. The type of relativity discussed below is known as Galilean or Newtonian relativity and is valid for physical situations in which velocities are relatively low. When velocities approach the speed of light, we must use Einstein's special relativity (see section V.10).

There are two general types of reference frames: inertial and non-inertial

Inertial: These frames translate at a constant vector velocity, meaning the velocity does not change direction or magnitude (i.e., travel in a straight line without acceleration).

Non-inertial: These frames include all other situations in which there is non-constant velocity, such as acceleration or rotation. Galilean relativity does not apply to non-inertial frames, as explained below.

Galilean relativity states that the laws of physics are the same in all inertial frames. That is, these same laws would apply to an experiment performed on the surface of the Earth and an experiment performed in a reference frame moving at constant velocity with respect to the earth. For instance, two baseball players can have the same game of catch either standing on the ground or in a moving bus (so long as the bus's motion has constant direction and magnitude).

It is true, however, that phenomenon will have different appearance depending on our frame of reference. Relative velocity is a useful concept to help us analyze such cases. We can understand relative velocity by again considering the game of catch being played on a bus:

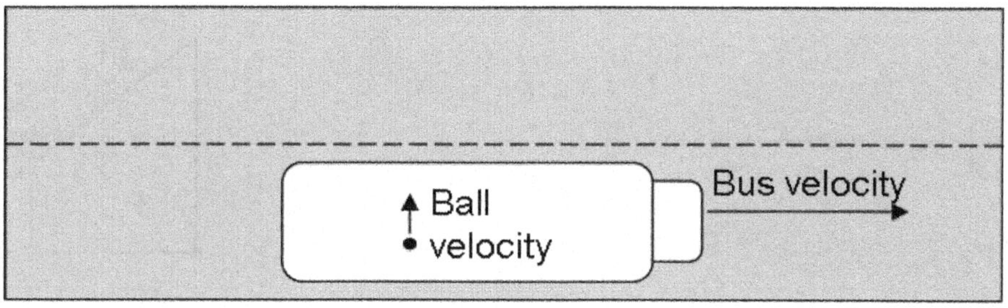

Inside the frame of reference of the bus, the ball travels at the velocity with which it was thrown and straight across the bus (shown by the ball velocity vector above). However, if we use stationary earth as our frame of reference, then the ball is not only moving across the bus, but down the road at the velocity with which the bus is driven. To determine the ball's velocity relative to the earth, then, we must add the ball's velocity relative to the bus and the bus's velocity relative to the earth. This can be performed with simple vector addition.

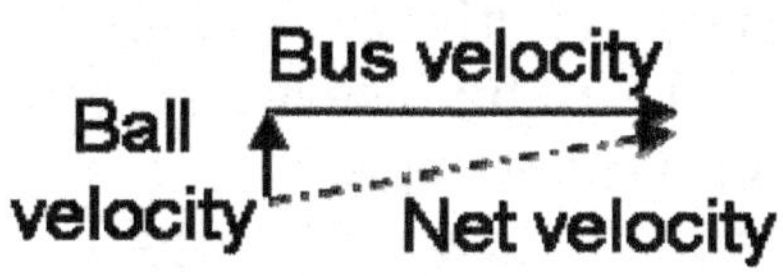

COMPETENCY 26.0 THE TEACHER UNDERSTANDS THE LAWS OF MOTION.

Skill 26.1 Identifies and analyzes the forces acting in a given situation and constructs a free-body diagram.

Free body diagrams are simple sketches that show all the objects and forces in a given physical situation. This makes them very useful for understanding and solving physical problems. These diagrams show relative positions, masses, and the direction in which forces are acting.

Below are two examples of free body diagrams. We can use them to help solve problems.

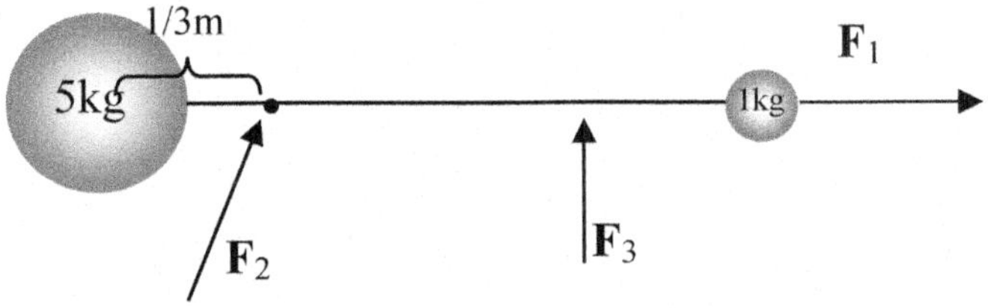

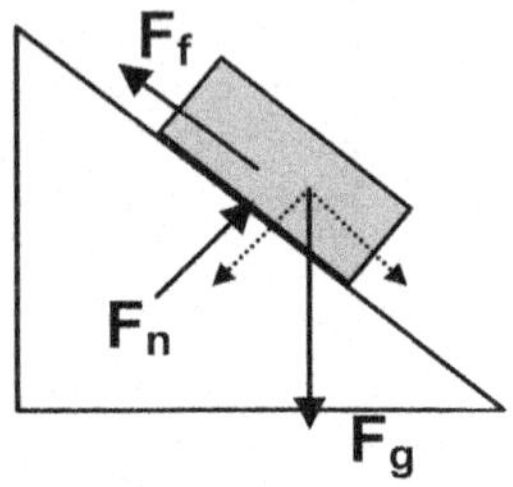

Some of the common forces that act on a body are the following:

Gravity
This is the force that pulls a body towards the center of the earth, i.e. downwards, and is also called the weight of the body. It is given by

$$W = mg$$

Where m is the mass of the body and g=9.81 m/s^2 is the acceleration due to gravity.

Normal force
When a body is pressed against a surface it experiences a reaction force that is perpendicular to the surface and in the direction away from the surface. For instance, an object resting on a table experiences an upward reaction force from the table that is equal and opposite to the force that the object exerts on the table. When the table is horizontal and no additional force is being applied to the object, the normal force is equal to the weight of the object.

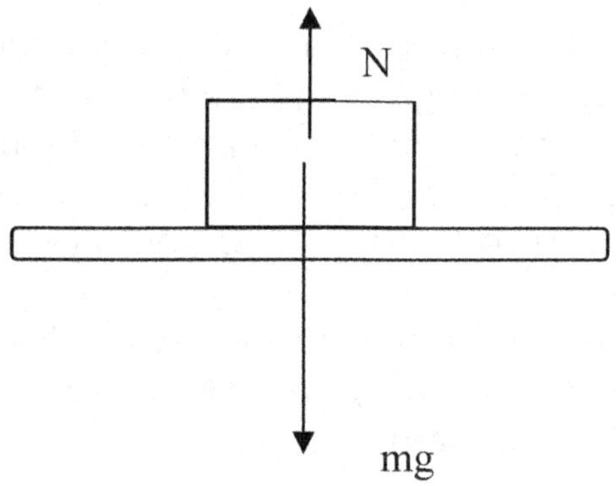

Friction

Friction is the force on a body that opposes its sliding over a surface. This force is due to the bonding between the two surfaces and is greater for rough surfaces. It acts in the direction opposite to the force attempting to move the object. When the object is at rest, the frictional force is known as **static friction**. The frictional force on an object in motion is known as **kinetic friction**.

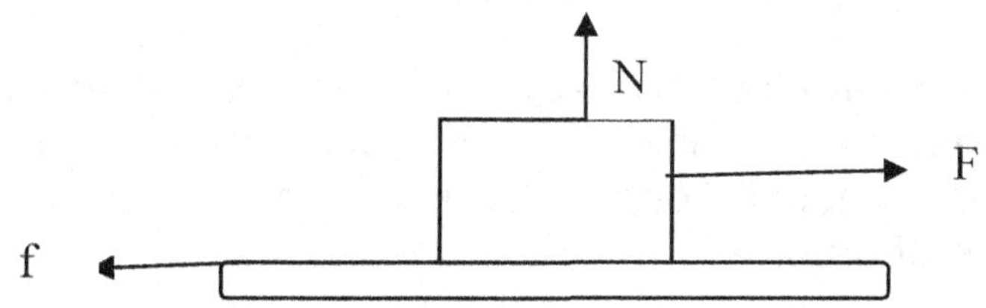

The frictional force is usually directly proportional to the normal force and can be calculated as $\mathbf{F_f} = \mu\ \mathbf{F_n}$ where μ is either the coefficient of static friction or kinetic friction depending on whether the object is at rest or in motion.

Tension/Compression

Tension is the force that acts in a rope, cable or rod that is attached to something and is being pulled. Tension acts along the cord. When a hand pulls a rope attached to a box, for instance, the tension T in the rope acts to pull the rope apart while it works on the box and the hand in the opposite direction as shown below:

Compression is the opposite of tension in that the force acts to shorten a rigid body instead of pulling it apart.

Net force

The forces that act on a body come from many different sources. Their effect on a body, however, is the same; a change in the state of motion of the body as given by Newton's laws of motion. Therefore, once we identify the magnitude and direction of each force acting on a body, we can combine the effect of all the forces together using vector addition and find the net force.

Skill 26.2 Solves problems involving the vector nature of force (e.g., resolving forces into components, analyzing static or dynamic equilibrium of a particle).

An object is said to be in a state of equilibrium when the forces exerted upon it are balanced. That is to say, forces to the left balance the forces exerted to the right, and upward forces are balanced by downward forces. The net force acting on the object is zero and the acceleration is 0 meters per second squared. This does not necessarily mean that the object is at rest. According to Newton's first law of motion, an object at equilibrium is either at rest and remaining at rest (**static equilibrium**), or in motion and continuing in motion with the same speed and direction (**dynamic equilibrium**).

Equilibrium of forces is often used to analyze situations where objects are in static equilibrium. One can determine the weight of an object in static equilibrium or the forces necessary to hold an object at equilibrium. The following are examples of each type of problem.

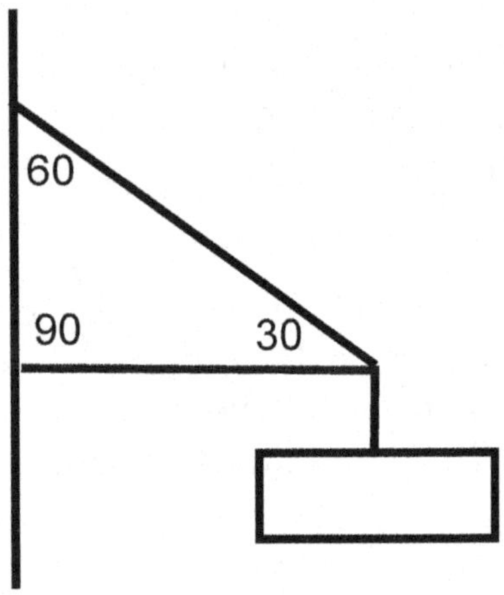

Problem: A sign hangs outside a building supported as shown in the diagram. The sign has a mass of 50 kg. Calculate the tension in the cable.

Solution: Since there is only one upward pulling cable it must balance the weight. The sign exerts a downward force of 490 N. Therefore, the cable pulls upwards with a force of 490 N. It does so at an angle of 30 degrees. To find the total tension in the cable:

$$F_{total} = 490 \text{ N} / \sin 30°$$

$$F_{total} = 980 \text{ N}$$

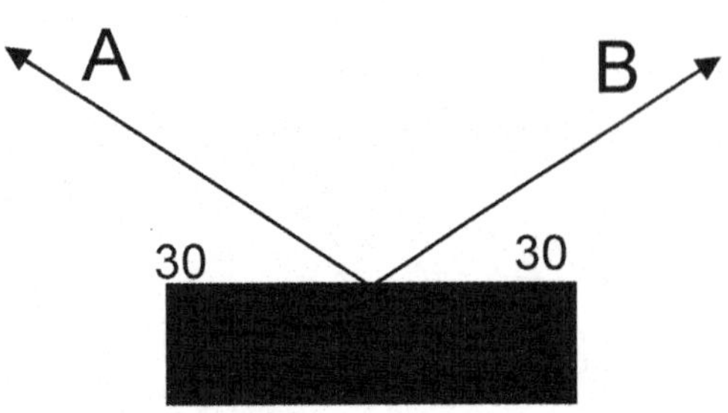

Problem: A block is held in static equilibrium by two cables. Suppose the tension in cables A and B are measured to be 50 Newtons each. The angle formed by each cable with the horizontal is 30 degrees. Calculate the weight of the block.

Solution: We know that the upward pull of the cable must balance the downward force of the weight of the block and the right pulling forces must balance the left pulling forces.

Using trigonometry we know that the y component of each cable can be calculated as:

$$F_y = 50 \text{ N} \sin 30°$$
$$F_y = 25 \text{ N}$$

Since there are two cables supplying an upward force of 25 N each, the overall downward force supplied by the block must be 50 N.

Equilibrium of moments (torques)

For an object to be in equilibrium the forces acting on it must be balanced. This applies to linear as well as rotational forces known as moments or torques. In the two dimensional example below, torque can only be applied in two directions; clockwise and counter clockwise. The convention is that positive rotation is counter clockwise and negative is clockwise. For the object to be in equilibrium, the sum of the applied torques must be zero, in addition to the sum of all forces being zero.

Let us consider a horizontal bar at equilibrium so that the bar experiences neither rotation nor translation. We can define rotation by choosing any point along the bar and labeling it A for axis.

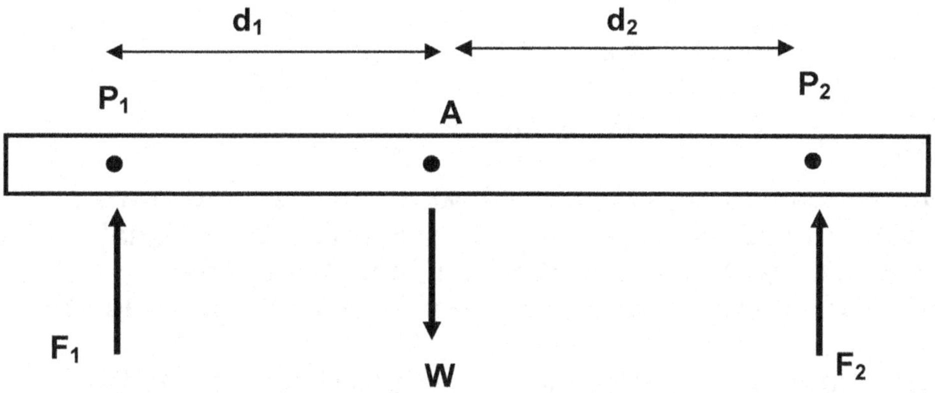

The bar experiences two point forces at either end, labeled F_1 and F_2. The torque applied to the bar by each of these forces is given by multiplying the force by the moment arm, the distance between the point where the force is applied and the axis. In the case of F_1 the torque is as follows (negative since rotation is clockwise):

$$\tau_1 = -F_1 d_1$$

The torque applied by F2 is given by (positive since rotation is counterclockwise):

$$\tau_2 = F_2 d_2$$

The other force acting on the bar is the force of gravity, or the weight of the bar. This is not a point force, but rather acts at all points along the bar. However, we can consider that the weight acts in the center of the bar, at a point called the center of mass. In the case of this example, we are taking the axis to be located at the center of mass. Since the axis is located at the center of mass, the torque exerted on the bar due to its weight is zero.

Suppose F_1=2 N, d_1=0.4 m, and d_2=0.5 m. Let us calculate F_2. We will use our knowledge that the sum of all torques must equal zero when that object is at equilibrium.

$$\tau_1 + \tau_2 = 0$$

$$-F_1 d_1 + F_2 d_2 = 0$$
$$(-2\ \text{N} \times 0.4\ \text{m}) + (F_2 \times 0.5\ \text{m}) = 0$$
$$-0.8 + 0.5\ F_2 = 0$$
$$F_2 = 1.6\ \text{N}$$

It is also possible to calculate the weight of the bar since we know that the sum of all forces must be zero. Since F1 and F2 act up but weight acts down we have:

$$2\ \text{N} + 1.6\ \text{N} - W = 0$$

$$W = 3.6\ \text{N}$$

Skill 26.3 Identifies and applies Newton's laws to analyze and solve a variety of practical problems (e.g., properties of frictional forces, acceleration of a particle on an inclined plane, displacement of a mass on a spring, forces on a pendulum).

Newton's first law of motion: "An object at rest tends to stay at rest and an object in motion tends to stay in motion with the same speed and in the same direction unless acted upon by an unbalanced force". This tendency of an object to continue in its state of rest or motion is known as **inertia**. Note that, at any point in time, most objects have multiple forces acting on them. If the vector addition of all the forces on an object results in a zero net force, then the forces on the object are said to be **balanced**. If the net force on an object is non-zero, an **unbalanced** force is acting on the object.

Prior to Newton's formulation of this law, being at rest was considered the natural state of all objects because at the earth's surface we have the force of gravity working at all times which causes nearly any object put into motion to eventually come to rest. Newton's brilliant leap was to recognize that an unbalanced force changes the motion of a body, whether that body begins at rest or at some non-zero speed.

We experience the consequences of this law everyday. For instance, the first law is why seat belts are necessary to prevent injuries. When a car stops suddenly, say by hitting a road barrier, the driver continues on forward due to inertia until acted upon by a force. The seat belt provides that force and distributes the load across the whole body rather than allowing the driver to fly forward and experience the force against the steering wheel.

Newton's second law of motion: "The acceleration of an object as produced by a net force is directly proportional to the magnitude of the net force, in the same direction as the net force, and inversely proportional to the mass of the object". In the equation form, it is stated as ***F=*** ma, force equals mass times acceleration. It is important, again, to remember that this is the net force and that forces are vector quantities. Thus if an object is acted upon by 12 forces that sum to zero, there is no acceleration. Also, this law embodies the idea of inertia as a consequence of mass. For a given force, the resulting acceleration is proportionally smaller for a more massive object because the larger object has more inertia.

The first two laws are generally applied together via the equation **F=ma.** The first law is largely the conceptual foundation for the more specific and quantitative second law. Newton's first law and second law are valid only in **inertial reference frames** (described in previous section).

The **weight** of an object is the result of the gravitational force of the earth acting on its mass. The acceleration due to Earth's gravity on an object is 9.81 m/s^2. Since force equals mass * acceleration, the magnitude of the gravitational force created by the earth on an object is

$$F_{Gravity} = m_{object} \cdot 9.81\, {}^{m}\!/_{s^2}$$

Example: For the arrangement shown, find the force necessary to overcome the 500 N force pushing to the left and move the truck to the right with an acceleration of 5 m/s^2.

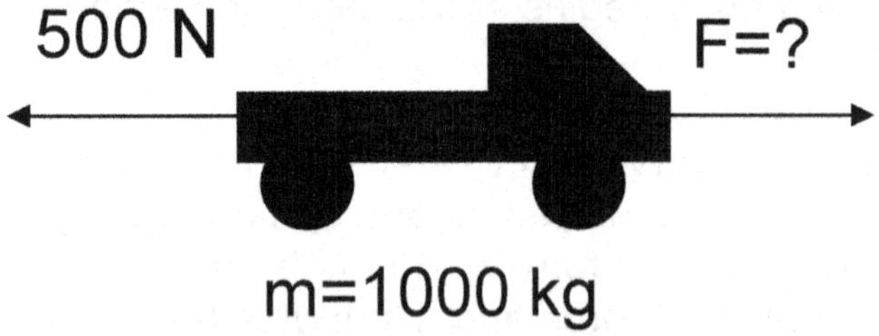

Solution: Since we know the acceleration and mass, we can calculate the net force necessary to move the truck with this acceleration. Assuming that to the right is the positive direction we sum the forces and get
F-500N = 1000kg x 5 m/s^2. Solving for F, we get 5500 N.

Newton's third law of motion: "For every action, there is an equal and opposite reaction". This statement means that, in every interaction, there is a pair of forces acting on the two interacting objects. The size of the force on the first object equals the size of the force on the second object. The direction of the force on the first object is opposite to the direction of the force on the second object.

1. The propulsion/movement of fish through water: A fish uses its fins to push water backwards. The water pushes back on the fish. Because the force on the fish is unbalanced the fish moves forward.
2. The motion of car: A car's wheels push against the road and the road pushes back. Since the force of the road on the car is unbalanced the car moves forward.
3. Walking: When one pushes backwards on the foot with the muscles of the leg, the floor pushes back on the foot. If the forces of the leg on the foot and the floor on the foot are balanced, the foot will not move and the muscles of the body can move the other leg forward.

In the real world, whenever an object moves its motion is opposed by a force known as friction. How strong the frictional force is depends on numerous factors such as the roughness of the surfaces (for two objects sliding against each other) or the viscosity of the liquid an object is moving through. Most problems involving the effect of friction on motion deal with sliding friction. This is the type of friction that makes it harder to push a box across cement than across a marble floor.

When you try and push an object from rest, you must overcome the maximum **static friction** force to get it to move. Once the object is in motion, you are working against **kinetic friction** which is smaller than the static friction force previously mentioned. Sliding friction is primarily dependent on two things, the **coefficient of friction (μ)** which is dependent on the roughness of the surfaces involved and the amount of force pushing the two surfaces together. This force is also known as the **normal force (F_n)**, the perpendicular force between two surfaces. When an object is resting on a flat surface, the normal force is pushing opposite to the gravitational force – straight up. When the object is resting on an incline, the normal force is less (because it is only opposing that portion of the gravitational force acting perpendicularly to the object) and its direction is perpendicular to the surface of incline but at an angle from the ground. Therefore, for an object experiencing no external action, the magnitude of the normal force is either equal to or less than the magnitude of the gravitational force **(F_g)** acting on it. The frictional force (**F_f**) acts perpendicularly to the normal force, opposing the direction of the object's motion.

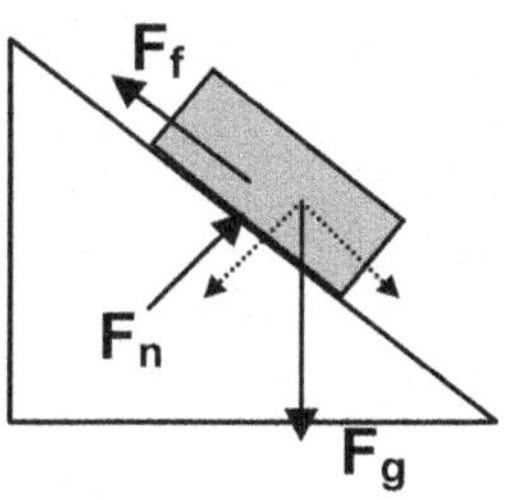

The frictional force is normally directly proportional to the normal force and, unless you are told otherwise, can be calculated as $F_f = \mu F_n$ where μ is either the coefficient of static friction or kinetic friction depending on whether the object starts at rest or in motion. In the first case, the problem is often stated as "how much force does it take to start an object moving" and the frictional force is given by $F_f > \mu_s F_n$ where μ_s is the coefficient of static friction. When questions are of the form "what is the magnitude of the frictional force opposing the motion of this object," the frictional force is given by $F_f = \mu_k F_n$ where μ_k is the coefficient of kinetic friction.

A static frictional force is needed in order to start a ball or a wheel rolling; without this force the object would just slide or spin. **Rolling friction** is the force that resists the rolling motion of an object such as a wheel once it is already in motion. Rolling friction arises from the roughness of the surfaces in contact and from the deformation of the rolling object or surface on which it is rolling. Rolling resistance $F_f = \mu_r F_n$ where μ_r is the coefficient of rolling friction.

There are several important things to remember when solving problems about friction.

1. The frictional force acts in opposition to the direction of motion.
2. The frictional force is proportional to, and acts perpendicular to, the normal force.
3. The normal force is perpendicular to the surface the object is lying on. If there is a force pushing the object against the surface, it will increase the normal force.

Problem:

A woman is pushing an 800N box across the floor. She pushes with a force of 1000 N in the direction indicated in the diagram below. The coefficient of kinetic friction is 0.50. If the box is already moving, what is the force of friction acting on the box?

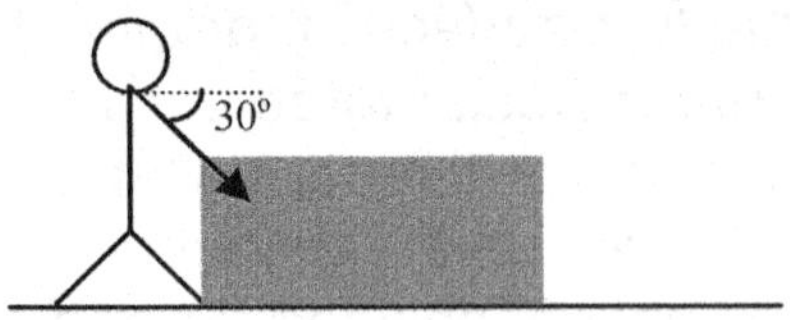

Solution:
First it is necessary to solve for the normal force.

$\mathbf{F_n}$= 800N + 1000N (sin 30°) = 1300N
Then, since $\mathbf{F_f} = \mu\ \mathbf{F_n}$ = 0.5*1300=650N

COMPETENCY 27.0 THE TEACHER UNDERSTANDS THE CONCEPTS OF GRAVITATIONAL AND ELECTROMAGNETIC FORCES IN NATURE.

Skill 27.1 Applies the Law of Universal Gravitation to solve a variety of problems (e.g., determining the gravitational fields of the planets, analyzing properties of satellite orbits).

Newton's universal law of gravitation states that any two objects experience a force between them as the result of their masses. Specifically, the force between two masses m_1 and m_2 can be summarized as

$$F = G\frac{m_1 m_2}{r^2}$$

where G is the gravitational constant ($G = 6.672 \times 10^{-11} Nm^2 / kg^2$), and r is the distance between the two objects.

Important things to remember:

1. The gravitational force is proportional to the masses of the two objects, but *inversely* proportional to the *square of the distance* between the two objects.
2. When calculating the effects of the acceleration due to gravity for an object above the earth's surface, the distance above the surface is ignored because it is inconsequential compared to the radius of the earth. The constant figure of 9.81 m/s^2 is used instead.

Problem: Two identical 4 kg balls are floating in space, 2 meters apart. What is the magnitude of the gravitational force they exert on each other?

Solution:

$$F = G\frac{m_1 m_2}{r^2} = G\frac{4\times 4}{2^2} = 4G = 2.67\times 10^{-10}\,N$$

For a satellite of mass m in orbit around the earth (mass M), the gravitational attraction of the earth provides the centripetal force that keeps the satellite in motion:

$$\frac{GMm}{r^2} = \frac{mv^2}{r} = mr\omega^2 = mr\left(\frac{2\pi}{T}\right)^2$$

Thus the period T of rotation of the satellite may be obtained from the equation

$$\frac{T^2}{r^3} = \frac{4\pi^2}{GM}$$

Johannes Kepler was a German mathematician who studied the astronomical observations made by Tyco Brahe. He derived the following three laws of planetary motion. Kepler's laws also predict the motion of comets.

First law

This law describes the shape of planetary orbits. Specifically, the orbit of a planet is an ellipse that has the sun at one of the foci. Such an orbit looks like this:

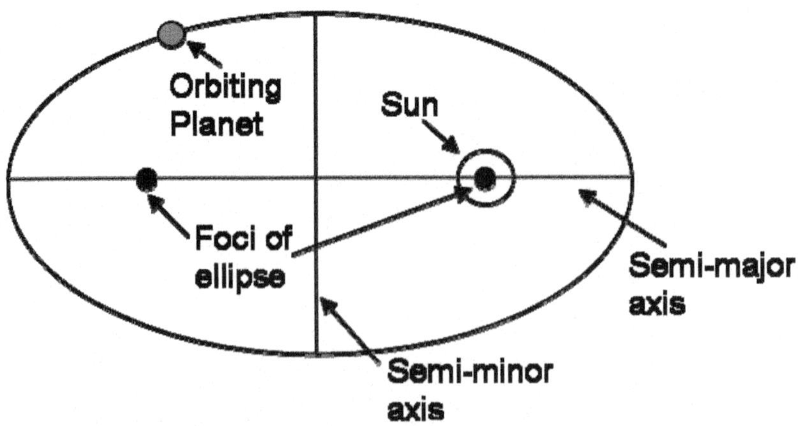

To analyze this situation mathematically, remember that the semi-major axis is denoted a, the semi-minor axis denoted b, and the general equation for an ellipse in polar coordinates is:

$$r = \frac{l}{1 + e\cos\theta}$$

Where r=radial coordinate
θ=angular coordinate
l= semi-latus rectum ($l = b^2/a$)
e=eccentricity (for an ellipse, e= $\sqrt{1 - \frac{b^2}{a^2}}$

Thus, we can also determine the planet's maximum and minimum distance from the sun.

The point at which the planet is closest to the sun is known as the perihelion and occurs when θ=0:

$$r_{\min} = \frac{l}{1+e}$$

The point at which the planet is farthest from the sun is known as the aphelion and occurs when θ=180°:

$$r_{\max} = \frac{l}{1-e}$$

Second Law

The second law pertains to the relative speed of a planet as it orbits. This law says that a line joining the planet and the Sun sweeps out equal areas in equal intervals of time. In the diagram below, the two shaded areas demonstrate equal areas.

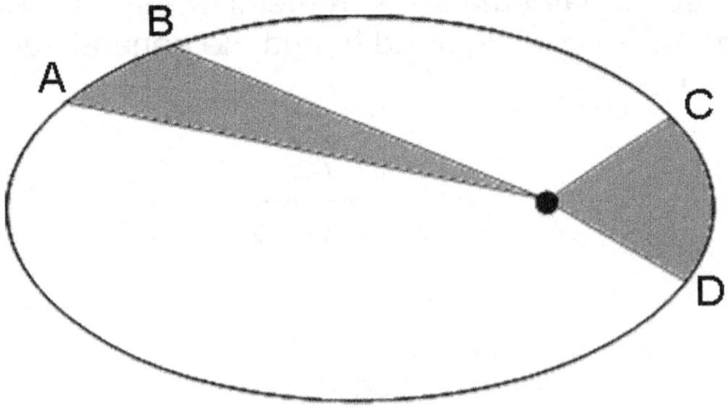

By Kepler's second law, we know that the planet will take the same amount of time to move between points A and B and between points C and D. Note that this means that the speed of the planet is inversely proportional to its distance from the sun (i.e., the plant moves fastest when it is closest to the sun). You can view an animation of this changing speed here:

http://home.cvc.org/science/kepler.gif

Kepler was only able to demonstrate the existence of this phenomenon but we now know that it is an effect of the Sun's gravity. The gravity of the Sun pulls the planet toward it thereby accelerating the planet as it nears. Using the first two laws together, Kepler was able to calculate a planet's position from the time elapsed since the perihelion.

Third law

The third law is also known as the harmonic law and it relates the size of a planet's orbital to the time needed to complete it. It states that the square of a planet's period is proportional to the cube of its mean distances from the Sun (this mean distance can be shown to be equal to the semi-major axis). So, we can state the third law as:

$$P^2 \propto a^3$$

where P=planet's orbital period (length of time needed to complete one orbit)
a=semi-major axis of orbit

Furthermore, for two planets A and B:

$$P_A^2 / P_B^2 = a_A^3 / a_B^3$$

The units for period and semi-major axis have been defined such that $P^2a^{-3}=1$ for all planets in our solar system. These units are sidereal years (yr) and astronomical units (AU). Sample values are given in the table below. Note that in each case $P^2 \sim a^3$

Planet	P (yr)	a (AU)	P^2	a^3
Venus	0.62	0.72	0.39	0.37
Earth	1.0	1.0	1.0	1.0
Jupiter	11.9	5.20	142	141

Skill 27.2 Calculates electrostatic forces, fields, and potentials.

Any point charge may experience force resulting from attraction to or repulsion from another charged object. The easiest way to begin analyzing this phenomenon and calculating this force is by considering two point charges. Let us say that the charge on the first point is Q_1, the charge on the second point is Q_2, and the distance between them is r. Their interaction is governed by **Coulomb's Law** which gives the formula for the force F as:

$$F = k\frac{Q_1Q_2}{r^2}$$

where $k = 9.0\times10^9 \frac{N\cdot m^2}{C^2}$ (known as Coulomb's constant)

The charge is a scalar quantity, however, the force has direction. For two point charges, the direction of the force is along a line joining the two charges. Note that the force will be repulsive if the two charges are both positive or both negative and attractive if one charge is positive and the other negative. Thus, a negative force indicates an attractive force.

When more than one point charge is exerting force on a point charge, we simply apply Coulomb's Law multiple times and then combine the forces as we would in any statics problem. Let's examine the process in the following example problem.

<u>Problem:</u> Three point charges are located at the vertices of a right triangle as shown below. Charges, angles, and distances are provided (drawing not to scale). Find the force exerted on the point charge A.

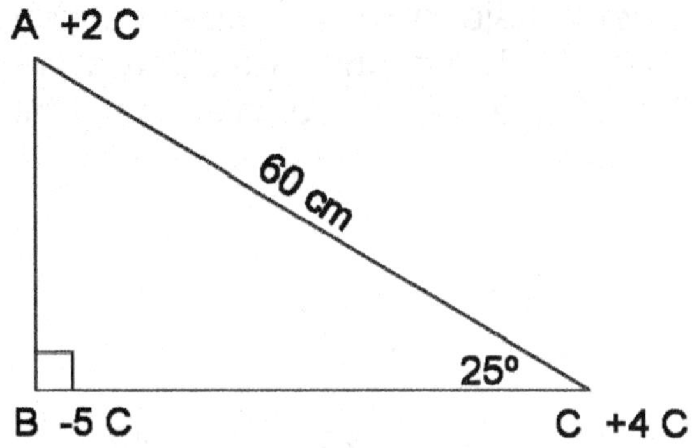

Solution: First we find the individual forces exerted on A by point B and point C. We have the information we need to find the magnitude of the force exerted on A by C.

$$F_{AC} = k\frac{Q_1Q_2}{r^2} = 9\times10^9\,\frac{N\cdot m^2}{C^2}\left(\frac{4C\times2C}{(0.6m)^2}\right) = 2\times10^{11}\,N$$

To determine the magnitude of the force exerted on A by B, we must first determine the distance between them.

$$\sin 25° = \frac{r_{AB}}{60cm}$$
$$r_{AB} = 60cm\times\sin 25° = 25cm$$

Now we can determine the force.

$$F_{AB} = k\frac{Q_1Q_2}{r^2} = 9\times10^9\,\frac{N\cdot m^2}{C^2}\left(\frac{-5C\times2C}{(0.25m)^2}\right) = -1.4\times10^{12}\,N$$

We can see that there is an attraction in the direction of B (negative force) and repulsion in the direction of C (positive force). To find the net force, we must consider the direction of these forces (along the line connecting any two point charges). We add them together using the law of cosines.

$$F_A^{\,2} = F_{AB}^{\,2} + F_{AC}^{\,2} - 2F_{AB}F_{AC}\cos 75°$$
$$F_A^{\,2} = (-1.4\times10^{12}\,N)^2 + (2\times10^{11}\,N)^2 - 2(-1.4\times10^{12}\,N)(2\times10^{11}\,N)^2\cos 75°$$
$$F_A = 1.5\times10^{12}\,N$$

This gives us the magnitude of the net force, now we will find its direction using the law of sines.

$$\frac{\sin\theta}{F_{AC}} = \frac{\sin 75^\circ}{F_A}$$

$$\sin\theta = F_{AC}\frac{\sin 75^\circ}{F_A} = 2\times10^{11}N\frac{\sin 75^\circ}{1.5\times10^{12}N}$$

$$\theta = 7.3^\circ$$

Thus, the net force on A is 7.3° west of south and has magnitude 1.5 x 10^{12}N. Looking back at our diagram, this makes sense, because A should be attracted to B (pulled straight south) but the repulsion away from C "pushes" this force in a westward direction.

An **electric field** exists in the space surrounding a charge. Electric fields have both direction and magnitude determined by the strength and direction in which they exhibit force on a test charge. The units used to measure electric fields are newtons per coulomb (N/C). **Electric potential** is simply the **potential energy** per unit of charge. Given this definition, it is clear that electric potential must be measured in joules per coulomb and this unit is known as a volt (J/C=V).

Within an electric field there are typically differences in potential energy. This **potential difference** may be referred to as **voltage**. The difference in electrical potential between two points is the amount of work needed to move a unit charge from the first point to the second point. Stated mathematically, this is:

$$V = \frac{W}{Q}$$

where V= the potential difference
W= the work done to move the charge
Q= the charge

We know from mechanics, however, that work is simply force applied over a certain distance. We can combine this with Coulomb's law to find the work done between two charges distance r apart.

$$W = F.r = k\frac{Q_1Q_2}{r^2}.r = k\frac{Q_1Q_2}{r}$$

Now we can simply substitute this back into the equation above for electric potential:

$$V_2 = \frac{W}{Q_2} = \frac{k\frac{Q_1Q_2}{r}}{Q_2} = k\frac{Q_1}{r}$$

Let's examine a sample problem involving electrical potential.

Problem: What is the electric potential at point A due to the 2 shown charges? If a charge of +2.0 C were infinitely far away, how much work would be required to bring it to point A?

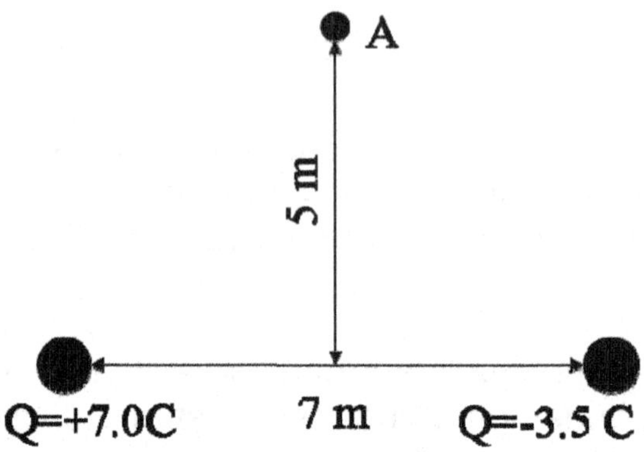

Solution: To determine the electric potential at point A, we simple find and add the potential from the two charges (this is the principle of superposition). From the diagram, we can assume that A is equidistant from each charge. Using the Pythagorean theorem, we determine this distance to be 6.1 m.

$$V = \frac{kq}{r} = k\left(\frac{7.0C}{6.1m} + \frac{-3.5C}{6.1m}\right) = 9 \times 10^9 \frac{N.m^2}{C^2}\left(0.57\frac{C}{m}\right) = 5.13 \times 10^9 V$$

Now, let's consider bringing the charged particle to point A. We assume that electric potential of these particle is initially zero because it is infinitely far away. Since now know the potential at point A, we can calculate the work necessary to bring the particle from V=0, i.e. the potential energy of the charge in the electrical field:

$$W = VQ = (5.13 \times 10^9) \times 2J = 10.26 \times 10^9 J$$

The large results for potential and work make it apparent how large the unit coulomb is. For this reason, most problems deal in microcoulombs (μC).

In any physical phenomenon, flux refers to rate of movement of a substance or energy through a certain area. Flux can be used to quantify the movement of mass, heat, momentum, light, molecules and other things. Flux depends on density of flow, area, and direction of the flow. To visualize this, imagine a kitchen sieve under a tap of flowing water. The water that passes through the sieve is the flux; the flux will decrease if we lower the water flow rate, decrease the size of the sieve, or tilt the sieve away from direction of the water's flow. Electric flux, then, is just the number of electric field lines that pass through a given area. It is given by the following equation:

$$\Phi=E(\cos\phi)A$$

where Φ=flux
E=the electric field
A=area
ϕ= the angle between the electric field and a vector normal to the surface A

Thus, if a plane is parallel to an electric field, no field lines will pass through that plane and the flux through it will be zero. If a plane is perpendicular to an electric field, the flux through it will be maximal.

Gauss's Law says that the electric flux through a surface is equal to the charge enclosed by a surface divided by a constant ε_0 (permittivity of free space). The simplest mathematical statement of this law is:

$$\Phi=Q_A/\varepsilon_0$$

where Q_A= the charge enclosed by the surface

Gauss's Law provides us with a useful and powerful method to calculate electric fields. For instance, imagine a solid conducting sphere with a net charge Q_s. We know from Gauss's Law that the electric field inside the sphere must be zero and all the excess charge lies on the outer surface of the sphere. The field produced by this sphere is the same a point charge of value Q_s. This conclusion is true whether the sphere is solid or hollow.

Skill 27.3 Understands the properties of magnetic materials and the molecular theory of magnetism.

Magnetism is a phenomenon in which certain materials, known as magnetic materials, attract or repel each other. A magnet has two poles, a south pole and a north pole. Like poles repel while unlike poles attract. Magnetic poles always occur in pairs known as **magnetic dipoles**. One cannot isolate a single magnetic pole. If a magnet is broken in half, opposite poles appear at both sides of the break point so that one now has two magnets each with a south pole and a north pole. No matter how small the pieces a magnet is broken into, the smallest unit at the atomic level is still a dipole.

A large magnet can be thought of as one with many small dipoles that are aligned in such a way that apart from the pole areas, the internal south and north poles cancel each other out. Destroying this long range order within a magnet by heating or hammering can demagnetize it. The dipoles in a **non-magnetic** material are randomly aligned while they are perfectly aligned in a preferred direction in **permanent** magnets. In a **ferromagnet**, there are domains where the magnetic dipoles are aligned, however, the domains themselves are randomly oriented. A ferromagnet can be magnetized by placing it in an external magnetic field that exerts a force to line up the domains.

Weber's theory of molecular magnetism assumes that all magnetic substances are made up of tiny molecular magnets. An unmagnetized material has no magnetic effect because adjacent molecular magnets neutralize the magnetic forces of its molecular magnets.

Most of the molecular magnets of a magnetized material line up so that each molecule's north pole points in one direction, and the south pole points in the opposite direction. When the molecules are aligned in this manner, the material has one effective north pole and one effective south pole.

The retention of magnetism in the fragments of a broken magnet suggests the existence of molecular magnets. Regardless of how many times you break a magnet, each fragment is still a magnet. Each fragment has a pole at the breaking point that is equally strong as, but opposite to, the pole at the breaking point of the other fragment. Reconstructing the magnet in the order in which it was broken restores its original properties. This suggests that the molecules of a magnet are magnets themselves.

Skill 27.4 Identifies the source of the magnetic field and calculates the magnetic field for various simple current distributions.

A magnet produces a magnetic field that exerts a force on any other magnet or current-carrying conductor placed in the field. Magnetic field lines are a good way to visualize a magnetic field. The distance between magnetic fields lines indicates the strength of the magnetic field such that the lines are closer together near the poles of the magnets where the magnetic field is the strongest. The lines spread out above and below the middle of the magnet, as the field is weakest at those points furthest from the two poles. The SI unit for magnetic field known as magnetic induction is Tesla(T) given by 1T = 1 N.s/(C.m) = 1 N/(A.m). Magnetic fields are often expressed in the smaller unit Gauss (G) (1 T = 10,000 G). Magnetic field lines always point from the north pole of a magnet to the south pole.

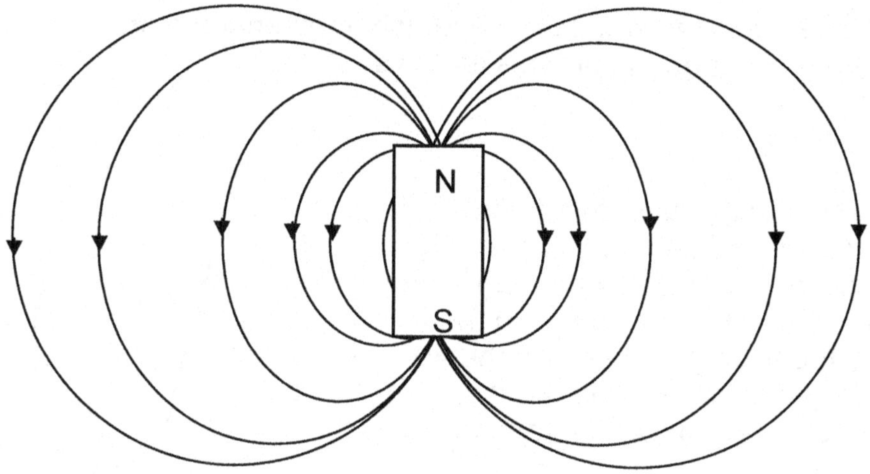

Magnetic field lines can be plotted with a magnetized needle that is free to turn in 3 dimensions. Usually a compass needle is used in demonstrations. The direction tangent to the magnetic field line is the direction the compass needle will point in a magnetic field. Iron filings spread on a flat surface or magnetic field viewing film which contains a slurry of iron filings are another way to see magnetic field lines.

The magnetic force exerted on a charge moving in a magnetic field depends on the size and velocity of the charge as well as the magnitude of the magnetic field. One important fact to remember is that only the velocity of the charge in a direction perpendicular to the magnetic field will affect the force exerted. Therefore, a charge moving parallel to the magnetic field will have no force acting upon it whereas a charge will feel the greatest force when moving perpendicular to the magnetic field.

Conductors through which electrical currents travel will produce magnetic fields: The magnetic field dB induced at a distance r by an element of current Idl flowing through a wire element of length dl is given by the **Biot-Savart** law

$$dB = \frac{\mu_0}{4\pi}\frac{Idl \times \hat{r}}{r^2}$$

where μ_0 is a constant known as the permeability of free space and $\hat{r}$ is the unit vector pointing from the current element to the point where the magnetic field is calculated.

An alternate statement of this law is **Ampere's law** according to which the line integral of $B.dl$ around any closed path enclosing a steady current I is given by

$$\oint_C B \cdot dl = \mu_0 I$$

The basis of this phenomenon is the same no matter what the shape of the conductor, but we will consider three common situations:

Straight Wire
Around a current-carrying straight wire, the magnetic field lines form concentric circles around the wire. The direction of the magnetic field is given by the right-hand rule: When the thumb of the right hand points in the direction of the current, the fingers curl around the wire in the direction of the magnetic field. Note the direction of the current and magnetic field in the diagram.

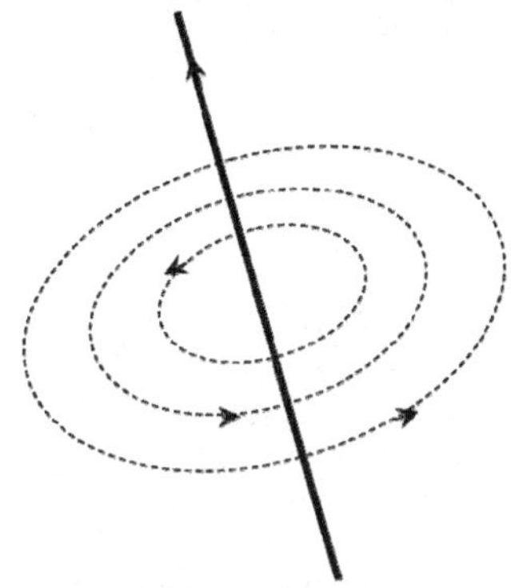

To find the magnetic field of an infinitely long (allowing us to disregarding end effects) we apply Ampere's Law to a circular path at a distance r around the wire:

$$B = \frac{\mu_0 I}{2\pi r}$$

where μ_0=the permeability of free space ($4\pi \times 10^{-7}$ T·m/A)
I=current
r=distance from the wire

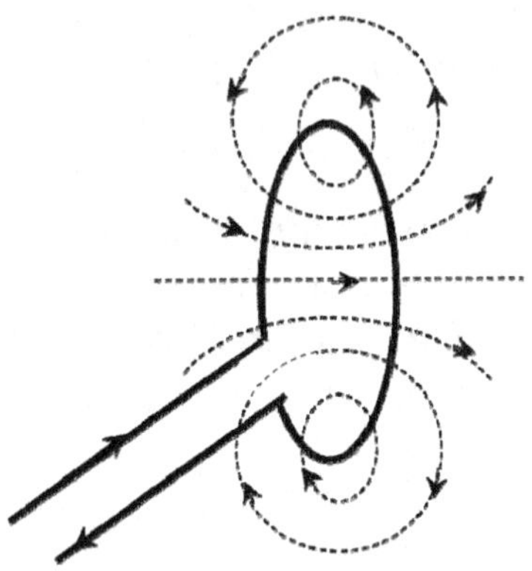

Loops
Like the straight wire from which it's been made, a looped wire has magnetic field lines that form concentric circles with direction following the right-hand rule. However, the field are additive in the center of the loop creating a field like the one shown. The magnetic field of a loop is found similarly to that for a straight wire.

In the center of the loop, the magnetic field is:

$$B = \frac{\mu_0 I}{2r}$$

Solenoids

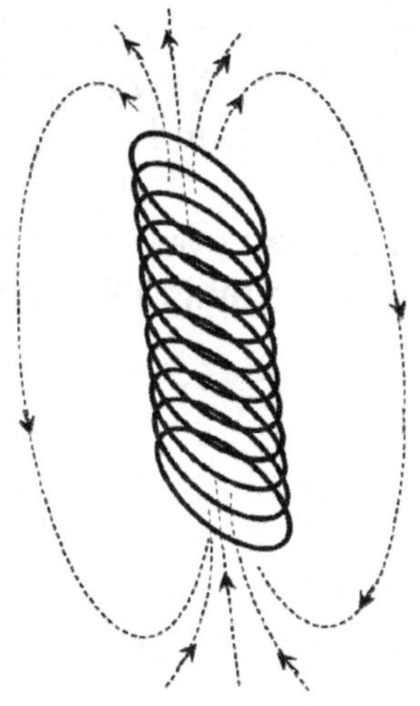

A solenoid is essentially a coil of conduction wire wrapped around a central object. This means it is a series of loops and the magnetic field is similarly a sum of the fields that would form around several loops, as shown.

The magnetic field of a solenoid can be found as with the following equation:

$$B = \mu_0 nI$$

In this equation, n is turn density, which is simply the number of turns divided by the length of the solenoid.

Displacement current

While Ampere's law works perfectly for a steady current, for a situation where the current varies and a charge builds up (e.g. charging of a capacitor) it does not hold. Maxwell amended Ampere's law to include an additional term that includes the displacement current. This is not a true current but actually refers to changes in the electric field and is given by

$$I_d = \varepsilon_0 \frac{d\varphi_e}{dt}$$

where φ_e is the flux of the electric field. Including the displacement current, Ampere's law is given by

$$\oint \boldsymbol{B}.d\boldsymbol{l} = \mu_0 I + \mu_0 \varepsilon_0 \frac{d\varphi_e}{dt}$$

The displacement current essentially indicates that changing electric flux produces a magnetic field.

Skill 27.5 Analyzes the magnetic force on charged particles and current-carrying conductors.

The direction of the magnetic force, or the magnetic component of the **Lorenz force** (force on a charged particle in an electrical and magnetic field), is always at a right angle to the plane formed by the velocity vector v and the magnetic field B and is given by applying the right hand rule - if the fingers of the right hand are curled in a way that seems to rotate the v vector into the B vector, the thumb points in the direction of the force. The magnitude of the force is equal to the cross product of the velocity of the charge with the magnetic field multiplied by the magnitude of the charge.

F=q (**v** X **B**) *or* F=q v Bsin (θ)

Where θ is the angle formed between the vectors of velocity of the charge and direction of magnetic field.

Problem: Assuming we have a particle of 1 x 10^{-6} kg that has a charge of -8 coulombs that is moving perpendicular to a magnetic field in a clockwise direction on a circular path with a radius of 2 m and a speed of 2000 m/s, let's determine the magnitude and direction of the magnetic field acting upon it.

Solution: We know the mass, charge, speed, and path radius of the charged particle. Combining the equation above with the equation for centripetal force we get

$$qvB = \frac{mv^2}{r} \quad \text{or} \quad B = \frac{mv}{qr}$$

Thus B= (1 x 10^{-6} kg) (2000m/s) / (-8 C)(2 m) = 1.25 x 10^{-4} Tesla

Since the particle is moving in a clockwise direction, we use the right hand rule and point our fingers clockwise along a circular path in the plane of the paper while pointing the thumb towards the center in the direction of the centripetal force. This requires the fingers to curl in a way that indicates that the magnetic field is pointing out of the page. However, since the particle has a negative charge we must reverse the final direction of the magnetic field into the page.

A **mass spectrometer** measures the mass to charge ratio of ions using a setup similar to the one described above. m/q is determined by measuring the path radius of particles of known velocity moving in a known magnetic field.

A **cyclotron**, a type of particle accelerator, also uses a perpendicular magnetic field to keep particles on a circular path. After each half circle, the particles are accelerated by an electric field and the path radius is increased. Thus the beam of particles moves faster and faster in a growing spiral within the confines of the cyclotron until they exit at a high speed near the outer edge. Its compactness is one of the advantages a cyclotron has over linear accelerators.

The **force on a current-carrying conductor** in a magnetic field is the sum of the forces on the moving charged particles that create the current. For a current I flowing in a straight wire segment of length l in a magnetic field B, this force is given by

$$\boldsymbol{F} = I\boldsymbol{l} \times \boldsymbol{B}$$

where $\boldsymbol{l}$ is a vector of magnitude l and direction in the direction of the current.

When a current-carrying loop is placed in a magnetic field, the net force on it is zero since the forces on the different parts of the loop act in different directions and cancel each other out. There is, however, a net torque on the loop that tends to rotate it so that the area of the loop is perpendicular to the magnetic field. For a current I flowing in a loop of area A, this torque is given by

$$\tau = IA\hat{\boldsymbol{n}} \times \boldsymbol{B}$$

where $\hat{\boldsymbol{n}}$ is the unit vector perpendicular to the plane of the loop.

Magnetic flux (Gauss's law of magnetism)

Carl Friedrich Gauss developed laws that related electric or gravitational flux to electrical charge or mass, respectively. Gauss's law, along with others, was eventually generalized by James Clerk Maxwell to explain the relationships between electromagnetic phenomena (Maxwell's Equations).

To understand Gauss's law for magnetism, we must first define magnetic flux. Magnetic flux is the magnetic field that passes through a given area. It is given by the following equation:

$$\Phi = B(\cos\phi)A$$

where Φ=flux
B=the magnetic field
A=area
ϕ= the angle between the electric field and a vector normal to the surface A

Thus, if a plane is parallel to a magnetic field, no magnetic field lines will pass through that plane and the flux will be zero. If a plane is perpendicular to a magnetic field, the flux will be maximal.

Now we can state Gauss's law of magnetism: the net magnetic flux out of any closed surface is zero. Mathematically, this may be stated as:

$$\vec{\nabla} \cdot \vec{B} = 0$$

where $\vec{\nabla}$ = the del operator
$\vec{B}$ = magnetic field

One of the most important implications of this law is that there are no magnetic monopoles (that is, magnets always have a positive and negative pole). This is because a magnetic monopole source would give a non-zero product in the equation above. For a magnetic dipole with closed surface, of course, the net flux will always be zero. This is because the magnetic flux directed inward toward the south pole is always equal to the magnetic flux outward from the north pole.

Skill 27.6 Understands induced electric and magnetic fields and analyzes the relationship between electricity and magnetism.

When the magnetic flux through a coil is changed, a voltage is produced which is known as induced electromagnetic force. Magnetic flux is a term used to describe the number of magnetic fields lines that pass through an area and is described by the equation:

$$\Phi = B\ A \cos\theta$$

Where Φ is the angle between the magnetic field B, and the normal to the plane of the coil of area A

By changing any of these three inputs, magnetic field, area of coil, or angle between field and coil, the flux will change and an EMF can be induced. The speed at which these changes occur also affects the magnitude of the EMF, as a more rapid transition generates more EMF than a gradual one. This is described by **Faraday's law** of induction:

$$\varepsilon = -N\ \Delta\Phi\ /\ \Delta t$$

where ε is emf induced, N is the number of loops in a coil, t is time, and Φ is magnetic flux

The negative sign signifies **Lenz's law** which states that induced emf in a coil acts to oppose any change in magnetic flux. Thus the current flows in a way that creates a magnetic field in the direction opposing the change in flux. The right hand rule for this is that if your fingers curl in the direction of the induced current, your thumb points in the direction of the magnetic field it produces through the loop.

Consider a coil lying flat on the page with a square cross section that is 10 cm by 5 cm. The coil consists of 10 loops and has a magnetic field of 0.5 T passing through it coming out of the page. Let's find the induced EMF when the magnetic field is changed to 0.8 T in 2 seconds.

First, let's find the initial magnetic flux: Φ_i

$\Phi_i = BA \cos\theta = (.5\text{ T})\ (.05\text{ m})\ (.1\text{m}) \cos 0° = 0.0025\text{ T m}^2$

And the final magnetic flux: Φ_f

$\Phi f= BA \cos \theta= (0.8\ T)\ (.05\ m)\ (.1m) \cos 0° = 0.004\ T\ m^2$

The induced emf is calculated then by

$\varepsilon = -N\ \Delta\Phi / \Delta t = -10\ (.004\ T\ m^2 - .0025\ T\ m^2) / 2\ s = -0.0075$ volts.

To determine the direction the current flows in the coil we need to apply the right hand rule and Lenz's law. The magnetic flux is being increased out of the page, with your thumb pointing up the fingers are coiling counterclockwise. However, Lenz's law tells us the current will oppose the change in flux so the current in the coil will be flowing clockwise.

Skill 27.7 Understands the electromagnetic spectrum and the production of electromagnetic waves.

The electromagnetic spectrum is measured using frequency (f) in hertz or wavelength (λ) in meters. The frequency times the wavelength of every electromagnetic wave equals the speed of light (3.0 x 10^8 meters/second).

Roughly, the range of wavelengths of the electromagnetic spectrum is:

	f	λ
Radio waves	10^{5} - 10^{-1} hertz	10^{3} - 10^{9} meters
Microwaves	$3x10^{9}$ - $3x10^{11}$ hertz	10^{-3} - 10^{-1} meters
Infrared radiation	$3x10^{11}$ - $4x10^{14}$ hertz	$7x10^{-7}$ - 10^{-3} meters
Visible light	$4x10^{14}$ - $7.5x10^{14}$ hertz	$4x10^{-7}$ - $7x10^{-7}$ meters
Ultraviolet radiation	$7.5x10^{14}$ - $3x10^{16}$ hertz	10^{-8} - $4x10^{-7}$ meters
X-Rays	$3x10^{16}$ - $3x10^{19}$ hertz	10^{-11} - 10^{-8} meters
Gamma Rays	$>3x10^{19}$ hertz	$<10^{-11}$ meters

Radio waves are used for transmitting data. Common examples are television, cell phones, and wireless computer networks. Microwaves are used to heat food and deliver Wi-Fi service. Infrared waves are utilized in night vision goggles. Visible light we are all familiar with as the human eye is most sensitive to this wavelength range. Light of different colors have different wavelengths. In the visible range, red light has the largest wavelength while violet light has the smallest. UV light causes sunburns and would be even more harmful if most of it were not captured in the Earth's ozone layer. X-rays aid us in the medical field and gamma rays are most useful in the field of astronomy.

A changing electric field produces a magnetic field, and a changing magnetic field produces an electric field. Electromagnetic waves are produced when an electric field connects with a magnetic field. The waves are produced by the motion of electrically charged particles. The electric and magnetic fields are perpendicular to each other and to the direction of the wave.

Radio waves are produced by antennae whose dimensions must be comparable to the wavelength in order for the antennae to work efficiently and by cosmic phenomena in deep space. Microwaves are produced by vacuum-based devices such as the magnetron, klystron, traveling wave tube, and gyrotron. Anything with a temperature greater than about 10 kelvin produces terahertz radiation; however, the emissions are very weak. The only effective sources of stronger terahertz radiation are the gyrotron, far infrared laser, quantum cascade laser, synchrotron light sources, the backward wave oscillator, the free electron laser, and single-cycle sources used in Terahertz time domain spectroscopy. Infrared radiation comes from the sun and from visible light that is absorbed and then re-radiated at longer wavelengths. Ultraviolet radiation is naturally transmitted by the sun. Artificial UV radiation is created in the form of vacuum UV and extreme UV. X-rays are produced by accelerating electrons in order to collide with a metal target. Gamma rays are produced by sub-atomic particle interaction, such as radioactive decay, but most are created by nuclear reactions in space.

COMPETENCY 28.0 THE TEACHER UNDERSTANDS APPLICATIONS OF ELECTRICITY AND MAGNETISM.

Skill 28.1 Analyzes common examples of electrostatics (e.g., a charged balloon attached to a wall, behavior of an electroscope, charging by induction).

Electrical current requires the free flow of electrons. Various materials allow different degrees of electron movement and are classified as conductors, insulators, or semiconductors (in certain, typically man-made environments, superconductors also exist). When charge is transferred to a mass of material, the response is highly dependent on whether that material is a conductor or insulator.

Conductors: Those materials that allow for free and easy movement of electrons are called conductors. Some of the best conductors are metal, especially copper and silver. This is because these materials are held together with metallic bonds, which involve de-localized electrons shared by atoms in a lattice. If a charge is transferred to a conductor, the electrons will flow freely and the charge will quickly distribute itself across the material in a manner dictated by the conductor's shape.

Insulators: Materials that do not allow conduction are call insulators. Good insulators include, glass, rubber, and wood. These materials have chemical structures in which the electrons are closely localized to the individual atoms. In contrast to a conductor, a charge transferred to an insulator will remain localized at the point where it was introduced. This is because the movement of electrons will be highly impeded.

Semiconductors: Materials with intermediate conduction properties are known as semiconductors. Their properties are similar to insulators in that they have few free electrons to carry the charge. However, these electrons can be thermally excited into higher energy states that allow them sufficient freedom to transmit electrical charge. The electrical properties of a semiconductor are often improved by introducing impurities, a procedure known as doping. Doping introduces extra electrons or extra electron acceptors to facilitate the movement of charge. *Temporary* changes in electrical properties can be induced by applying an electrical field.

Charge can be transferred to materials in a few different ways:

Conduction: In the most general sense, electrical conduction is the movement of charged particles (electrons) through a medium. As explained above, in conducting materials, electrons loosely attached to atoms are capable of carrying an electrical current. This requires that the atoms of the conducting material be brought into physical contact with the charge source. For example, if a conductor is brought in contact with another charged conductor or a current source (such as a battery), the charge will be transferred to that conductor.

Friction: When two materials are brought into contact (or rubbed against each other), electrons may be transferred between them. If the materials are then separated, one will be left with a negative charge and the other will have a positive charge. This phenomenon is known as the triboelectric effect. Both the polarity and strength of the resultant charge depends upon the materials, surface roughness, temperature, and strain. Some common examples of combinations that produce significant charges are glass with silk and rubber with fur.

Induction: Electromagnetic induction is the production of voltage that occurs when a conductor interacts with a magnetic field. A conductor can be charged either by moving it through a static electric field or by placing it in a changing magnetic field. Induction was discovered by Michael Faraday and Faraday's Law governs this phenomenon.

Skill 28.2 Understands electric current, resistance and resistivity, potential difference, capacitance, and electromotive force in conductors and circuits.

Semiconductors, conductors and superconductors are differentiated by how "easily" current can flow in the presence of an applied electric field. Dielectrics (insulators) conduct little or no current in the presence of an applied electric field, as the charged components of the material (for example, atoms and their associated electrons) are tightly bound and are not free to move within the material. In the case of materials with some amount of conductivity, so-called valence (higher energy level) electrons are only loosely bound and may move among positive charge centers (i.e., atoms or ions).

Conductors, such as metals like aluminum and iron, have numerous valence electrons distributed among the atoms that compose the material. These electrons can move freely, especially under the influence of an applied electric field. In the absence of any applied field, these electrons move randomly, resulting in no net current. If a static electric field is applied, the mobile electrons reorient themselves to minimize the energy of the system and create an equipotential on the surface of the metal (in the ideal case of infinite conductivity).The highly mobile charge carriers do not allow a net field inside the metal; thus conductors can "shield" electromagnetic fields.

Semiconductors, such as silicon and germanium, are materials that can neither be described as conductors, nor as insulators**. A certain number of charge carriers are mobile in the semiconductor**, but this number is nowhere near the free charge populations of conductors such as metals. "Doping" of a semiconductor by adding so-called donor atoms or acceptor atoms to the intrinsic (or pure) semiconductor can increase the conductivity of the material. Furthermore, combination of a number of differently doped semiconductors (such as donor-doped silicon and acceptor-doped silicon) can produce a device with beneficial electrical characteristics (such as the diode). The conductivity of such devices can be controlled by applying voltages across specific portions of the device.

A useful way to visualize the relationship of conductors and semiconductors is by way of energy band diagrams. For purposes of comparison, an example of an insulator is included here as well.

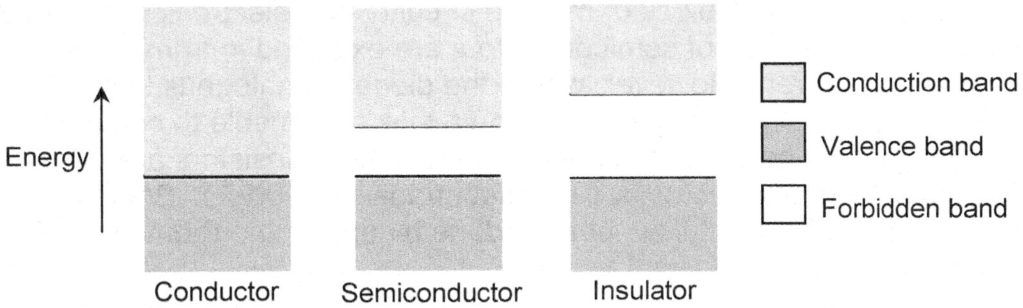

Prior to the 1960s most electronic apparatuses relied on either vacuum tubes or mechanical devices, typically relays. As semiconductors became increasingly available, however, vacuum tubes were replaced with solid state devices. Vacuum tubes conduct electrons through a heated vacuum but semiconductors allow electrons to flow through them while still solid. Solid state has come to mean any circuit that does not contain the aforementioned vacuum tubes and, in short, operates with "no moving parts". The lack of these elements make solid state devices more resistant to physical stressors, such as vibration, and more durable in general, since they are less susceptible to wear. A familiar example of this are the solid state "flash cards" that are popular for data storage. Previously, hard disk players were primarily used for a similar function but their moving parts make them of limited durability and less practical to transport.

Semiconductors are extremely useful because their conductive properties can be controlled. This is done by "doping" or introducing impurities. The impurity or dopant is used to introduce extra electrons or extra free orbital space that can be filled with electrons. This allows much freer movement of electron and, therefore, flow of current through the material. Semiconductors doped to contain extra electrons are known as N-type, while those doped to contain extra "holes" are known a P-type. Typically, either type of semiconductor can be made from the same base material. For instance, silicon can be doped with boron to create a P-type semiconductor or with phosphorus to create an N-type semiconductor. Two of the most common and important semiconductor devices today are diodes and transistors.

Diodes: Diodes restrict the direction of current (electron) flow in a given direction. Somewhat analogous to a check valve, they allow the electrical current to flow in one direction, but not in the other. Thus diodes are found in almost all circuits where single direction current flow is required. While early diodes were made from vacuum tubes, today most are made from semiconductors. This is often done by designing a diode with P-type semiconducting material on one side and N-type on the other. The current flows into the P-type side and on to the N-type side, but cannot flow in the opposite direction. Alternatively, the diode can be made from a conducting metal and a semiconductor and function in a similar fashion.

Transistors: Just like the vacuum tubes they replaced, transistors control current flow. They serve a variety of functions in circuits and can act as amplifiers, switches, voltage regulators, signal modulators, or oscillators. They are perhaps the most important building block of modern circuitry and electronics. As in solid state diodes, the properties of semiconductors are exploited in transistors to control the flow of current. However, where the diode is analogous to a check valve, a transistor function more like a tap on a sink and is able to control the rate of current flow or eliminate it all together. Today, most transistors are either bipolar junction transistors (BJT) or field effect transistors (FET). Both accomplish control of current flow, but it is done by applying current in the BJTs and voltage in the FETs.

Ohm's Law is the most important tool we posses to analyze electrical circuits. Ohm's Law states that the current passing through a conductor is directly proportional to the voltage drop and inversely proportional to the resistance of the conductor. Stated mathematically, this is:

$$V=IR$$

Problem:
The circuit diagram at right shows three resistors connected to a battery in series. A current of 1.0A flows through the circuit in the direction shown. It is known that the equivalent resistance of this circuit is 25 Ω. What is the total voltage supplied by the battery?

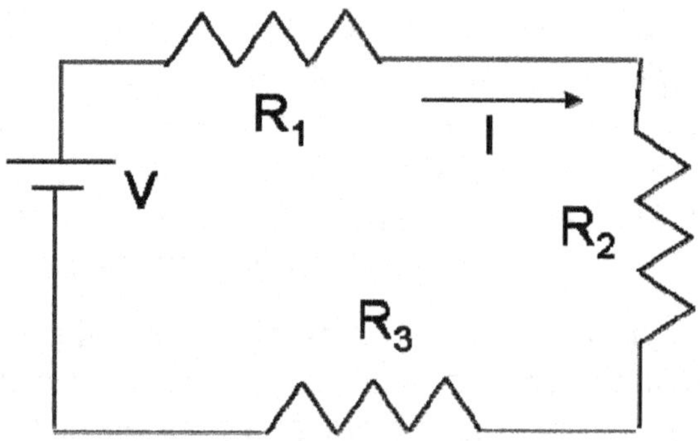

Solution:
To determine the battery's voltage, we simply apply Ohm's Law:

$$V = IR = 1.0A \times 25\Omega = 25V$$

Conductors are those materials which allow for the free passage of electrical current. However, all materials exhibit a certain opposition to the movement of electrons. This opposition is known as **resistivity** (ρ). Resistivity is determined experimentally by measuring the resistance of a uniformly shaped sample of the material and applying the following equation:

$$\rho = R\frac{A}{l}$$

where ρ = static resistivity of the material
R = electrical resistance of the material sample
A = cross-sectional area of the material sample
L = length of the material sample

The temperature at which these measurements are taken is important as it has been shown that resistivity is a function of temperature. For conductors, resistivity increases with increasing temperature and decreases with decreasing temperature. At extremely low temperatures resistivity assumes a low and constant value known as residual resistivity (ρ_0). Residual resistivity is a function of the type and purity of the conductor.

The following equation allows us to calculate the resistivity ρ of a material at any temperature given the resistivity at a reference temperature, in this case at $20^0 C$:

$$\rho = \rho_{20}[1 + a(t - 20)]$$

where ρ_{20} = resistivity at $20^0 C$
a= proportionality constant characteristic of the material
t=temperature in Celsius

Problem: The tungsten filament in a certain light bulb is a wire 8 μm in diameter and 10 mm long. Given that, for tungsten, $\rho_{20} = 5.5 \times 10^{-8}$ Ω·m and a=$4.5 \times 10^{-3} K^{-1}$, what will the resistance of the filament be at 45°C?

Solution: First we must find the resistivity of the tungsten at 45°C:

$$\rho = 5.5 \times 10^{-8}\left(1 + 4.5 \times 10^{-3}(45 - 20)\right) = 6.1 \times 10^{-8}\Omega.m$$

Now we can rearrange the equation defining resistivity and solve for the resistance of the filament:

$$R = \rho\frac{l}{A} = 6.1 \times 10^{-8} \times 0.01 / (\pi(4 \times 10^{-6})^2) = 12.1\Omega$$

Capacitance (C) is a measure of the stored electric charge per unit electric potential. The mathematical definition is:

$$C = \frac{Q}{V}$$

It follows from the definition above that the units of capacitance are coulombs per volt, a unit known as a farad (F=C/V). In circuits, devices called parallel plate capacitors are formed by two closely spaced conductors. The function of capacitors is to store electrical energy. When a voltage is applied, electrical charges build up in both the conductors (typically referred to as plates). These charges on the two plates have equal magnitude but opposite sign. The capacitance of a capacitor is a function of the distance d between the two plates and the area A of the plates:

$$C \approx \frac{\varepsilon A}{d}; A \gg d^2$$

Capacitance also depends on the permittivity of the non-conducting matter between the plates of the capacitor. This matter may be only air or almost any other non-conducting material and is referred to as a **dielectric**. The permittivity of empty space ε_0 is roughly equivalent to that for air, ε_{air}=8.854x10^{-12} C ^{2}/N•m^2. For other materials, the dielectric constant, κ, is the permittivity of the material in relation to air ($\kappa=\varepsilon/\varepsilon_{air}$). The make-up of the dielectric is critical to the capacitor's function because it determines the maximum energy that can be stored by the capacitor. This is because an overly strong electric field will eventually destroy the dielectric.

In summary, a capacitor is "charged" as electrical energy is delivered to it and opposite charges accumulate on the two plates. The two plates generate electric fields and a voltage develops across the dielectric. The energy stored in the capacitor, then, is equal to the amount of work necessary to create this voltage. The mathematical statement of this is:

$$E_{\text{stored}} = \frac{1}{2}CV^2 = \frac{1}{2}\frac{Q^2}{C} = \frac{1}{2}VQ$$

The work per unit volume or the **electric field energy density** within a capacitor can be shown to be $\eta = \frac{1}{2}\varepsilon E^2$. This result is generally valid for the energy per unit volume of any electrostatic field, not only for a constant field within a capacitor.

<u>Problem</u>: Imagine that a parallel plate capacitor has an area of 10.00 cm ^{2} and a capacitance of 4.50 pF. The capacitor is connected to a 12.0 V battery. The capacitor is completely charged and then the battery is removed. What is the separation of the plates in the capacitor? How much energy is stored between the plates? We've assumed that this capacitor initially had no dielectric (i.e., only air between the plates) but now imagine it has a Mylar dielectric that fully fills the space. What will the new capacitance be? (for Mylar, □=3.5)

<u>Solution</u>: To determine the separation of the plates, we use our equation for a capacitor:

$$C = \frac{\varepsilon_0 A}{d}$$

We can simply solve for d and plug in our values:

$$d = \varepsilon_0 \frac{A}{C} = \left(8.854 \times 10^{-12} \frac{\mathrm{C}}{\mathrm{N} \cdot \mathrm{m}^2}\right) \frac{10 \times 10^{-4} m^2}{4.5 \times 10^{-12} F} = 1.97 \times 10^{-3} m = 1.97 mm$$

Similarly, to find stored energy, we simply employ the equation above:

$$E_{stored} = \frac{1}{2}QV$$

But we don't yet know the charge Q, so we must first find it from the definition of capacitance:

$$C = \frac{Q}{V}$$

$$Q = CV = (4.5 \times 10^{-12}) \times (12V) = 5.4 \times 10^{-11} C$$

Now we can find the stored energy:

$$E_{stored} = \frac{1}{2}QV = \frac{1}{2}(5.4 \times 10^{-11} C)(12V) = 3.24 \times 10^{-10} J$$

To find the capacitance with a Mylar dielectric, we again use the equation for capacitance of a parallel plate capacitor. Note that the new capacitance can be found by multiplying the original capacitance by κ:

$$C = \frac{\kappa_{Mylar}\varepsilon_0 A}{d} = \kappa_{Mylar} C_0 = 3.5 \times 4.5 pF = 15.75 pF$$

Skill 28.3 Analyzes series and parallel DC circuits in terms of current, resistance, voltage, and power.

Often resistors and capacitors are used together in series or parallel. Two components are in series if one end of the first element is connected to one end of the second component. The components are in parallel if both ends of one element are connected to the corresponding ends of another. A series circuit has a single path for current flow through all of its elements. A parallel circuit is one that requires more than one path for current flow in order to reach all of the circuit elements. Below is a diagram demonstrating a simple circuit with resistors in parallel (on right) and in series (on left). Note the symbols used for a battery (noted V) and the resistors (noted R).

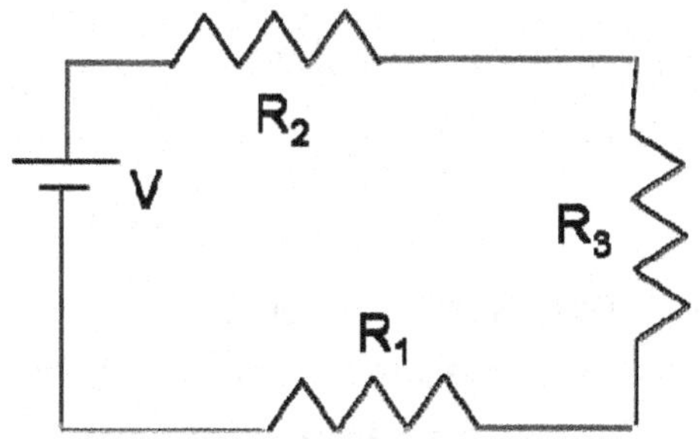

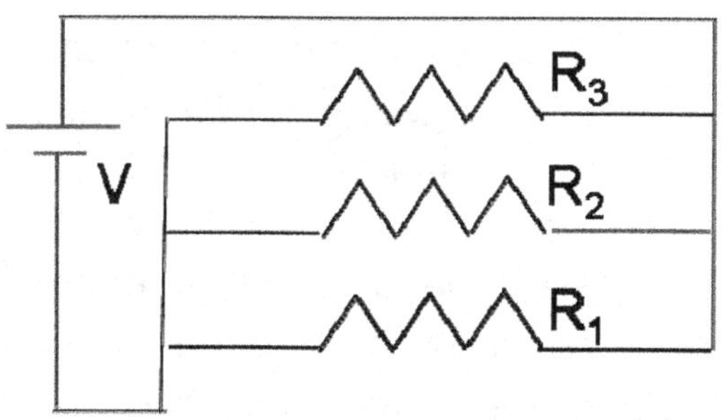

Thus, when the resistors are placed in series, the current through each one will be the same. When they are placed in parallel, the voltage through each one will be the same. To understand basic circuitry, it is important to master the rules by which the equivalent resistance (R_{eq}) or capacitance (C_{eq}) can be calculated from a number of resistors or capacitors:

Resistors in parallel: $\frac{1}{R_{eq}} = \frac{1}{R_1} + \frac{1}{R_2} + \cdots + \frac{1}{R_n}$

Resistors in series: $R_{eq} = R_1 + R_2 + \cdots + R_n$

Capacitors in parallel: $C_{eq} = C_1 + C_2 + \cdots + C_n$

Capacitors in series: $\frac{1}{C_{eq}} = \frac{1}{C_1} + \frac{1}{C_2} + \cdots + \frac{1}{C_n}$

Skill 28.4 Identifies basic components and characteristics of AC circuits.

Alternating current (AC) is a type of electrical current with cyclically varying magnitude and direction. This is differentiated from direct current (DC), which has constant direction. AC is the type of current delivered to businesses and residences.

Though other waveforms are sometimes used, the vast majority of AC current is sinusoidal. Thus we can use wave terminology to help us describe AC current. Since AC current is a function of time, we can express it mathematically as:

$$v(t) = V_{peak} \cdot \sin(\omega t)$$

where V_{peak}= the peak voltage; the maximum value of the voltage
ω=angular frequency; a measure of rotation rate
t=time

The instantaneous power or energy transmission per unit time is given by $I^2{}_{peak} R \sin^2(\omega t)$, a value that varies over time. In order to asses the overall rate of energy transmission, however, we need some kind of average value. The **root mean square value (V_{rms}, I_{rms})** is a specific type of average given by the following formulae:

$$V_{rms} = \frac{V_{peak}}{\sqrt{2}} \; ; \; I_{rms} = \frac{I_{peak}}{\sqrt{2}} \; ; \; I_{rms} = \frac{V_{rms}}{R}$$

V_{rms} is useful because an AC current will deliver the same power as a DC current if its $V_{rms}=V_{DC}$, i.e. average power or average energy transmission per unit time is given by $P_{av} = V_{rms} I_{rms}$.

Skill 28.5 Understands the operation of an electromagnet.

The movement of electric charges (i.e., a current density **J**) results in a magnetic field **H**, as described by the Maxwell equation based on Ampere's law:

$$\nabla \times \mathbf{H}(r,t) = \frac{\partial \mathbf{D}(r,t)}{\partial t} + \mathbf{J}(r,t)$$

For the case where the time derivative of the electric flux density **D** is zero, this reduces to the simpler Ampere's law:

$$\nabla \times \mathbf{H}(r,t) = \mathbf{J}(r,t)$$

Electromagnets take advantage of this relationship between the current and the magnetic field by, for example, coiling a current-carrying wire (a solenoid), sometimes around a ferromagnetic or paramagnetic material. This creates a magnetic dipole with a strength that varies depending on a number of factors.

The direction of the dipole relative to the direction of current flow is determined by the right hand rule: the fingers curl in the direction of the current, and the thumb points in the (north) direction of the dipole.

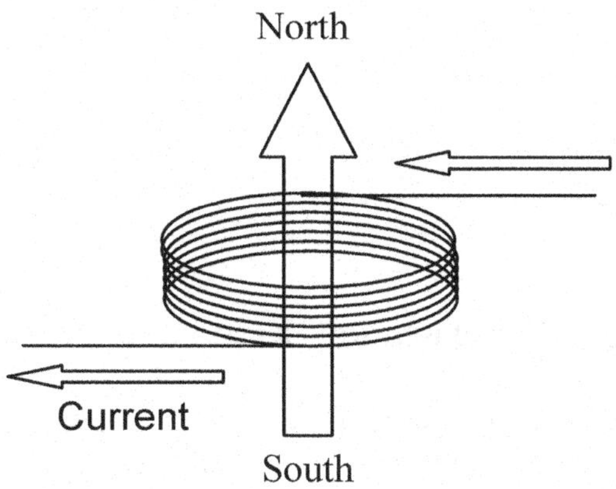

The strength of the electromagnet (i.e., the induced magnetic dipole) can be increased by adding to the number of turns (or loops) of current-carrying wire. The principle of superposition applies here and the induced magnetic field varies proportionally according to the number of turns. Also, by varying the current, the strength of the electromagnet can be increased or decreased. Imperfections in the solenoid, such as imperfectly packed wire loops, which can cause magnetic flux leakage, can adversely affect the strength of the electromagnet. Also, the spatial extent of the solenoid (or other form of electromagnet, such as a toroid) must be considered if exact numerical calculations are pursued.

Another critical factor that affects the electromagnet is the material used inside the coil; specifically, the magnetic properties of the material. Magnetizable materials with permeabilities (μ) greater than unity can be used to increase the magnitude of the induced magnetic dipole. Ferromagnetic materials, for example, can have very large permeabilities. One such example is iron. A ferromagnetic core is magnetized when the magnetic field induced by the current-carrying wire is applied resulting in a stronger magnetic dipole than would have been produced if a non-magnetic material (such as air) had been used. The ability of a magnetic material core to increase the strength of the electromagnet is limited, however. In the case of a ferromagnetic material, once all the magnetic domains have been aligned with the magnetic field of the electromagnet, saturation has been reached and the material cannot be magnetized further.

Skill 28.6 Understands the operation of electric meters, motors, generators, and transformers.

Electric Meters

Electrical meters function by utilizing the following familiar equations:

Across a resistor (Resistor R):

$$V_R = IR_R$$

Across a capacitor (Capacitor C):

$$V_C = IX_C$$

Across an inductor (Inductor L):

$$V_L = IX_L$$

Where V=voltage, I=current, R=resistance, X=reactance.

Ammeter: An ammeter placed in series in a circuit measures the current through the circuit. An ammeter typically has a very small resistance so that the current in the circuit is not changed too much by insertion of the ammeter.

Voltmeter: A voltmeter is used to measure potential difference. The potential difference across a resistor is measured by a voltmeter placed in parallel across it. An ideal voltmeter has very high resistance so that it does not appreciably alter circuit resistance and therefore the voltage drop it is measuring.

Galvanometer: A galvanometer is a device that measures current and is a component of an ammeter or a voltmeter. A typical galvanometer consists of a coil of wire, an indicator and a scale that is designed to be proportional to the current in the galvanometer. The principle that a current-carrying wire experiences a force in a magnetic field is used in the construction of a galvanometer. In order to create a voltmeter from a galvanometer, resistors are added in series to it. To build an ammeter using a galvanometer, a small resistance known as a shunt resistor is placed in parallel with it.

Potentiometer: A potentiometer is a variable resistance device in which the user can vary the resistance to control the current and voltage applied to a circuit. Since the potentiometer can be used to control what fraction of the emf of a battery is applied to a circuit, it is also known as a voltage divider. It can be used to measure an unknown voltage by comparing it with a known value.

Multimeter: A common electrical meter, typically known as a multimeter, is capable of measuring voltage, resistance, and current. Many of these devices can also measure capacitance (farads), frequency (hertz), duty cycle (a percentage), temperature (degrees), conductance (siemens), and inductance (henrys).

Motors

Electric motors are found in many common appliances such as fans and washing machines. The operation of a motor is based on the principle that a magnetic field exerts a force on a current carrying conductor. This force is essentially due to the fact that the current carrying conductor itself generates a magnetic field; the basic principle that governs the behavior of an electromagnet. In a motor, this idea is used to convert electrical energy into mechanical energy, most commonly rotational energy. Thus the components of the simplest motors must include a strong magnet and a current-carrying coil placed in the magnetic field in such a way that the force on it causes it to rotate.

Motors may be run using DC or AC current and may be designed in a number of ways with varying levels of complexity. A very basic DC motor consists of the following components:

- A **field magnet**
- An **armature** with a coil around it that rotates between the poles of the field magnet
- A **power supply** that supplies current to the armature
- An **axle** that transfers the rotational energy of the armature to the working parts of the motor
- A set of **commutators** and **brushes** that reverse the direction of power flow every half rotation so that the armature continues to rotate

Generators

Generators are devices that are the opposite of motors in that they convert **mechanical energy into electrical energy**. The mechanical energy can come from a variety of sources; combustion engines, blowing wind, falling water, or even a hand crank or bicycle wheel. Most generators rely upon electromagnetic induction to create an electrical current. These generators basically consist of magnets and a coil. The magnets create a magnetic field and the coil is located within this field. Mechanical energy, from whatever source, is used to spin the coil within this field. As stated by Faraday's Law, this produces a voltage.

Transformers

Electromagnetic induction is used in a transformer, a device that magnetically couples two circuits together to allow the transfer of energy between the two circuits without requiring motion. Typically, a transformer consists of a couple of coils and a magnetic core. A changing voltage applied to one coil (the primary) creates a flux in the magnetic core, which induces voltage in the other coil (the secondary). All transformers operate on this simple principle though they range in size and function from those in tiny microphones to those that connect the components of the US power grid.

One of the most important functions of transformers is that they allow us to "step-up" and "step-down" between vastly different voltages. To determine how the voltage is changed by a transformer, we employ any of the following relationships:

$$\frac{V_s}{V_p} = \frac{n_s}{n_p} = \frac{I_p}{I_s}$$

where V_s=secondary voltage
V_p=primary voltage
n_s=number of turns on secondary coil
n_p=number of turns on primary coil
I_p=primary current
I_s=secondary current

Problem: If a step-up transformer has 500 turns on its primary coil and 800 turns on its secondary coil, what will be the output (secondary) voltage be if the primary coil is supplied with 120 V?

Solution:

$$\frac{V_s}{V_p} = \frac{n_s}{n_p}$$

$$V_s = \frac{n_s}{n_p} \times V_p = \frac{800}{500} \times 120V = 192V$$

COMPETENCY 29.0 THE TEACHER UNDERSTANDS THE CONSERVATION OF ENERGY AND MOMENTUM.

Skill 29.1 Understands the concept of work.

In physics, work done by a constant force is defined as force times distance $W = F \cdot s$. Work is a scalar quantity, it does not have direction, and it is usually measured in Joules ($N \cdot m$). It is important to remember, when doing calculations about work, that the only part of the force that contributes to the work is the part that acts in the direction of the displacement. Therefore, sometimes the equation is written as $W = F \cdot s \cos\theta$, where θ is the angle between the force and the displacement.

When the force applied to an object varies over the distance moved, the total work done in moving from a point x_1 to a point x_2 is given by

$$W = \int_{x_1}^{x_2} F_x dx$$

where F_x is the force applied to the object, in the direction of movement, at any point x.

Problem: A man uses a constant 6N force to pull a 10kg block, as shown below, over a distance of 3 m. How much work did he do?

Solution:

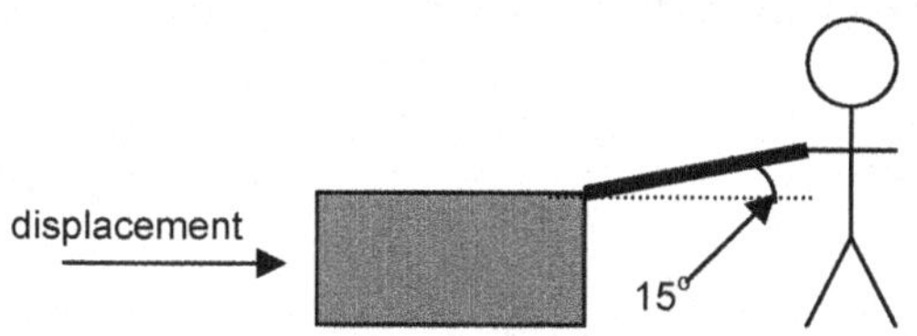

$$W = F \cdot s \cos\theta$$
$$W = 6 \cdot 3 \cos 15 = 17.4J$$

Notice that you did not actually need the mass of the box in order to calculate the work done.

Skill 29.2 Understands the relationships among work, energy, and power.

Power is defined in relationship to work. It is the rate at which work is done, or, in other words, the amount of work done in a certain period of time: $P = W/t$. There are many different units for power, but the one most commonly seen in physics problems is the Watt which is measured in Joules per second. Another commonly discussed unit of power is horsepower, and 1hp=746 W.

Problem: A woman standing in her 4th story apartment raises a 10kg box of groceries from the ground using a rope. She is pulling at a constant rate, and it takes her 5 seconds to raise the box one meter. How much power is she using to raise the box?

Solution:

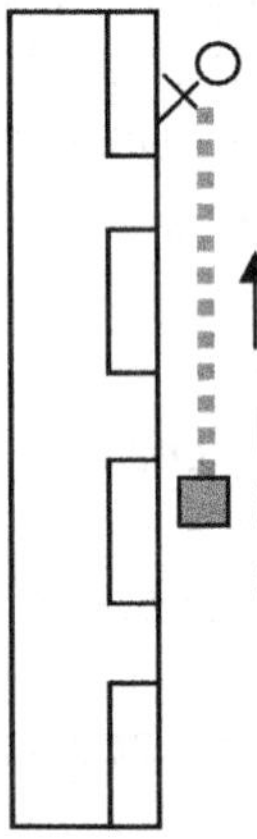

$$P = W/t$$

$$P = \frac{F \cdot s}{t} = \frac{mgh}{t} = \frac{10*9.8*1}{5} = 19.6W$$

Notice that, because she is pulling at a constant rate, you don't need to know the actual distance she has raised the box. 2 meters in 10 seconds would give you the same result as 5 meters in 25 seconds.

Energy is also defined, in relation to work, as the ability of an object to do work. As such, it is measured in the same units as work, usually Joules. Most problems relating work to energy are looking at two specific kinds of energy. The first, kinetic energy, is the energy of motion. The heavier an object is and the faster it is going, the more energy it has resulting in a greater capacity for work. The equation for kinetic energy is : $KE = \frac{1}{2}mv^2$.

Problem:

A 1500 kg car is moving at 60m/s down a highway when it crashes into a 3000kg truck. In the moment before impact, how much kinetic energy does the car have?

Solution:

$$KE = \frac{1}{2}mv^2 = \frac{1}{2} \cdot 1500 \cdot 60^2 = 2.7 \times 10^6 J$$

The other form of energy frequently discussed in relationship to work is gravitational potential energy, or potential energy, the energy of position. Potential energy is calculated as $PE = mgh$ where h is the distance the object is capable of falling.

Problem:

Which has more potential energy, a 2 kg box held 5 m above the ground or a 10 kg box held 1 m above the ground?

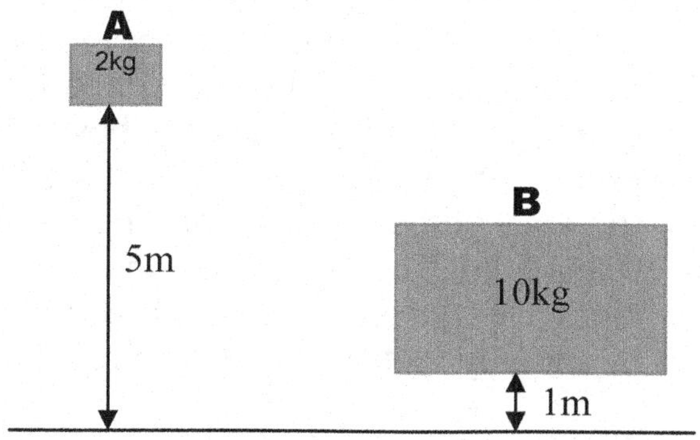

Solution:

$$PE_A = mgh = 2 \cdot g \cdot 5 = 10g$$
$$PE_B = mgh = 10 \cdot g \cdot 1 = 10g$$
$$PE_A = PE_B$$

Skill 29.3 Solves problems using the conservation of mechanical energy in a physical system (e.g., determining potential energy for conservative forces, analyzing the motion of a pendulum).

The principle of conservation of energy states that an isolated system maintains a constant total amount of energy despite the fact that the energy may change forms. To put it another way, energy cannot be created or destroyed but can be changed from one form to another. For example, friction can turn kinetic energy into thermal energy. Other forms of energy include electrical energy, chemical energy, and mechanical energy.

A **conservative** force is one that conserves mechanical energy (kinetic + potential energy), i.e. there is no change in mechanical energy when a conservative force acts on an object. Consider a mass on a spring on a frictionless surface. This is a closed loop system. If conservative forces alone act on the mass during each cycle, the velocity of the mass at the beginning and the end of the cycle must be the same for the mechanical energy to have been conserved. In this way, the force has done no work. At any point in the cycle of motion, the total mechanical energy of the system remains constant even though the energy moves back and forth between kinetic and potential forms. If work is done on the mass, then the forces acting on the mass are **nonconservative**. In a real system there will be some dissipative forces that will convert some of the mechanical energy to thermal energy. Conservative forces are independent of path the object takes, while nonconservative forces are path dependent.

Gravity is a conservative force. This can be illustrated by imagining an object tossed into the air. On the upward journey the work done by gravity is the negative product of mass, acceleration, and height. On the downward journey, the work done by gravity is the positive value of this amount. Thus for the total loop the work is zero.

Friction is a nonconservative force. If a box is pushed along a rough surface from one side of the room to the other and back, friction opposes the movement in both directions; so the work done by friction cannot be equal to zero. This example also helps illustrate how nonconservative forces are path dependent. More work is done by friction if the path is tortuous rather than straight, even if the start and end points are the same. Let's try an example with a small box of mass 5 kg. The box moves in a circle 2 meters in diameter. The coefficient of kinetic friction between the box and the surface it rests on is 0.2. How much work is done by friction during one revolution?

The force exerted by friction is calculated by

$$\mathbf{F_k = \mu_k\, F_n = (0.2)\ (5\ kg)\ (9.8\ m/s^2) = 9.8\ N}$$

The force opposes the movement of the box during the entire distance of one revolution, or approximately 6.3 meters ($2\pi r$).

The total work done by friction is

$$\mathbf{W = F\ x\ \cos\theta = (9.8\ N)\ (6.3\ m)\ (\cos 180) = -61.7\ Joules}$$

As expected, the work is not zero since friction is not a conservative force. Since it does negative work on an object, it reduces the mechanical energy of the object and is a **dissipative** force.

Harmonic motion or harmonic oscillation is seen in any system that follows **Hooke's Law**. Hooke's law simply predicts the behavior of certain bodies as they return to equilibrium following a displacement. It is given by the following equation:

$$F = -kx$$

Where F=restoring force
x=displacement
k=a positive constant

From this equation we can see that the harmonic motion is neither damped nor driven. Harmonic motion is observed as sinusoidal oscillations about an equilibrium point. Both the amplitude and the frequency are constant. Further, the amplitude is always positive and is a function of the original force that disrupted the equilibrium. If an object's oscillation is governed solely by Hooke's law, it is a simple harmonic oscillator. Below are examples of simple harmonic oscillators:

Pendula: A pendulum is a mass on the end of a rigid rod or a string. An initial push will cause the pendulum to swing back and forth. This motion will be harmonic as long as the pendulum moves through an angle of less than 15°.

Masses connected to springs: A spring is simply the familiar helical coil of metal that is used to store mechanical energy. In a typical system, one end of a spring is attached to a mass and the other to a solid surface (a wall, ceiling, etc). If the spring is then stretched or compressed (i.e., removed from equilibrium) it will oscillate harmonically.

Vibrating strings: A string or rope tied tightly at both ends will oscillate harmonically when it is struck or plucked. This is often the mechanism used to generate sound in string-based instruments such as guitars and pianos.

Several simple harmonic oscillations maybe superimposed to create complex harmonic motion. The best-known example of complex harmonic motion is a musical chord.

View an animation of harmonic oscillation here:
http://en.wikipedia.org/wiki/Image:Simple_harmonic_motion_animation.gif

The displacement of a simple harmonic oscillator varies sinusoidally with time and is given by

$$x = A\cos(\omega t + \delta)$$

where A is the maximum displacement or amplitude, ω is the angular frequency and δ is the phase constant. We can see from the equation that the displacement goes through a full cycle at time intervals given by the period $T = 2\pi / \omega$. The figure below displays a graphical representation of the displacement of a simple harmonic oscillator.

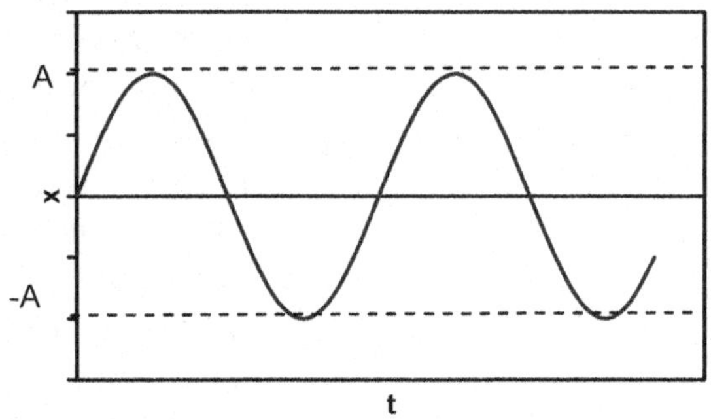

In the absence of dissipative forces, the total energy of a simple harmonic oscillator remains constant; however, the proportion of kinetic to potential energy varies. At x = A and x = -A all of the energy is potential. At x = 0 all of the energy is kinetic.

Skill 29.4 Applies the work-energy theorem to analyze and solve a variety of practical problems (e.g., finding the speed of an object given its potential energy, determining the work done by frictional forces on a decelerating car).

The work-energy theorem states that the amount of work done on an object is equal to its change in mechanical energy (kinetic or potential energy). Specifically, in systems where multiple forces are at work, the energy change of the system is the work done by the *net* force on the object. Problems dealing with the work-energy theorem may look at changes in kinetic energy, changes in potential energy, or some combination of the two. It is also important to remember that only external forces can cause changes in an object's total amount of mechanical energy. Internal forces, such as spring force or gravity, only lead to conversions between kinetic and potential energy rather than changes in the total level of mechanical energy.

<u>Example</u>:

1. A woman driving a 2000 kg car along a level road at 30 m/s takes her foot off the gas to see how far her car will roll before it slows to a stop. She discovers that it takes 150m. What is the average force of friction acting on the car?

$$W = \Delta KE$$
$$f \cdot s\cos\theta = \tfrac{1}{2}mv^2_{final} - \tfrac{1}{2}mv^2_{initial}$$
$$f \cdot 150 \cdot (-1) = \tfrac{1}{2} \cdot 2000 \cdot 0^2 - \tfrac{1}{2} \cdot 2000 \cdot 30^2$$
$$-150f = -900000$$
$$f = 6000\,N$$

According to the work-energy theorem, the amount of work done on the car is equal to the change in its mechanical energy which in this case is its change in kinetic energy. Since the only force acting on the car is friction, all the work can be attributed to the frictional force. It is important to realize, for this problem, that the force of friction is against the direction of motion, and thus $\cos\theta = -1$.

2. A 20kg child lifts his .5 kg ball off the floor to put it away on his bookshelf 1.5 meters above the ground. How much work has he done?

$$W = \Delta PE$$
$$W = mgh_{final} - mgh_{initial}$$
$$W = mg(h_{final} - h_{initial}) = .5 \cdot 9.8 \cdot 1.5 = 7.35J$$

Skill 29.5 Understands linear and angular momentum.

Linear motion is measured in rectangular coordinates. Rotational motion is measured differently, in terms of the angle of displacement. There are three common ways to measure rotational displacement; degrees, revolutions, and radians. Degrees and revolutions have an easy to understand relationship, one revolution is 360°. Radians are slightly less well known and are defined as

$\frac{arc\ length}{radius}$. Therefore 360°=2π radians and 1 radian = 57.3°.

The major concepts of linear motion are duplicated in rotational motion with linear displacement replaced by **angular displacement**.

Angular velocity ω = rate of change of angular displacement.
Angular acceleration α = rate of change of angular velocity.

Also, the linear velocity v of a rolling object can be written as $v = r\omega$ and the linear acceleration as $a = r\alpha$.

One important difference in the equations relates to the use of mass in rotational systems. In rotational problems, not only is the mass of an object important but also its location. In order to include the spatial distribution of the mass of the object, a term called **moment of inertia** is used, $I = m_1 r_1^2 + m_2 r_2^2 + \cdots + m_n r_n^2$. The moment of inertia is always defined with respect to a particular axis of rotation.

<u>Example</u>:

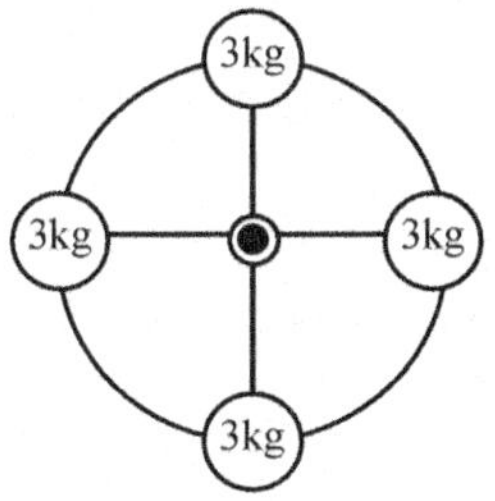

If the radius of the wheel on the left is 0.75m, what is its moment of inertia about an axis running through its center perpendicular to the plane of the wheel?

$$I = 3 \cdot 0.75^2 + 3 \cdot 0.75^2 + 3 \cdot 0.75^2 + 3 \cdot 0.75^2 = 6.75$$

Note: $I_{Sphere} = \frac{2}{5} mr^2$, $I_{Hoop/Ring} = mr^2$, $I_{disk} = \frac{1}{2} mr^2$

The rotational analog of Newton's second law of motion is given in terms of **torque** τ, moment of inertia I, and angular acceleration α :

$$\tau = I\alpha$$

where the torque τ is the rotational force on the body. In simple terms, the torque τ produced by a force F acting at a distance r from the point of rotation is given by the product of r and the component of the force that is perpendicular to the line joining the point of rotation to the point of action of the force.

A concept related to the moment of inertia is the **radius of gyration** (*k*), which is the average distance of the mass of an object from its axis of rotation, i.e., the distance from the axis where a point mass m would have the same moment of inertia.

$k_{Sphere} = \sqrt{\frac{2}{5}}r,\ k_{Hoop/Ring} = r,\ k_{disk} = \frac{r}{\sqrt{2}}$. As you can see $I = mk^2$

This is analogous to the concept of center of mass, the point where an equivalent mass of infinitely small size would be located, in the case of linear motion.

Angular momentum (*L*), and **rotational kinetic energy (KE_r),** are therefore defined as follows: $L = I\omega, \quad KE_r = \frac{1}{2}I\omega^2$

As with all systems, energy is conserved unless the system is acted on by an external force. This can be used to solve problems such as the one below.

Example:

A uniform ball of radius *r* and mass *m* starts from rest and rolls down a frictionless incline of height *h*. When the ball reaches the ground, how fast is it going?

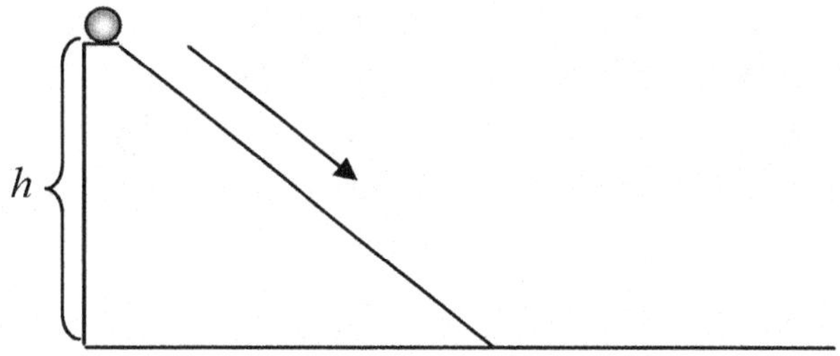

$$PE_{initial} + KE_{rotational/initial} + KE_{linear/initial} = PE_{final} + KE_{rotational/final} + KE_{linear/final}$$

$$mgh + 0 + 0 = 0 + \frac{1}{2}I\omega_{final}^{2} + \frac{1}{2}mv_{final}^{2} \rightarrow mgh = \frac{1}{2}\cdot\frac{2}{5}mr^{2}\omega_{final}^{2} + \frac{1}{2}mv_{final}^{2}$$

$$mgh = \frac{1}{5}mr^{2}\left(\frac{v_{final}}{r}\right)^{2} + \frac{1}{2}mv_{final}^{2} \rightarrow mgh = \frac{1}{5}mv_{final}^{2} + \frac{1}{2}mv_{final}^{2}$$

$$gh = \frac{7}{10}v_{final}^{2} \rightarrow v_{final} = \sqrt{\frac{10}{7}gh}$$

Similarly, unless a net torque acts on a system, the angular momentum remains constant in both magnitude and direction. This can be used to solve many different types of problems including ones involving satellite motion.

Example:
A planet of mass *m* is circling a star in an orbit like the one below. If its velocity at point A is 60,000m/s, and r_B=8 r_A, what is its velocity at point B?

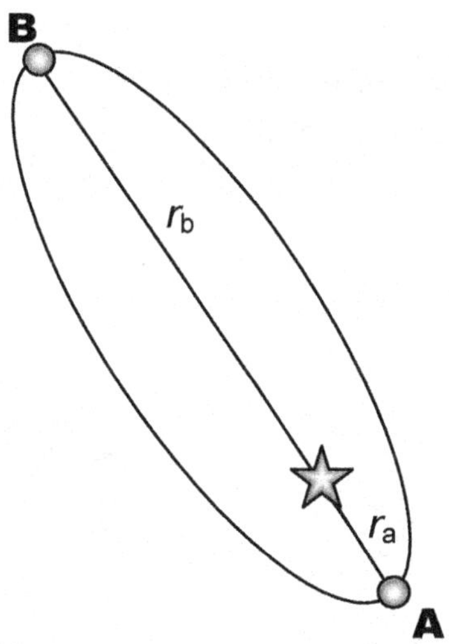

$$I_B\omega_B = I_A\omega_A$$

$$mr_B^2\omega_B = mr_A^2\omega_A$$

$$r_B^2\omega_B = r_A^2\omega_A$$

$$r_B^2\frac{v_B}{r_B} = r_A^2\frac{v_A}{r_A}$$

$$r_Bv_B = r_Av_A$$

$$8r_Av_B = r_Av_A$$

$$v_B = \frac{v_A}{8} = 7500m/s$$

Skill 29.6 Solves a variety of problems (e.g., collisions) using the conservation of linear and angular momentum.

A collision may be **elastic** or **inelastic**. In a totally elastic collision, the kinetic energy is conserved along with the momentum. In a totally inelastic collision, on the other hand, the kinetic energy associated with the center of mass remains unchanged but the kinetic energy relative to the center of mass is lost. An example of a totally inelastic collision is one in which the bodies stick to each other and move together after the collision. Most collisions are neither perfectly elastic nor perfectly inelastic and only a portion of the kinetic energy relative to the center of mass is lost.

Imagine two carts rolling towards each other as in the diagram below

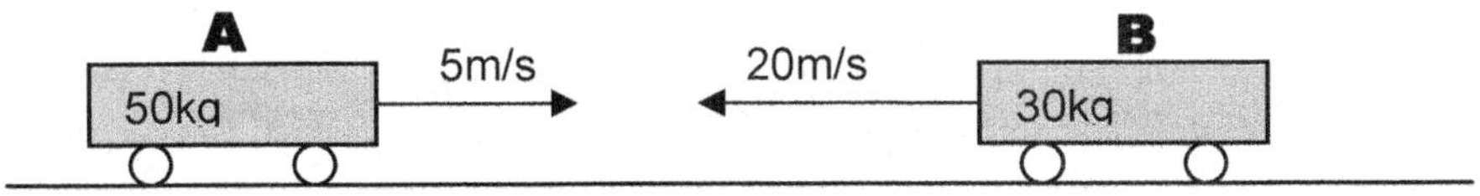

Before the collision, cart **A** has 250 kg m/s of momentum, and cart **B** has –600 kg m/s of momentum. In other words, the system has a total momentum of –350 kg m/s of momentum.

After the inelastic collision, the two cards stick to each other, and continue moving. How do we determine how fast, and in what direction, they go?

We know that the new mass of the cart is 80kg, and that the total momentum of the system is –350 kg m/s. Therefore, the velocity of the two carts stuck together must be $\frac{-350}{80} = -4.375\,m/s$

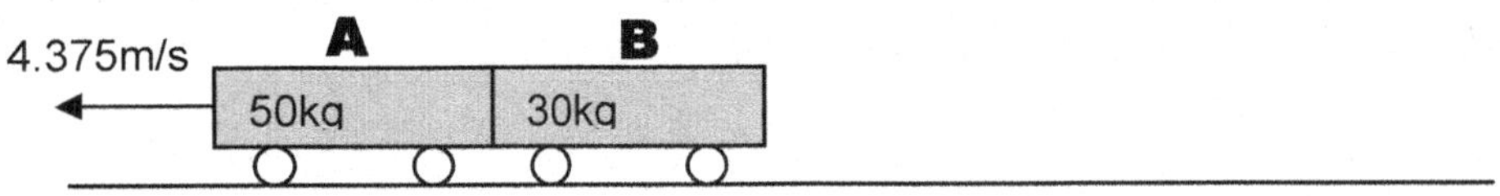

Conservation of momentum works the same way in two dimensions, the only change is that you need to use vector math to determine the total momentum and any changes, instead of simple addition.

Imagine a pool table like the one below. Both balls are 0.5 kg in mass.

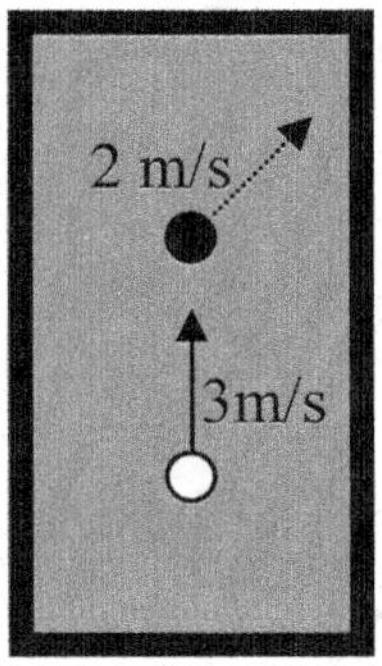

Before the collision, the white ball is moving with the velocity indicated by the solid line and the black ball is at rest.
After the collision the black ball is moving with the velocity indicated by the dashed line (a 135° angle from the direction of the white ball).

With what speed, and in what direction, is the white ball moving after the collision?

$p_{white/before} = .5 \cdot (0,3) = (0,1.5) \quad p_{black/before} = 0 \quad p_{total/before} = (0,1.5)$

$p_{black/after} = .5 \cdot (2\cos 45, 2\sin 45) = (0.71, 0.71)$

$p_{white/after} = (-0.71, 0.79)$

i.e. the white ball has a velocity of $v = \sqrt{(-.71)^2 + (0.79)^2} = 1.06 m/s$

and is moving at an angle of $\theta = \tan^{-1}\left(\frac{0.79}{-0.71}\right) = -48°$ from the horizontal

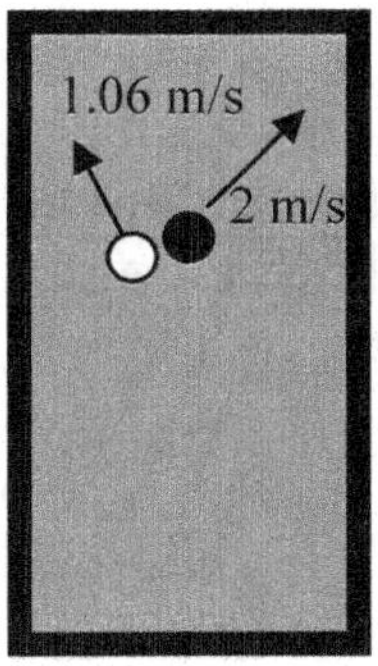

The **impulse-momentum theorem** states that any impulse acting on a system changes the momentum of that system. When considering the impulse-momentum theorem, there are several factors that need to be taken into account. The first factor is that momentum is a vector quantity $p = m \cdot v$. It has both magnitude and direction. Therefore, any action that causes either the speed or the direction of an object to change causes a change in its momentum. An impulse is defined as a force acting over a period of time (integral of force over time), and any impulse acting on the system is equivalent to a change in its momentum, as you can see from the equations below:

$$F = m \cdot a \rightarrow F = m \cdot \frac{\Delta v}{t} \rightarrow F \cdot t = m \cdot \Delta v$$

i.e. Forces acting over time cause a change in momentum.

Sample Problems:

1. A 1 kg ball is rolled towards a wall at 4 m/s. It hits the wall, and bounces back off the wall at 3 m/s.

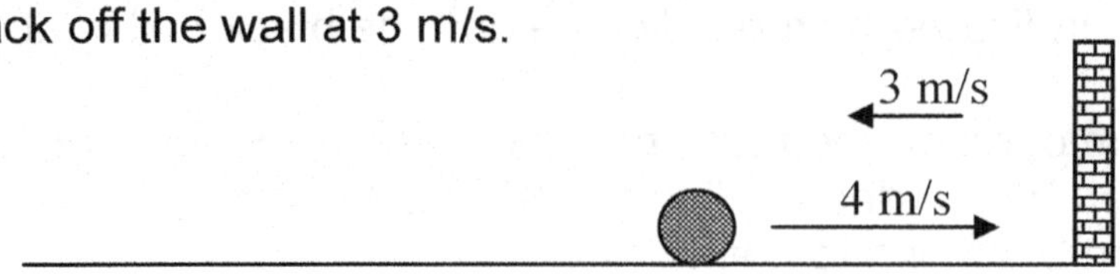

   a. What is the change in velocity?

      The velocity goes from +4m/s to –3m/s, a net change of -7m/s.

At what point does the impulse occur?

The impulse occurs when the ball hits the wall.

2. A 30kg woman is in a car accident. She was driving at 50m/s when she had to hit the brakes to avoid hitting the car in front of her.
   a. The automatic tensioning device in her seatbelt slows her down to a stop over a period of one half second. How much force does it apply?

$$F = m \cdot \frac{\Delta v}{t} \rightarrow F = 30 \cdot \frac{50}{.5} = 3000N$$

   b. If she hadn't been wearing a seatbelt, the windshield would have stopped her in .001 seconds. How much force would have been applied there?

$$F = m \cdot \frac{\Delta v}{t} \rightarrow F = 30 \cdot \frac{50}{.001} = 1500000N$$

COMPETENCY 30.0 THE TEACHER UNDERSTANDS THE LAWS OF THERMODYNAMICS.

Skill 30.1 Understands methods of heat transfer (i.e., convection, conduction, radiation).

All heat transfer is the movement of thermal energy from hot to cold matter. This movement down a thermal gradient is a consequence of the second law of thermodynamics. The three methods of heat transfer are listed and explained below.

Conduction: Electron diffusion or photo vibration is responsible for this mode of heat transfer. The bodies of matter themselves do not move; the heat is transferred because adjacent atoms that vibrate against each other or as electrons flow between atoms. This type of heat transfer is most common when two solids come in direct contact with each other. This is because molecules in a solid are in close contact with one another and so the electrons can flow freely. It stands to reason, then, that metals are good conductors of thermal energy. This is because their metallic bonds allow the freest movement of electrons. Similarly, conduction is better in denser solids. Examples of conduction can be seen in the use of copper to quickly convey heat in cooking pots, the flow of heat from a hot water bottle to a person's body, or the cooling of a warm drink with ice.

The amount of heat transferred by conduction through a material depends on several factors. It is directly proportional to the temperature difference ΔT between the surface from which the heat is flowing and the surface to which it is transferred. Heat flow *H* increases with the area *A* through which the flow occurs and also with the time duration *t*. The thickness of the material reduces the flow of heat. The relationship between all these variables is expressed as

$$H = \frac{k.t.A.\Delta T}{d}$$

where the proportionality constant *k* is known as the **thermal conductivity**, a property of the material. Thermal conductivity of a good conductor is close to 1 (0.97 cal/cm.s.^0C for silver) while good insulators have thermal conductivity that is nearly zero (0.0005 cal/cm.s.^0C for wood).

Problem: A glass window pane is 50 cm long and 30 cm wide. The glass is 1 cm thick. If the temperature indoors is 15^0C higher than it is outside, how much heat will be lost through the window in 30 minutes? The thermal conductivity of glass is 0.0025 cal/cm.s.^0C.

Solution: The window has area *A* = 1500 sq. cm and thickness *d* = 1 cm. Duration of heat flow is 1800 s and the temperature difference $\Delta T = 15^0C$. Therefore heat loss through the window is given by

$$H = (0.0025 \times 1800 \times 1500 \times 15)/1 = 101250 \text{ calories}$$

Convection: Convection involves some conduction but is distinct in that it involves the movement of warm particles to cooler areas. Convection may be either natural or forced, depending on how the current of warm particles develops. Natural convection occurs when molecules near a heat source absorb thermal energy (typically via conduction), become less dense, and rise. Cooler molecules then take their place and a natural current is formed. Forced convection, as the name suggests, occurs when liquids or gases are moved by pumps, fans, or other means to be brought into contact with warmer or cooler masses. Because the free motion of particles with different thermal energy is key to this mode of heat transfer, convection is most common in liquid and gases. Convection can, however, transfer heat between a liquid or gas and a solid. Forced convection is used in “forced air” home heating systems and is common in industrial manufacturing processes. Additionally, natural convection is responsible for ocean currents and many atmospheric events. Finally, natural convection often arises in association with conduction, for instance in the air near a radiator or the water in a pot on the stove. The mathematical analysis of heat transfer by convection is far more complicated than for conduction or radiation and will not be addressed here.

Radiation: This method of heat transfer occurs via electromagnetic radiation. All matter warmer than absolute zero (that is, all known matter) radiates heat. This radiation occurs regardless of the presence of any medium. Thus, it occurs even in a vacuum. Since light and radiant heat are both part of the EM spectrum, we can easily visualize how heat is transferred via radiation. For instance, just like light, radiant heat is reflected by shiny materials and absorbed by dark materials. Common examples of radiant heat include the way sunlight travels from the sun to warm the earth, the use of radiators in homes, and the warmth of incandescent light bulbs.

The amount of energy radiated by a body at temperature T and having a surface area A is given by the Stefan-Boltzmann law expressed as

$$I = e\sigma AT^4$$

where I is the radiated power in watts, e (a number between 0 and 1) is the **emissivity** of the body and σ is a universal constant known as **Stefan's constant** that has a value of $5.6703 \times 10^{-8} W/m^2.K^4$. Black objects absorb and radiate energy very well and have emissivity close to 1. Shiny objects that reflect energy are not good absorbers or radiators and have emissivity close to zero.

A body not only radiates thermal energy but also absorbs energy from its surroundings. The net power radiation from a body at temperature T in an environment at temperature T_0 is given by

$$I = e\sigma A(T^4 - T_0^{\,4})$$

Problem: Calculate the net power radiated by a body of surface area 2 sq. m, temperature 30^0C and emissivity 0.5 placed in a room at a temperature of 15^0C.

Solution: $I = 0.5 \times 5.67 \times 10^{-8} \times 2(303^4 - 288^4) = 88\ \text{W}$

Skill 30.2 Understands the molecular interpretation of temperature and heat.

Heat is the thermal energy a body has due to the kinetic and potential energy of its atoms and molecules and is measured in the same units as any other form of energy, the SI unit being Joule. The traditional unit for the measurement of heat is the calorie that is related to the Joule through the relationship **1 calorie = 4.184 Joule**. Many other forms of energy, mechanical energy when you rub your hands together or electrical energy from a light bulb for instance, can be converted into heat energy. For a detailed discussion of heat energy see section IV.4.

Heat is generally measured in terms of temperature, a measure of the average internal energy of a material. Temperature is an **intensive property**, meaning that it does not depend on the amount of material. Heat content is an **extensive property** because more material at the same temperature will contain more heat. The relationship between the change in heat energy of a material and the change in its temperature is given by $\Delta Q = mC\Delta T$, where ΔQ is the change in heat energy, m is the mass of the material, ΔT is the change in temperature and C is the specific heat which is characteristic of a particular material. Section IV.4 provides a detailed discussion of heat, temperature and specific heat.

There are four generally recognized temperature scales, Celsius, Fahrenheit, Kelvin and Rankine. The Kelvin and Rankine scales are absolute temperature scales corresponding to the Celsius and Fahrenheit scales, respectively. Absolute temperature scales have a zero reading when the temperature reaches absolute zero (the theoretical point at which no thermal energy exists). The absolute temperature scales are useful for many calculations in chemistry and physics.

To convert between Celsius and Fahrenheit, use the following relationship:

$$x\ °F = (5/9)(x - 32)\ °C$$

To convert to the absolute temperature scales, use the appropriate conversion below:

x °F=x+ 459.67 °R

x °C= x+273.15 K

Note that the size of each degree on the Fahrenheit/ Rankine scale is smaller than the size of a degree on the Celsius/Kelvin scale.

Skill 30.3 Solves problems involving thermal expansion, heat capacity, and the relationship between heat and other forms of energy.

Most solid and liquid materials expand when heated with a change in dimension proportional to the change in temperature. A notable exception to this is water between 0^0C and 4^0C.

If we consider a long rod of length *L* that increases in length by ΔL when heated, the fractional change in length $\Delta L / L$ is directly proportional to the change in temperature ΔT.

$$\Delta L / L = \alpha . \Delta T$$

The constant of proportionality α is known as the **coefficient of linear expansion** and is a property of the material of which the rod is made.

Problem: The temperature of an iron rod 10 meters long changes from -3^0C to 12^0C. If iron has a coefficient of linear expansion of 0.000011 per ^0C, by how much does the rod expand?

Solution: The length of the rod *L* = 10 meters.
Change in temperature $\Delta T = 12^0C - (-3^0C) = 15^0C$
Change in length of the rod $\Delta L = 0.000011 \times 10 \times 15 = .00165$ meters

If instead of a rod, we consider an area *A* that increases by ΔA when heated, we find that the fractional change in area is proportional to the change in temperature ΔT. The proportionality constant in this case is known as the **coefficient of area expansion** and is related to the coefficient of linear expansion as demonstrated below.

If *A* is a rectangle with dimensions L_1 and L_2, then

$$\begin{aligned} A + \Delta A &= (L_1 + \Delta L_1)(L_2 + \Delta L_2) \\ &= (L_1 + \alpha L_1 \Delta T)(L_2 + \alpha L_2 \Delta T) \\ &= L_1 L_2 + 2\alpha L_1 L_2 \Delta T + \alpha^2 (\Delta T)^2 \end{aligned}$$

Ignoring the higher order term for small changes in temperature, we find that

$$\Delta A = 2\alpha A \Delta T = \gamma A \Delta T$$

Thus the coefficient of area expansion $\gamma = 2\alpha$.

Following the same procedure as above, we can show that the change in volume of a material when heated may be expressed as

$$\Delta V = 3\alpha V \Delta T = \beta V \Delta T$$

where the **coefficient of volume expansion** $\beta = 3\alpha$.

Problem: An aluminum sphere of radius 10cm is heated from $0^0 C$ to $25^0 C$. What is the change in its volume? The coefficient of linear expansion of aluminum is 0.000024 per $^0 C$.

Solution: Volume *V* of the sphere = $\frac{4}{3}\Pi r^3 = \frac{4}{3} \times 3.14 \times 1000 cm^3 = 4186.67 cm^3$

Change in volume of the sphere = $3 \times 0.000024 \times 4186.67 \times 25 = 7.54 cm^3$

The internal energy of a material is the sum of the total kinetic energy of its molecules and the potential energy of interactions between those molecules. Total kinetic energy includes the contributions from translational motion and other components of motion such as rotation. The potential energy includes energy stored in the form of resisting intermolecular attractions between molecules.

The enthalpy (*H*) of a material is the sum of its internal energy and the mechanical work it can do by driving a piston. A change in the enthalpy of a substance is the total energy change caused by adding/removing heat at constant pressure.

When a material is heated and experiences a phase change, thermal energy is used to break the intermolecular bonds holding the material together. Similarly, bonds are formed with the release of thermal energy when a material changes its phase during cooling. Therefore, the energy of a material increases during a phase change that requires heat and decreases during a phase change that releases heat. For example, the energy of H_2O increases when ice melts and decreases when water freezes.

Heat capacity and specific heat

A substance's molar heat capacity is the heat required to change the temperature of one mole of the substance by one degree. Heat capacity has units of joules per mol- kelvin or joules per mol- °C. The two units are interchangeable because we are only concerned with differences between one temperature and another. A Kelvin degree and a Celsius degree are the same size.

The specific heat of a substance (also called specific heat capacity) is the heat required to change the temperature of one gram or kilogram by one degree. Specific heat has units of joules per gram-°C or joules per kilogram-°C.

These terms are used to solve problems involving a change in temperature by applying the formula:

$q = n \times C \times \Delta T$ where $q \Rightarrow$ heat added (positive) or evolved (negative)

$n \Rightarrow$ amount of material

$C \Rightarrow$ molar heat capacity if n is in moles, specific heat if n is a mass

$\Delta T \Rightarrow$ change in temperature $T_{final} - T_{initial}$

Example:

What is the change in energy of 10 g of gold at 25 °C when it is heated beyond its melting point to 1300 °C. You will need the following data for gold:

Solid heat capacity: 28 J/mol-K

Molten heat capacity: 20 J/mol-K

Enthalpy of fusion: 12.6 kJ/mol

Melting point: 1064 °C

Solution: First determine the number of moles used: $10\text{ g} \times \dfrac{1\text{ mol}}{197\text{ g}} = 0.051\text{ mol}$.

There are then three steps. 1) Heat the solid. 2) Melt the solid. 3) Heat the liquid. All three require energy so they will be positive numbers.

1) Heat the solid:

$$q_1 = n \times C \times \Delta T = 0.051\text{ mol} \times 28\ \frac{\text{J}}{\text{mol-K}} \times (1064\text{ °C} - 25\text{ °C})$$

$$= 1.48 \times 10^3\text{ J} = 1.48\text{ kJ}$$

2) Melt the solid: $q_2 = n \times \Delta H_{fusion} = 0.051\text{ mol} \times 12.6\ \dfrac{\text{kJ}}{\text{mol}}$

$$= 0.64\text{ kJ}$$

3) Heat the liquid:

$$q_3 = n \times C \times \Delta T = 0.051 \text{ mol} \times 20 \frac{\text{J}}{\text{mol-K}} \times (1300\ °\text{C} - 1064\ °\text{C})$$

$$= 2.4 \times 10^2 \text{ J} = 0.24 \text{ kJ}$$

The sum of the three processes is the total change in energy of the gold:

$$q = q_1 + q_2 + q_3 = 1.48 \text{ kJ} + 0.64 \text{ kJ} + 0.24 \text{ kJ} = 2.36 \text{ kJ}$$

$$= 2.4 \text{ kJ}$$

A **temperature vs. heat graph** can demonstrate these processes visually. One can also calculate the specific heat or latent heat of phase change for the material by studying the details of the graph.

Example: The plot below shows heat applied to 1g of ice at -40C. The horizontal parts of the graph show the phase changes where the material absorbs heat but stays at the same temperature. The graph shows that ice melts into water at 0C and the water undergoes a further phase change into steam at 100C.

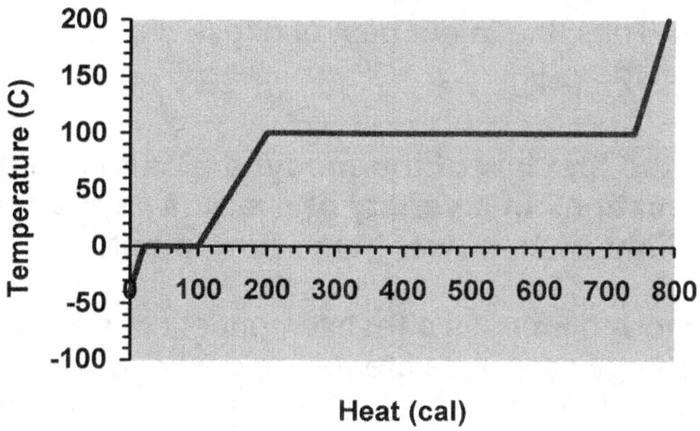

The specific heat of ice, water and steam and the latent heat of fusion and vaporization may be calculated from each of the five segments of the graph.

For instance, we see from the flat segment at temperature 0C that the ice absorbs 80 cal of heat. The latent heat L of a material is defined by the equation $\Delta Q = mL$ where ΔQ is the quantity of heat transferred and m is the mass of the material. Since the mass of the material in this example is 1g, the latent heat of fusion of ice is given by $L = \Delta Q / m = 80$ cal/g.

The next segment shows a rise in the temperature of water and may be used to calculate the specific heat C of water defined by $\Delta Q = mC\Delta T$, where ΔQ is the quantity of heat absorbed, m is the mass of the material and ΔT is the change in temperature. According to the graph, ΔQ = 200-100 =100 cal and ΔT = 100-0=100C. Thus, C = 100/100 = 1 cal/gC.

Problem: The plot below shows the change in temperature when heat is transferred to 0.5g of a material. Find the initial specific heat of the material and the latent heat of phase change.

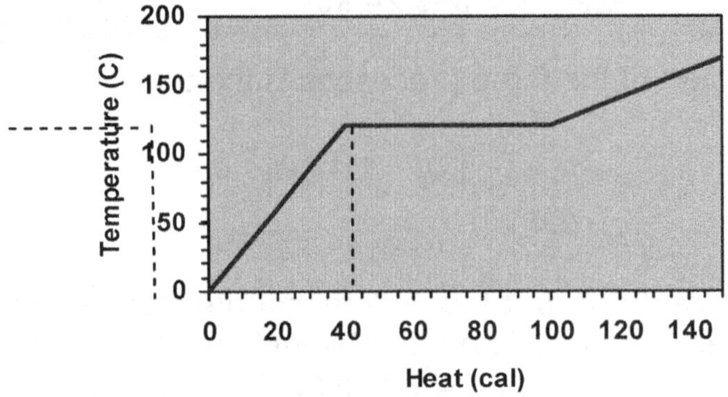

Solution: Looking at the first segment of the graph, we see that $\Delta Q = 40$ cal and $\Delta T = 120$ C. Since the mass m = 0.5g, the specific heat of the material is given by $C = \Delta Q / (m \Delta T)$ = 40/(0.5 X120) = 0.67 cal/gC.

The flat segment of the graph represents the phase change. Here $\Delta Q = 100 - 40 = 60$ cal. Thus, the latent heat of phase change is given by $L = \Delta Q / m$ = 60/(0.5) = 120 cal/g.

Skill 30.4 Applies the first law of thermodynamics to analyze energy transformations in a variety of everyday situations (e.g., electric light bulb, power generating plant).

The first law of thermodynamics is a restatement of conservation of energy, i.e. the principle that energy cannot be created or destroyed. It also governs the behavior of a system and its surroundings. The change in heat energy supplied to a system (Q) is equal to the sum of the change in the internal energy (U) and the change in the work (W) done by the system against internal forces.

The internal energy of a material is the sum of the total kinetic energy of its molecules and the potential energy of interactions between those molecules. Total kinetic energy includes the contributions from translational motion and other components of motion such as rotation. The potential energy includes energy stored in the form of resisting intermolecular attractions between molecules. Mathematically, we can express the relationship between the heat supplied to a system, its internal energy and work done by it as

$$\Delta Q = \Delta U + \Delta W$$

The first law of thermodynamics can be observed in an electric light bulb. Work is done on the system as an electric current is driven through the tungsten filament. The system becomes hotter and there is a positive change in energy. Once the filament becomes hot enough, it begins to glow. At that point, it does work. A fluorescent light bulb converts about 20% of the electric current into visible light energy, whereas an incandescent light bulb converts only about 5%. In both cases, the remaining current is converted to waste heat.

In an example of a power generating plant, the feed-pump works to supply water to a boiler. Heat is supplied to the system through the boiler and drives the steam turbine. The steam turbine produces work and heat. The heat is rejected from the system by the condenser. Overall, we have work and heat going in and work and heat going out.

Now let us examine a sample problem that relies upon this law.

Problem: A closed tank has a volume of 40.0 m^3 and is filled with air at 25°C and 100 kPa. We desire to maintain the temperature in the tank constant at 25°C as water is pumped into it. How much heat will have to be removed from the air in the tank to fill the tank ½ full?

Solution: The problem involves isothermal compression of a gas, so $\Delta U_{gas}=0$. Consulting the equation above, $\Delta Q = \Delta U + \Delta W$, it is clear that the heat removed from the gas must be equal to the work done by the gas.

$$Q_{gas} = W_{gas} = P_{gas}V_1 \ln\left(\frac{V_2}{V_T}\right) = P_{gas}V_T \ln\left(\frac{\frac{1}{2}V_T}{V_T}\right) = P_{gas}V_T \ln \frac{1}{2}$$

$$= (100kPa)(40.0m^3)(-0.69314) = -2772.58kJ$$

Thus, the gas in the tank must lose 2772.58 kJ to maintain its temperature.

Many forms of energy exist all around us. Energy is defined as the ability to do work. If you are able to measure how much work an object does, or how much heat is exchanged, you can determine the amount of energy that is in a system. Energy and work are measured in Joules. As the law of conservation of energy states, energy can be neither created nor destroyed however, it can be transformed from one form to another. Energy cannot be truly "lost" then, although energy may be wasted or not used to perform work.

Some typical forms of energy are mechanical, heat, sound, electrical, light, chemical, nuclear, and magnetic. Energy can be transformed from mechanical to heat by friction. Additionally, mechanical kinetic energy can combine with magnetic energy, to transform into electrical energy, as when a magnet is spun inside a metal coil. Electrical energy is transformed into light and heat energy when light bulb is turned on and the filament begins to glow. A firefly uses phosphorescence to transform chemical energy into light energy. Within mechanical energy, energy can transform between potential and kinetic repeatedly as is the case with a pendulum.

Energy can also be transformed into matter and vice versa. The equation $E=mc^2$ quantifies the relationship between matter and energy. The conversion of mass to other forms of energy can liberate vast amounts of energy, as shown by nuclear reactors and weapons. However, the mass equivalent of a unit of energy is very small, which is why energy loss is not typically measured by weight.

Energy transformations are classified as thermodynamically reversible or irreversible. A reversible process is one in which no energy is dissipated into empty quantum states, or states of energy with increased disorder. The easiest way to explain the concept is to consider a roller coaster car on a track. A reversible energy transformation occurs as the car travels up and down converting potential energy into kinetic and back. Without friction, the transformation is 100% efficient and no energy is wasted, and the transformation is reversible. However, we know that friction generates heat, and that the heat generated cannot be completely recovered as usable energy, which results in the transformation being irreversible.

Skill 30.5 Understands the concept of entropy and its relationship to the second law of thermodynamics.

To understand the second law of thermodynamics, we must first understand the concept of entropy. Entropy is the transformation of energy to a more disordered state and is the measure of how much energy or heat is available for work. The greater the entropy of a system, the less energy is available for work. The simplest statement of the second law of thermodynamics is that the entropy of an isolated system not in equilibrium tends to increase over time. The entropy approaches a maximum value at equilibrium. Below are several common examples in which we see the manifestation of the second law.

- The diffusion of molecules of perfume out of an open bottle
- Even the most carefully designed engine releases some heat and cannot convert all the chemical energy in the fuel into mechanical energy
- A block sliding on a rough surface slows down
- An ice cube sitting on a hot sidewalk melts into a little puddle; we must provide energy to a freezer to facilitate the creation of ice

When discussing the second law, scientists often refer to the “arrow of time”. This is to help us conceptualize how the second law forces events to proceed in a certain direction. To understand the direction of the arrow of time, consider some of the examples above; we would never think of them as proceeding in reverse. That is, as time progresses, we would never see a puddle in the hot sun spontaneously freeze into an ice cube or the molecules of perfume dispersed in a room spontaneously re-concentrate themselves in the bottle. The above-mentioned examples are **spontaneous** as well as **irreversible**, both characteristic of increased entropy. Entropy change is zero for a **reversible process**, a process where infinitesimal quasi-static changes in the absence of dissipative forces can bring a system back to its original state without a net change to the system or its surroundings. All real processes are irreversible. The idea of a reversible process, however, is a useful abstraction that can be a good approximation in some cases.

The second law of thermodynamics may also be stated in the following ways:

1. No machine is 100% efficient.
2. Heat cannot spontaneously pass from a colder to a hotter object.

If we consider a **heat engine** that absorbs heat Q_h from a hot reservoir at temperature T_h and does work W while rejecting heat Q_c to a cold reservoir at a lower temperature T_c, $Q_h - Q_c = \mathrm{W}$. The efficiency of the engine is the ratio of the work done to the heat absorbed and is given by

$$\varepsilon = \frac{W}{Q_h} = \frac{Q_h - Q_c}{Q_h} = 1 - \frac{Q_c}{Q_h}$$

It is impossible to build a heat engine with 100% efficiency, i.e. one where $Q_c = 0$.

Carnot described an ideal reversible engine, the **Carnot engine**, that works between two heat reservoirs in a cycle known as the **Carnot cycle** which consists of two isothermal (12 and 34) and two adiabatic processes (23 and 41) as shown in the diagram below.

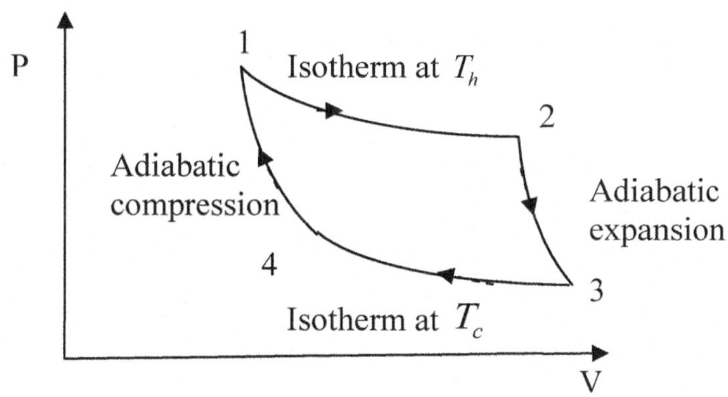

The efficiency of a Carnot engine is given by $\varepsilon = 1 - \frac{T_c}{T_h}$ where the temperature values are absolute temperatures. This is the highest efficiency that any engine working between T_c and T_h can reach.

According to **Carnot's theorem**, no engine working between two heat reservoirs can be more efficient than a reversible engine. All such reversible engines have the same efficiency.

COMPETENCY 31.0 THE TEACHER UNDERSTANDS THE CHARACTERISTICS AND BEHAVIOR OF WAVES.

Skill 31.1 Understands interrelationships among wave characteristics such as velocity, frequency, wavelength, and amplitude and relates them to properties of sound and light (e.g., pitch, color).

To fully understand waves, it is important to understand many of the terms used to characterize them.

Wave velocity: Two velocities are used to describe waves. The first is phase velocity, which is the rate at which a wave propagates. For instance, if you followed a single crest of a wave, it would appear to move at the phase velocity. The second type of velocity is known as group velocity and is the speed at which variations in the wave's amplitude shape propagate through space. Group velocity is often conceptualized as the velocity at which energy is transmitted by a wave. Phase velocity is denoted v_p and group velocity is denoted v_g.

Crest: The maximum value that a wave assumes; the highest point.

Trough: The lowest value that a wave assumes; the lowest point.

Nodes: The points on a wave with minimal amplitude.

Antinodes: The farthest point from the node on the amplitude axis; both the crests and the troughs are antinodes.

Amplitude: The distance from the wave's highest point (the crest) to the equilibrium point. This is a measure of the maximum disturbance caused by the wave and is typically denoted by A.

Wavelength: The distance between any two sequential troughs or crests denoted λ and representing a complete cycle in the repeated wave pattern.

Period: The time required for a complete wavelength or cycle to pass a given point. The period of a wave is usually denoted T.

Frequency: The number of periods or cycles per unit time (usually a second). The frequency is denoted f and is the inverse of the wave's period (that is, $f=1/T$).

Phase: This is a given position in the cycle of the wave. It is most commonly used in discussing a "being out of phase" or a "phase shift", an offset between waves.

Polarization: A property of transverse waves that describes the plane perpendicular to the direction of travel in which the oscillation occurs. Note that longitudinal waves are not polarized because they can oscillate only in one direction, the direction of travel.

Pitch: The frequency of a sound wave as perceived by the human ear. A high pitch sound corresponds to a high frequency sound wave and a low pitch sound corresponds to a low frequency sound wave.

Sound power: This is the sonic energy of a sound wave per unit time.

Sound intensity: This is simply the sound power per unit area.

We can visualize several of these terms on the following diagram of a simple, periodic sine wave on a scale of distance displacement (x-axis) vs. (y-axis):

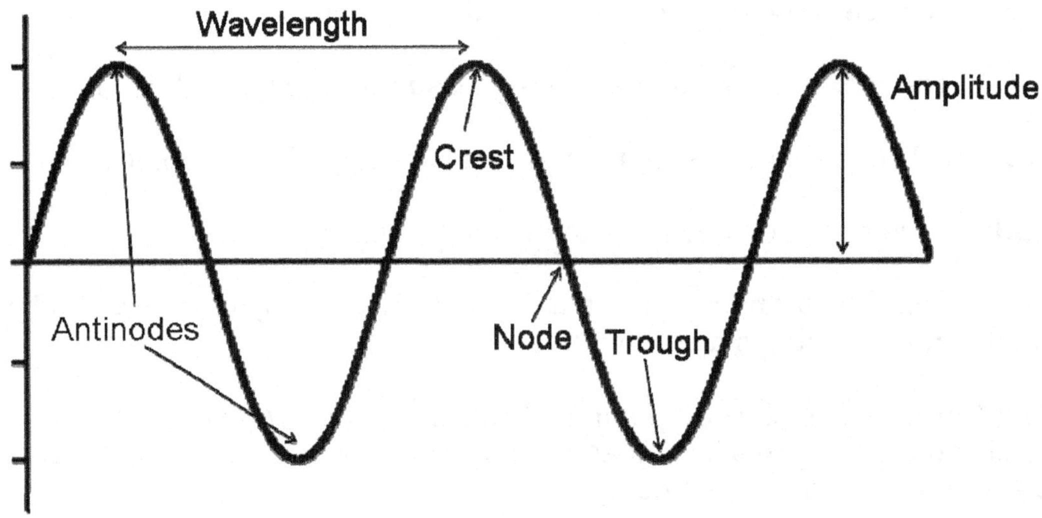

The phase velocity of a wave is related to its wavelength and frequency. Taking light waves, for instance, the speed of light c is equal to the distance traveled divided by time taken. Since the light wave travels the distance of one wavelength λ in the period of the wave T,

$$c = \frac{\lambda}{T}$$

The frequency of a wave, f, is the number of completed periods in one second. In general,

$$f = \frac{1}{T}$$

So the formula for the speed of light can be rewritten as

$$c = \lambda f$$

Thus the phase velocity of a wave is equal to the wavelength times the frequency.

Skill 31.2 Compares and contrasts transverse and longitudinal waves.

Transverse waves: Waves in which the oscillations are perpendicular to the direction in which in the wave travels.

Longitudinal waves: Waves in which the oscillations are in the direction in which the wave travels.

Polarization: A property of transverse waves that describes the plane perpendicular to the direction of travel in which the oscillation occurs. Note that longitudinal waves are not polarized because they can oscillate only in one direction, the direction of travel. In unpolarized light, the transverse oscillation occurs in all planes perpendicular to the direction of travel. Polarized light (created, for instance, by using polarizing filters that absorb light oscillating in other planes) oscillates in only a selected plane. An everyday example of polarization is found in polarized sunglasses which reduce glare.

The distance that a particle in a medium moves as a wave is transmitted through it is measured by particle displacement. Particle displacement can be measured in both transverse and longitudinal waves.

In a longitudinal wave, particle displacement is parallel to the wave's direction of travel. Thus, if we imagine a longitudinal waveform moving down a tube, particles will move back and forth parallel to the sides of the tube. Sound is one of the most important types of longitudinal waves and the described phenomenon is manifest as sound travels through air or water.

Transverse waves, on the other hand, oscillate in a direction perpendicular to the direction of wave travel. So let's imagine the same tube, this time with a transverse wave traveling down it. In this case, the particles oscillate up and down within the tube. Particle displacement in a transverse wave can also be easily visualized in the vibration of a taut string.

Excellent videos that demonstrate the movement of individual particles are available at the following web address:
http://www.kettering.edu/~drussell/Demos/waves/wavemotion.html

Finally, it is important to note that, in either type of wave, particle displacement depends on the particle velocity of the wave in the medium which is not the same as the velocity of the wave itself.

Skill 31.3 Describes how various waves are propagated through different media.

Mechanical waves differ from electromagnetic waves in that they need a medium through which to travel. This medium can be a solid, as in the case of the waves in a Slinky toy, a gas, such as a sound wave traveling to your ear, or liquid, such waves in the ocean. It follows, then, that mechanical waves cannot exist in a vacuum.

Mechanical waves rely on the local oscillation of each atom in a medium, but the material itself does not move; only the energy is transferred from atom to atom. Therefore the material through which the mechanical wave is traveling greatly affects the wave's propagation and speed. In particular, a material's elastic constant and density affect the speed at which a wave travels through it. Both of these properties of a medium can predict the extent to which the atoms will vibrate and pass along the energy of the wave. The general relationship between these properties and the speed of a wave in a solid is given by the following equation:

$$V = \sqrt{\frac{C_{ij}}{\rho}}$$

where V is the speed of sound, C_{ij} is the elastic constant, and ρ is the material density. It is worth noting that the elastic constant differs depending on direction in anisotropic materials and the ij subscripts indicate that this directionality must be taken into account.

A mechanical wave requires an initial energy input to be created. Once this initial energy is added, the wave will travel through the medium until all the energy has been dissipated. This initial amount of energy upon creation of the wave will also affect the extent of its propagation throughout the medium.

The type of mechanical wave is another factor that will determine propagation. There are three types of mechanical waves. Transverse waves are waves that cause the medium to vibrate at a 90 degree angle to the direction of travel of the wave. The next type is a longitudinal wave which causes the medium to vibrate in the same direction as the direction of travel of the wave. When the particles of the medium the longitudinal wave is traveling through are drawn close together it is called compression. When the particles of the medium it is traveling through are spread apart, it is called rarefaction. The final type of wave is a surface wave.

This type of wave travels along a surface that is between two mediums. An example of a surface wave would be waves in a pool, or in an ocean.

Skill 31.4 Applies properties of reflection and refraction to analyze optical phenomena (e.g., mirrors, lenses, fiber-optic cable).

Light interacts with matter by **reflection, absorption** or **transmission**. In a transparent material such as a lens, most of the light is transmitted through. Opaque objects such as rocks or cars partially absorb and partially reflect light.

The image you see in a mirrored surface is the result of the reflection of the light waves off the surface. Light waves follow the "law of reflection," i.e. the angle at which the light wave approaches a flat reflecting surface is equal to the angle at which it leaves the surface. Scattering is a form of reflection where the reflection happens in multiple directions. **Rayleigh scattering** is the scattering of an electromagnetic wave by particles that are much smaller than its wavelength. The amount of scattering is inversely related to the wavelength of the wave. Thus, the greater scattering of blue light compared to red light by particles in the atmosphere results in the sky appearing blue.

Absorption is the transfer of energy from a light wave to particles of matter. In the absorption process, a material converts some of the light energy into heat. Some of the energy may be radiated at a different frequency. When light crosses the boundary between two different media, its path is bent, or refracted.

Wave **refraction** is a change in direction of a wave due to a change in its speed. This most commonly occurs when a wave passes from one material to another, such as a light ray passing from air into water or glass. However, light is only one example of refraction; any type of wave can undergo refraction. Another example would be physical waves passing from water into oil. At the boundary of the two media, the wave velocity is altered, the direction changes, and the wavelength increases or decreases. However, the frequency remains constant.

Snell's Law describes how light bends, or refracts, when traveling from one medium to the next. It is expressed as

$$n_1 \sin\theta_1 = n_2 \sin\theta_2$$

where n_i represents the index of refraction in medium i, and θ_i represents the angle the light makes with the normal in medium i.

Problem: The index of refraction for light traveling from air into an optical fiber is 1.44. (a) In which direction does the light bend? (b) What is the angle of refraction inside the fiber, if the angle of incidence on the end of the fiber is 22°?

Solution: (a) The light will bend toward the normal since it is traveling from a rarer region (lower n) to a denser region (higher n).

(b) Let air be medium 1 and the optical fiber be medium 2:

$$n_1 \sin\theta_1 = n_2 \sin\theta_2$$
$$(1.00)\sin 22° = (1.44)\sin\theta_2$$
$$\sin\theta_2 = \frac{1.00}{1.44}\sin 22° = (.6944)(.3746) = 0.260$$
$$\theta_2 = \sin^{-1}(0.260) = 15°$$

The angle of refraction inside the fiber is $15°$.

Light travels at different speeds in different media. The speed of light in a vacuum is represented by

$$c = 2.99792458x10^8 \, m/s$$

but is usually rounded to

$$c = 3.00x10^8 \, m/s.$$

Light will never travel faster than this value. The **index of refraction**, n, is the amount by which light slows in a given material and is defined by the formula

$$n = \frac{c}{v}$$

where v represents the speed of light through the given material.

Problem: The speed of light in an unknown medium is measured to be $1.24x10^8 \, m/s$. What is the index of refraction of the medium?

Solution:

$$n = \frac{c}{v}$$
$$n = \frac{3.00x10^8}{1.24x10^8} = 2.42$$

Referring to a standard table showing indices of refraction, we would see that this index corresponds to the index of refraction for diamond.

Reflection is the change in direction of a wave at an interface between two dissimilar media such that the wave returns into the medium from which it originated. The most common example of this is light waves reflecting from a mirror, but sound and water waves can also be reflected. The law of reflection states that the angle of incidence is equal to the angle of reflection.

Reflection may occur whenever a wave travels from a medium of a given refractive index to another medium with a different index. A certain fraction of the light is reflected from the interface and the remainder is refracted. However, when the wave is moving from a dense medium into one less dense, that is the refractive index of the first is greater than the second, a critical angle exists which will create a phenomenon known as **total internal reflection**. In this situation all of the wave incident at an angle greater than the critical angle is reflected. When a wave reflects off a more dense material (higher refractive index) than that from which it originated, it undergoes a 180° phase change. In contrast, a less dense, lower refractive index material will reflect light in phase.

Fiber optics makes use of the phenomenon of total internal reflection. The light traveling through a fiber reflects off the walls at angles greater than the critical angle and thus keeps the wave confined to the narrow fiber.

Thin Lenses

A lens is a device that causes electromagnetic radiation to converge or diverge. The most familiar lenses are made of glass or plastic and designed to concentrate or disperse visible light. Two of the most important parameters for a lens are its thickness and it's focal length. Focal length is a measure of how strongly light is concentrated or dispersed by a lens. For a convex or converging lens, the focal length is the distance at which a beam of light will be focused to a single spot. Conversely, for a concave or diverging lens, the focal length is the distance to the point from which a beam appears to be diverging.

A thin lens in one in which focal length is much greater than lens thickness. For problems involving thin lenses, we can disregard any optical effects of the lens itself. Additionally, we can assume that the light that interacts with the lens makes a small angle with the optical axis of the system and so the sine and tangent values of the angle are approximately equal to the angle itself. This paraxial approximation, along with the thin lens assumptions, allows us to state:

$$\frac{1}{s}+\frac{1}{s'}=\frac{1}{f}$$

Where s=distance from the lens to the object (object location)
s'=distance from the lens to the image (image location)
f=focal length of the lens

Most lenses also cause some magnification of the object. Magnification is defined as:

$$m=\frac{y'}{y}=-\frac{s'}{s}$$

Where m=magnification
y'=image height
y=object height

The images produced by lenses can be either virtual or real. A virtual image is one that is created by rays of light that appear to diverge from a certain point. Virtual images cannot be seen on a screen because the light rays do not actually meet at the point where the image is located. If an image and object appear on the same side of a converging lens, that image is defined as virtual. For virtual images, the image location will be negative and the magnification positive. Real images, on the other hand, are formed by light rays actually passing through the image. Thus, real images are visible on a screen. Real images created by a converging lens are inverted and have a positive image location and negative magnification.

Sign conventions will make it easier to understand thin lens problems:

Focal length: positive for a converging lens; negative for a diverging lens
Object location: positive when in front of the lens; negative when behind the lens
Image location: positive when behind the lens; negative when in front of the lens
Image height: positive when upright; negative when upside-down.
Magnification: positive for an erect, virtual image; negative for an inverted, real image

Example:
A converging lens has a focal length of 10.00 cm and forms a 2.0 cm tall image of a 4.00 mm tall real object to the left of the lens. If the image is erect, is the image real or virtual? What are the locations of the object and the image?

Solution:

We begin by determining magnification:

$$m = \frac{y'}{y} = \frac{0.02m}{0.004m} = 5$$

Since the magnification is positive and the image is erect, we know the image must be virtual.

To find the locations of the object and image, we first relate them by using the magnification:

$$m = -\frac{s'}{s}$$

$$s' = -ms$$

Then we substitute into the thin lens equation, creating one variable in one unknown:

$$\frac{1}{s} + \frac{1}{s'} = \frac{1}{f}$$

$$\frac{1}{s} - \frac{1}{5s} = \frac{1}{10cm}$$

$$\frac{5-1}{5s} = \frac{1}{10cm}$$

$$s = \frac{40cm}{5} = 8cm \rightarrow \rightarrow s' = -5 \times 8cm = -40cm$$

Thus the object is located 8 cm to the left of the lens and the image is 40 cm to the left of the lens.

Mirrors
Plane mirrors

Plane mirrors form virtual images. In other words, the image is formed behind the mirror where light does not actually reach. The image size is equal to the object size and object distance is equal to the image distance; i.e. the image is the same distance behind the mirror as the object is in front of the mirror. Another characteristic of plane mirrors is left-right reversal.

Example: Suppose you are standing in front of a mirror with your right hand raised. The image in the mirror will be raising its left hand.

Problem: If a cat creeps toward a mirror at a rate of 0.20 m/s, at what speed will the cat and the cat's image approach each other?

Solution: In one second, the cat will be 0.20 meters closer to the mirror. At the same time, the cat's image will be 0.20 meters closer to the cat. Therefore, the cat and its image are approaching each other at the speed of 0.40 m/s.

Problem: If an object that is two feet tall is placed in front of a plane mirror, how tall will the image of the object be?

Solution: The image of the object will have the same dimensions as the actual object, in this case, a height of two feet. This is because the magnification of an image in a plane mirror is 1.

Curved mirrors

Curved mirrors are usually sections of spheres. In a concave mirror the inside of the spherical surface is silvered while in a convex mirror it is the outside of the spherical surface that is silvered.

Terminology associated with spherical mirrors:

Principal axis: The line joining the center of the sphere (of which we imagine the mirror is a section) to the center of the reflecting surface.

Center of curvature: The center of the sphere of which the mirror is a section.

Vertex: The point on the mirror where the principal axis meets the mirror or the geometric center of the mirror.

Focal point: The point at which light rays traveling parallel to the principal axis will meet after reflection in a concave mirror. For a convex mirror, it is the point from which light rays traveling parallel to the principal axis will appear to diverge from after reflection. The focal point is midway between the center of curvature and the vertex.

Focal length: The distance between the focal point and the vertex.

Radius of curvature: The distance between the center of curvature and the vertex, i.e. the radius of the sphere of which the mirror is a section. The radius of curvature is twice the focal length.

The relationship between the object distance from vertex s, the image distance from vertex s', and the focal length f is given by the equation

$$\frac{1}{s}+\frac{1}{s'}=\frac{1}{f}$$

Magnification is defined as:

$$m=\frac{y'}{y}=-\frac{s'}{s}$$

where m=magnification, y'=image height, y=object height

Image characteristics for **concave mirrors**:

1) If the object is located **beyond the center of curvature**, the image will be **real, inverted, smaller** and located between the focal point and center of curvature.
2) If the object is located **at the center of curvature**, the image will be **real, inverted, of the same height** and also located at the center of curvature.
3) If the object is located **between the center of curvature and focal point**, the image will be **real, inverted, larger** and located beyond the center of curvature.
4) If the object is located **at the focal point no image is formed**.
5) If the object is located **between the focal point and vertex**, the image will be **virtual, upright, larger** and located on the opposite side of the mirror.
6) If the object is located **at infinity** (very far away), the image is **real, inverted, smaller** and located at the focal point.

For **convex mirrors**, the image is always **virtual, upright, reduced in size** and formed on the opposite side of the mirror.

Problem: A concave mirror collects light from a star. If the light rays converge at 50 cm, what is the radius of curvature of the mirror?

Solution: The focal length, in this case, 50 cm, is the distance from the focal point to the mirror. Since the focal point is the midpoint of the line from the vertex to the center of curvature, or focal length, the focal length would be one-half the radius of curvature. Since the focal length in this case is 50 cm, the radius of curvature would be 100 cm.

Problem: An image of an object in a mirror is upright and reduced in size. In what type of mirror is this image being viewed, plane, concave, or convex?

Solution: The image in a plane mirror would be the same size as the object. The image in a concave mirror would be magnified if upright. Only a convex mirror would produce a reduced upright image of an object.

Skill 31.5 Applies principles of wave interference to analyze wave phenomena, including acoustical (e.g., harmonics) and optical phenomena (e.g., patterns created by thin films and diffraction gratings).

According to the **principle of linear superposition**, when two or more waves exist in the same place, the resultant wave is the sum of all the waves, i.e. the amplitude of the resulting wave at a point in space is the sum of the amplitudes of each of the component waves at that point.

Interference occurs when two or more waves are superimposed. Usually, interference is observed in coherent waves, well-correlated waves that have very similar frequencies or even come from the same source. Superposition of waves may result in either constructive or destructive interference. Constructive interference occurs when the crests of the two waves meet at the same point in time. Conversely, destructive interference occurs when the crest of one wave and the trough of the other meet at the same point in time. It follows, then, that constructive interference increases amplitude and destructive interference decreases it. We can also consider interference in terms of wave phase; waves that are out of phase with one another will interfere destructively while waves that are in phase with one another will interfere constructively. In the case of two simple sine waves with identical amplitudes, for instance, amplitude will double if the waves are exactly in phase and drop to zero if the waves are exactly 180° out of phase.

Additionally, interference can create a **standing wave**, a wave in which certain points always have amplitude of zero. Thus, the wave remains in a constant position. Standing waves typically results when two waves of the same frequency traveling in opposite directions through a single medium are superposed. View an animation of how interference can create a standing wave at the following URL:

http://www.glenbrook.k12.il.us/GBSSCI/PHYS/mmedia/waves/swf.html

A **standing wave** typically results from the interference between two waves of the same frequency traveling in opposite directions. The result is a stationary vibration pattern. One of the key characteristics of standing waves is that there are points in the medium where no movement occurs. The points are called nodes and the points where motion is maximal are called antinodes. This property allows for the analysis of various typical standing waves.

Vibrating string

Imagine a string of length L tied tightly at its two ends. We can generate a standing wave by plucking the string. The waves traveling along the string are reflected at the fixed end points and interfere with each other to produce standing waves. There will always be two nodes at the ends where the string is tied. Depending on the frequency of the wave that is generated, there may also be other nodes along the length of the string. In the diagrams below, examples are given of strings with 0, 1, or 2 additional nodes. These vibrations are known as the 1st, 2nd, and 3rd **harmonics**. The higher order harmonics follow the same pattern although they are not diagramed here.

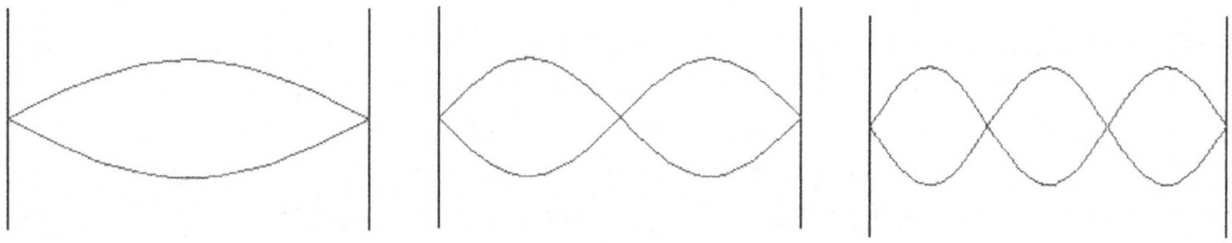

Since we know the length of the string (L) in each case, we can calculate the wavelength (λ) and frequency (f) for any harmonic using the following formula, where n=the harmonic order (n=1,2,3...) and v is the phase velocity of the wave.

$$\lambda_n = \frac{2L}{n} \qquad f_n = \frac{v}{\lambda_n} = n\frac{v}{2L}$$

Waves in a tube

Just as on a string, standing waves can propagate in gaseous or liquid medium inside a tube. In these cases we will also observe harmonic vibrations, but their nature will depend on whether the ends of the tube are closed or open. Specifically, an antinode will be observed at an open end and a node will appear at a closed end. Then, just as in the string example above, we can derive formulas that allow us to predict the wavelength and frequency of the harmonic vibrations that occur in a tube. Below, only frequencies are given; wavelength can be found by applying the formula $f=v/\lambda$.

For a tube with two closed ends or two open ends (note that this is the same as for the string described above):

$$f_n = \frac{nv}{2L}$$

where n = 1,2,3,4...

When only one end of a tube is closed, that end become a node and wave exhibits odd harmonics. For a tube with one close end and one open end:

$$f_n = \frac{nv}{4L}$$

where n = 1,3,5,7...

An animation at the following URL may be helpful in visualizing these various standing waves:

http://www.physics.smu.edu/~olness/www/05fall1320/applet/pipe-waves.html

When two sound waves with slightly different frequencies interfere with each other, **beats** result. We hear a beat as a periodic variation in volume with a rate that depends on the difference between the two frequencies. You may have observed this phenomenon when listening to two instruments being tuned to match; beating will be heard as the two instruments approach the same note and disappear when they are perfectly in tune.

All wavelengths in the EM spectrum can experience interference but it is easy to comprehend instances of interference in the spectrum of visible light. One classic example of this is **Thomas Young's double-slit experiment**. In this experiment a beam of light is shone through a paper with two slits and a striated wave pattern results on the screen. The light and dark bands correspond to the areas in which the light from the two slits has constructively (bright band) and destructively (dark band) interfered. Light from any source can be used to obtain interference patterns. For example, Newton's rings can be produced with sun light. However, in general, white light is less suited for producing clear interference patterns as it is a mix of a full spectrum of colors. Sodium light is close to monochromatic and is thus more suitable for producing interference patterns. The most suitable is laser light as it is almost perfectly monochromatic.

Problem: The interference maxima (location of bright spots created by constructive interference) for double-slit interference are given by

$$\frac{n\lambda}{d} = \frac{x}{D} = \sin\theta \quad n=1,2,3...$$

where λ is the wavelength of the light, *d* is the distance between the two slits, *D* is the distance between the slits and the screen on which the pattern is observed and *x* is the location of the nth maximum. If the two slits are 0.1mm apart, the screen is 5m away from the slits, and the first maximum beyond the center one is 2.0 cm from the center of the screen, what is the wavelength of the light?

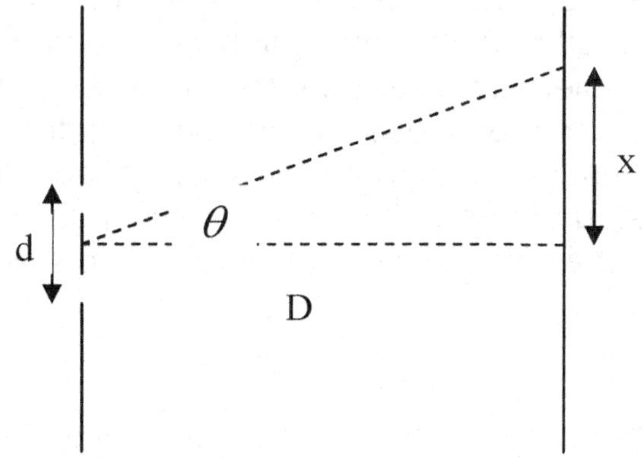

Solution: λ = xd/(Dn) = 0.02 x 0.0001/ (5 x1) = 400 nanometers

Thin-film interference occurs when light waves reflecting off the top surface of a film interfere with the waves reflecting from the bottom surface. We see colors in soap bubbles or in oil films floating on water since the criteria for constructive or destructive interference depend on the wavelength of the light. Non-reflective coatings on materials make use of destructive thin-film interference.

Diffraction is an important characteristic of waves. This occurs when part of a wave front is obstructed. Diffraction and interference are essentially the same physical process. Diffraction refers to various phenomena associated with wave propagation such as the bending, spreading, and interference of waves emerging from an aperture. It occurs with any type of wave including sound waves, water waves, and electromagnetic waves such as light and radio waves.

Here, we take a close look at important phenomena like single-slit diffraction, double-slit diffraction, diffraction grating, other forms of diffraction and lastly interference.

1. Single-slit diffraction: The simplest example of diffraction is single-slit diffraction in which the slit is narrow and a pattern of semi-circular ripples is formed after the wave passes through the slit.

2. Double-slit diffraction: These patterns are formed by the interference of light diffracting through two narrow slits.

3. Diffraction grating: Diffraction grating is a reflecting or transparent element whose optical properties are periodically modulated. In simple terms, diffraction gratings are fine parallel and equally spaced grooves or rulings on a material surface. When light is incident on a diffraction grating, light is reflected or transmitted in discrete directions, called diffraction orders. Because of their light dispersive properties, gratings are commonly used in monochromators and spectrophotometers. Gratings are usually designated by their groove density, expressed in grooves/millimeter. A fundamental property of gratings is that the angle of deviation of all but one of the diffracted beams depends on the wavelength of the incident light.

4. Other forms of diffraction:
i) Particle diffraction: It is the diffraction of particles such as electrons, which is used as a powerful argument for quantum theory. It is possible to observe the diffraction of particles such as neutrons or electrons and hence we are able to infer the existence of wave particle duality.
ii) Bragg diffraction: This is diffraction from a multiple slits, and is similar to what occurs when waves are scattered from a periodic structure such as atoms in a crystal or rulings on a diffraction grating. Bragg diffraction is used in X-ray crystallography to deduce the structure of a crystal from the angles at which the X-rays are diffracted from it.

Dispersion is the separation of a wave into its constituent wavelengths due to interaction with a material occurring in a wavelength-dependent manner (as in thin-film interference for instance).

Skill 31.6 Identifies and interprets how wave characteristics and behaviors are used in medical, industrial, and other real-world applications.

Ultrasound waves are high frequency, longitudinal waves used in medical imaging. The waves penetrate the human body well and some of the waves reflect off tissue boundaries. The pulses sent back are used to create a picture of the inside of the body.

In echo–sounding or echolocation, a system called sonar transmits sound waves to detect and locate objects underwater or measure the distance to the floor of a body of water. The sound wave pulses are reflected off the object or floor, picked up and timed. Since their speed is known, the distance can be calculated. From the pulses, a picture can be built.

Patterns of seismic waves picked up by seismometers can be used to locate distant earthquakes and major explosions and to research the internal structure of the earth. For example, the "shadows" of S-waves prove that the core of the earth is molten.

Radar uses electromagnetic waves to identify the speed, altitude, and direction of both fixed objects and objects in motion. Radio waves are discharged by a transmitter. A receiver detects the waves reflected off the object. Even though the waves are usually very weak, they can be amplified, which makes them more useful than sound waves or visible light. Radar is used in meteorology to detect precipitation, in air traffic control to monitor the locations of planes, in police work to detect speeders, and in military operations.

X-rays are a form of electromagnetic radiation. They are produced by accelerating electrons in order to collide with a metal target. X-rays behave more like a particle than a wave because of their short wavelength. X-rays do not pass through dense materials such as bone but do pass through softer tissues. Where the X-rays pass through soft tissue and strike the photographic plate, the plate turns black, creating a contrast between those areas not penetrated by the X-rays and those that are. X-rays are used in diagnostic imaging for teeth, the skeletal system, and detection of lung cancer, pulmonary edema, and pneumonia, and certain abdominal disorders. X-rays are also used in microscopic analysis to produce images of very small objects and in art to examine the underlying layers of paintings.

Monochromators are optical devices that send a narrow band of wavelengths of light selected from a wider range of wavelengths. They are often used in absorption spectrophotometers to supply light to a sample. Absorption spectrophotometers are used to study enzymes, create optical thermometers, and measure the performance of sunglasses, laser protective glasses, and other optical filters, among other things.

Microwaves are used in microwave ovens to cook food by passing microwave radiation through the food. They are used in broadcasting and telecommunications transmissions because of their short wavelength. A typical example is transmitting television news from a specially equipped van at a remote location to a television station. Bluetooth technology, cable TV, and Internet access on co-ax cable use microwaves. A maser works in a manner similar to a laser but uses microwaves. Most radio astronomy uses microwaves.

COMPETENCY 32.0 THE TEACHER UNDERSTANDS THE FUNDAMENTAL CONCEPTS OF QUANTUM PHYSICS.

Skill 32.1 Interprets wave-particle duality.

The dual wave and particle nature of light has long been considered. In 1924 **Louis de Broglie** suggested that not only light but all matter, particularly electrons, may exhibit wave as well as particle behavior. He proposed that the frequency f and wavelength λ of electron waves are given by the equations

$$f = \frac{E}{h}; \lambda = \frac{h}{p}$$

where p is the momentum of the electron, E is its energy and h is Planck's constant. These are the same relations that Planck proposed for photons. Using deBroglie's equations and considering electrons as standing waves in a circular Bohr orbit, the discrete energy states of an electron could be explained and led to the same set of energy levels found by Bohr. Schrodinger developed these ideas into **wave mechanics**, a general method for finding the quantization condition for a system.

Wave-particle duality is also expressed by **Heisenberg's uncertainty principle** which places a limit on the accuracy with which one can measure the properties of a physical system. This limit is not due to the imperfections of measuring instruments or experimental methods but arises from the fundamental wave-particle duality inherent in quantum systems.

One statement of the uncertainty principle is made in terms of the position and momentum of a particle. If Δx is the uncertainty in the position of a particle in one dimension and Δp the uncertainty in its momentum in that dimension, then according to the uncertainty principle

$$\Delta x \Delta p \geq \hbar / 2$$

where the reduced Planck's constant $\hbar = h / 2\pi = 1.05457168 \times 10^{-34} J.s$

Thus if we measure the position of a particle with greater and greater accuracy, at some point the accuracy in the measurement of its momentum will begin to fall. A simple way to understand this is by considering the wave nature of a subatomic particle. If the wave has a single wavelength, then the momentum of the particle is also exactly known using the DeBroglie momentum-wavelength relationship. The position of the wave, however, extends through all space. If waves of several different wavelengths are superposed, the position of the wave becomes increasingly localized as more wavelengths are added. The increased spread in wavelength, however, then results in an increased momentum spread.

An alternate statement of the uncertainty principle may be made in terms of energy and time.

$$\Delta E \Delta t \geq \hbar / 2$$

Thus, for a particle that has a very short lifetime, the uncertainty in the determination of its energy will be large.

Problem: If a proton is confined to a nucleus that is approximately $10^{-15} m$ in diameter, estimate the minimum uncertainty in its momentum in any one dimension.

Solution: The uncertainty of the position of the proton in any dimension cannot be greater than $10^{-15} m$. Using the uncertainty principle we find that the approximate uncertainty in its momentum in any one dimension must be greater than $\Delta p = \hbar / (2\Delta x) \approx 10^{-19} Kg.m/s$

Skill 32.2 Identifies examples and consequences of the Uncertainty Principle.

Heisenberg's uncertainty principle places a limit on the accuracy with which one can measure the properties of a physical system. This limit is not due to the imperfections of measuring instruments or experimental methods but arises from the fundamental wave-particle duality inherent in quantum systems.

One statement of the uncertainty principle is made in terms of the position and momentum of a particle. If Δx is the uncertainty in the position of a particle in one dimension and Δp the uncertainty in its momentum in that dimension, then according to the uncertainty principle

$$\Delta x \Delta p \geq \hbar / 2$$

where the reduced Planck's constant $\hbar = h / 2\pi = 1.05457168 \times 10^{-34} J.s$

Thus if we measure the position of a particle with greater and greater accuracy, at some point the accuracy in the measurement of its momentum will begin to fall. A simple way to understand this is by considering the wave nature of a subatomic particle. If the wave has a single wavelength, then the momentum of the particle is also exactly known using the DeBroglie momentum-wavelength relationship. The position of the wave, however, extends through all space. If waves of several different wavelengths are superposed, the position of the wave becomes increasingly localized as more wavelengths are added. The increased spread in wavelength, however, then results in an increased momentum spread.

An alternate statement of the uncertainty principle may be made in terms of energy and time.

$$\Delta E \Delta t \geq \hbar / 2$$

Thus, for a particle that has a very short lifetime, the uncertainty in the determination of its energy will be large.

Problem: If a proton is confined to a nucleus that is approximately $10^{-15} m$ in diameter, estimate the minimum uncertainty in its momentum in any one dimension.

Solution: The uncertainty of the position of the proton in any dimension cannot be greater than $10^{-15} m$. Using the uncertainty principle we find that the approximate uncertainty in its momentum in any one dimension must be greater than
$\Delta p = \hbar / (2\Delta x) \approx 10^{-19} Kg.m/s$

Skill 32.3 Understands the photoelectric effect.

The **photoelectric effect** occurs when **light shining on a clean metal surface causes the surface to emit electrons**. The energy of an absorbed photon is transferred to an electron as shown to the right. If this energy is greater than the binding energy holding the electron close to nearby nuclei then the electron will move. A high energy (high frequency, low wavelength) photon will not only dislodge an electron from the "electron sea" of a metal but it will also impart kinetic energy to the electron, making it move rapidly. These electrons in motion will produce an electric current if a circuit is present.

When the metal surface on which light is incident is a cathode with the anode held at a higher potential V, an electric current flows in the external circuit. It is observed that current flows only for light of higher frequencies. Also there is a threshold negative potential, the **stopping potential** V_0 below which no current will flow in the circuit.

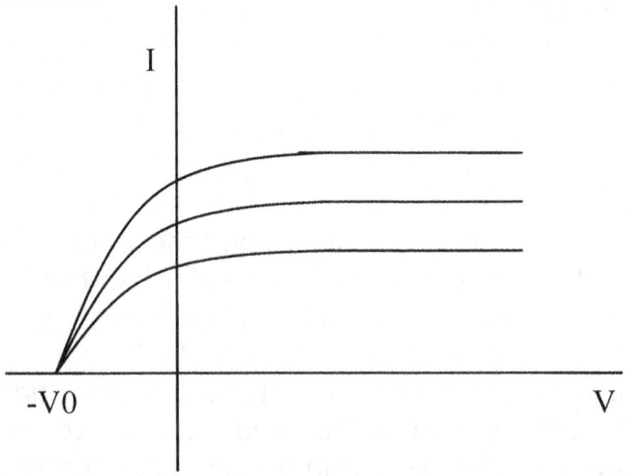

The figure displayed above shows current flow vs. potential for three different intensities of light. It shows that the maximum current flow increases with increasing light intensity but the stopping potential remains the same.

All these observations are counter-intuitive if one considers light to be a wave but may be understood in terms of light particles or photons. According to this interpretation, each photon transfers its energy to a single electron in the metal. Since the energy of a photon depends on its frequency, only a photon of higher frequency can transfer enough energy to an electron to enable it to pass the stopping potential threshold.

When V is negative, only electrons with a kinetic energy greater than $|eV|$ can reach the anode. The maximum kinetic energy of the emitted electrons is given by eV_0. This is expressed by Einstein's photoelectric equation as

$$(\tfrac{1}{2}mv^2)_{\max} = eV_0 = hf - \varphi$$

where the **work function** φ is the energy needed to release an electron from the metal and is characteristic of the metal.

Problem: The work function for potassium is 2.20eV. What is the stopping potential for light of wavelength 400nm?

Solution:

$$eV_0 = hf - \varphi = hc/\lambda - \varphi = 4.136\times 10^{-15} \times 3\times 10^{8} / (400\times 10^{-9}) - 2.20 = 3.10 - 2.20 = 0.90\text{eV}$$

Thus stopping potential $V_0 = 0.90\text{V}$

Skill 32.4 Uses the quantum model of the atom to describe and analyze absorption and emission spectra (e.g., line spectra, blackbody radiation).

As quantum theory was developed and popularized (primarily by Max Planck and Albert Einstein), chemists and physicists began to consider how it might apply to atomic structure. Niels Bohr put forward a model of the atom in which electrons could only orbit the nucleus in circular orbitals with specific distances from the nucleus, energy levels, and angular momentums. In this model, electrons could only make instantaneous "quantum leaps" between the fixed energy levels of the various orbitals. The Bohr model of the atom was altered slightly by Arnold Sommerfeld in 1916 to reflect the fact that the orbitals were elliptical instead of round.

Though the Bohr model is still thought to be largely correct, it was discovered that electrons do not truly occupy neat, cleanly defined orbitals. Rather, they exist as more of an "electron cloud." The work of Louis de Broglie, Erwin Schrödinger, and Werner Heisenberg showed that an electron can actually be located at any distance from the nucleus. However, we can find the *probability* that the electrons exists at given energy levels (i.e., in particular orbitals) and those probabilities will show that the electrons are most frequently organized within the orbitals originally described in the Bohr model.

The quantum structure of the atom describes electrons in discrete energy levels surrounding the nucleus. When an electron moves from a high energy orbital to a lower energy orbital, a quantum of electromagnetic radiation is emitted, and for an electron to move from a low energy to a higher energy level, a quantum of radiation must be absorbed. The particle that carries this electromagnetic force is called a **photon**. The quantum structure of the atom predicts that only photons corresponding to certain wavelengths of light will be emitted or absorbed by atoms. These distinct wavelengths are measured by **atomic spectroscopy**.

In **atomic absorption spectroscopy**, a continuous spectrum (light consisting of all wavelengths) is passed through the element. The frequencies of absorbed photons are then determined as the electrons increase in energy. An **absorption spectrum** in the visible region usually appears as a rainbow of color stretching from red to violet interrupted by a few black lines corresponding to distinct wavelengths of absorption.

In **atomic emission spectroscopy**, the electrons of an element are excited by heating or by an electric discharge. The frequencies of emitted photons are then determined as the electrons release energy. An **emission spectrum** in the visible region typically consists of lines of light at certain colors corresponding to distinct wavelengths of emission. The bands of emitted or absorbed light at these wavelengths are called **spectral lines**. **Each element has a unique line spectrum**. Light from a star (including the sun) may be analyzed to determine what elements are present.

The constituents of an atom include **protons** which have a positive charge, **neutrons** which have no charge, and **electrons** which have a negative charge. Atoms have no net charge and thus have an equal number of protons and electrons. Protons and neutrons are contained in a small volume at the center of the atom called the **nucleus**. Electrons move in the remaining space of the atom and have very little mass—about 1/1800 of the mass of a proton or neutron. Electrons are prevented from flying away from the nucleus by the attraction that exists between opposite electrical charges. This force is known as **electrostatic** or **coulomb** attraction. **Nuclear force**, which holds the nucleus together, is a byproduct of **strong interactions** and is clearly far stronger than electrostatic forces which would otherwise cause the protons in the nucleus to repel one another.

The identity of an **element** depends on the **number of protons** in the nucleus of the atom. This value is called the **atomic number** and it is sometimes written as a subscript before the symbol for the corresponding element. Atoms and ions of a given element that differ in number of neutrons have a different mass and are called **isotopes**. A nucleus with a specified number of protons and neutrons is called a **nuclide**, and a nuclear particle, either a proton or neutron, may be called a **nucleon**. The total number of nucleons is called the **mass number** and may be written as a superscript before the atomic symbol.

$^{14}_{6}C$ represents an atom of carbon with 6 protons and 8 neutrons.

The **number of neutrons** may be found by **subtracting the atomic number from the mass number**. For example, uranium- 235 has 235–92=143 neutrons because it has 235 nucleons and 92 protons. Different isotopes have different natural abundances and have different nuclear properties. Some nuclei are unstable and emit particles and electromagnetic radiation. These emissions from the nucleus are known as **radioactivity**; the unstable isotopes are known as **radioisotopes**; and the nuclear reactions that spontaneously alter them are known as **radioactive decay**. (For a discussion of nuclear reactions and binding energy see section V.7). Beta decay occurs as a consequence of the **weak interaction** force acting within the atomic nucleus.

Quantum #	Radius
$n \to \infty$	$r_\infty \to \infty$
$\vdots$	$\vdots$
$n = 5$	$r_5 = 25a_0$
$n = 4$	$r_4 = 16a_0$
$n = 3$	$r_3 = 9a_0$
$n = 2$	$r_2 = 4a_0$
$n = 1$	$r_1 = a_0$
$\oplus$ (H nucleus)	

An electron may exist at distinct radial distances (r_n) from the nucleus. These distances are proportional to the square of the **principal quantum number**, n. For a hydrogen atom (shown at left), the proportionality constant is called the **Bohr radius** ($a_0 = 5.29\times10^{-11}$ m). This value is the mean distance of an electron from the nucleus at the ground state of $n = 1$. The distances of other electron shells are found by the formula:

$$r_n = a_0 n^2 .$$

As $n\to\infty$, the electron is no longer part of the hydrogen atom. Ionization occurs and the atom become an H^+ ion.

A quantum of energy (ΔE) emitted from or absorbed by an electron transition is directly proportional to the frequency of radiation. The proportionality constant between them is **Planck's constant** ($h = 6.63\times10^{-34}$ J·s):

$$\Delta E = h\nu \text{ and } \Delta E = \frac{hc}{\lambda}.$$

The energy of an electron (E_n) is inversely proportional to its radius from the nucleus. For a hydrogen atom (shown below left), only the principle quantum number determines the energy of an electron by the **Rydberg constant** ($R_H = 2.18\times10^{-18}$ J):

$$E_n = -\frac{R_H}{n^2}.$$

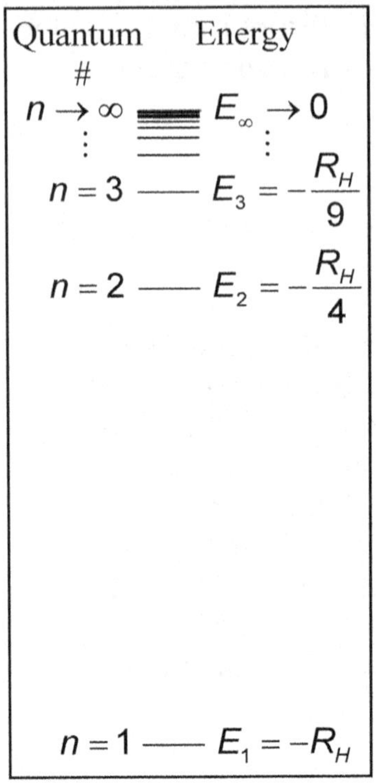

The Rydberg constant is used to determine the energy of a photon emitted or absorbed by an electron transition from one shell to another in the H atom:

$$\Delta E = R_H\left(\frac{1}{n_{initial}^2} - \frac{1}{n_{final}^2}\right).$$

When a photon is absorbed, n_{final} is greater than $n_{initial}$, resulting in positive values corresponding to an endothermic process. Ionization occurs when sufficient energy is added for the atom to lose its electron from the ground state. This corresponds to an electron transition from $n_{initial} = 1$ to $n_{final}\to\infty$. The Rydberg constant is the energy required to ionize one atom of hydrogen. Photon emission causes negative values corresponding to an exothermic process because $n_{initial}$ is greater than n_{final}.

Planck's constant and the speed of light are often used to express the Rydberg constant in units of s^{-1} or length. The formulas below determine the photon frequency or wavelength corresponding to a given electron transition:

$$\nu_{photon} = \left(\frac{R_H}{h}\right)\left|\frac{1}{n_{initial}^2} - \frac{1}{n_{final}^2}\right| \text{ and } \lambda_{photon} = \frac{1}{\left(\frac{R_H}{hc}\right)\left|\frac{1}{n_{initial}^2} - \frac{1}{n_{final}^2}\right|}.$$

These formulas **relate observed lines in the hydrogen spectrum to individual transitions** from one quantum state to another.

A simple optical spectroscope separates visible light into distinct wavelengths by passing the light through a prism or diffraction grating. When electrons in hydrogen gas are excited inside a discharge tube, the emission spectroscope shown below detects photons at four visible wavelengths.

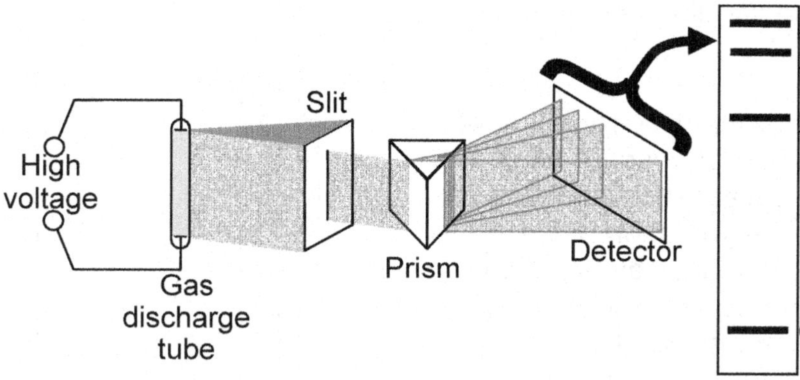

Visible spectral lines for hydrogen	Wavelength λ
Purple	411 nm
Blue	434 nm
Green	486 nm
Orange- red	656 nm

Every line in the hydrogen spectrum corresponds to a transition between electron energy levels. The four spectral lines from hydrogen emission spectroscopy in the visible range correspond to electron transitions from n = 3, 4, 5, and 6 to n =2 as shown in the table below.

Radiation type	Wavelength ×(nm)	Frequency ×(s^{-1})	Energy change ×E (J)	Electron transition $n_{initial} \to n_{final}$
Ultraviolet	≤397	≥7.55×10^{14}	≤-5.00×10^{-19}	∞→1, ... 2→1 ∞→2, ... 7→2
Purple	411	7.31×10^{14}	-4.84×10^{-19}	6→2
Blue	434	6.90×10^{14}	-4.58×10^{-19}	5→2
Green	486	6.17×10^{14}	-4.09×10^{-19}	4→2
Orange-red	656	4.57×10^{14}	-3.03×10^{-19}	3→2
Infrared and beyond	≥821	≤3.65×10^{14}	≥-2.42×10^{-19}	∞→3, ... 4→3 ∞→4, ... 5→4 ⋮

Electron transitions in the hydrogen atom

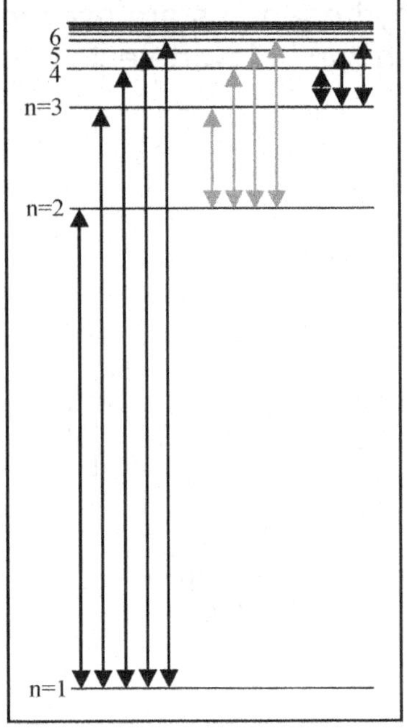

Most lines in the hydrogen spectrum are not at visible wavelengths. Larger energy transitions produce ultraviolet radiation and smaller energy transitions produce infrared or longer wavelengths of radiation. Transitions between the first three and the first six energy levels of the hydrogen atom are shown in the diagram to the right. The energy transitions producing the four visible spectral lines are colored grey.

Blackbody radiation is the characteristic radiation of an ideal blackbody, i.e. a body that absorbs all the radiation incident upon it. Theoretical calculations of the frequency distribution of this radiation using classical physics showed that the energy density of this wave should increase as frequency increases. This result agreed with experiments at shorter wavelengths but failed at large wavelengths where experiment shows that that the energy density of the radiation actually falls back to zero.

In trying to resolve this impasse and derive the spectral distribution of blackbody radiation, Max Planck proposed that an atom can absorb or emit energy only in chunks known as quanta. The energy E contained in each quantum depends on the frequency of the radiation and is given by $E = hf$ where Planck's constant $h = 6.626 \times 10^{-34}\, J.s = 4.136 \times 10^{-15} eV.s$. Using this quantum hypothesis, Planck was able to provide an explanation for blackbody radiation that matched experiment.

Einstein extended Planck's idea further to suggest that quantization is a fundamental property of electromagnetic radiation which consists of quanta of energy known as **photons**. The energy of each photon is hf where h is Planck's constant.

Skill 32.5 Explores real-world applications of quantum phenomena (e.g., lasers, photoelectric sensors, semiconductors, superconductivity).

The names "laser" and "maser" are acronyms for "light amplification by stimulated emission of radiation" and "microwave amplification by stimulated emission of radiation." Thus, as is evident by their names, these two devices are based on the principle of stimulated emission.

According to quantum theory, an atom in an excited state has a certain probability in a given time frame for relaxing to a lower state through, for example, the emission of a photon of the same energy as the energy difference between the two states. Stimulated emission, however, may take place when the excited atom is perturbed by a passing photon of the same energy as the excitation. In such a case, the atom relaxes to a lower energy by emitting a new photon, resulting in a total of two photons of equal frequency (and, thus, energy) and equal phase. That is to say, the photons are coherent.

In order to produce a significant level of stimulated emission for application in typical lasers and masers, population inversion is required. Population inversion is a situation in which there are more atoms in a particular excited state than in a particular lower-energy state. In order to achieve this condition, the material must be "pumped," which can be performed using electromagnetic fields (light). Pumping to produce population inversion often requires three or more atomic energy levels.

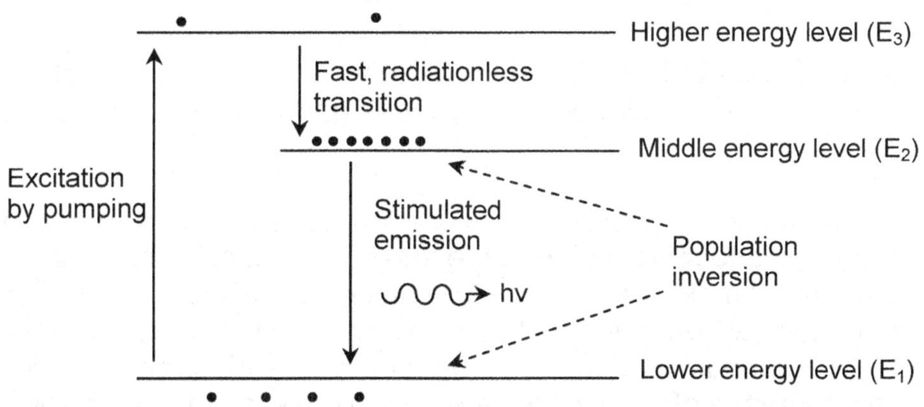

The above energy level diagram is for a so-called three-level laser (or maser). Other lasers may use more levels, depending on the material that is being pumped and the desired frequency output. The transition between the energy levels E_3 and E_2 is noted as being fast and radiationless; that is, almost as soon as the electron(s) of an atom are pumped to E_3, they decay to E_2 without radiating light. The energy difference can be emitted in the form of a phonon (vibration mode in the material, or heat).

The abundance of excited atoms due to population inversion allow for a "chain reaction" to form when photons of the proper frequency (proportional by Planck's constant to the energy difference between the excited state and lower-energy state) are incident. This results in a cascading increase in photon intensity through stimulated emission, thus producing the coherent, high-power light that is used by the laser or maser. This process has its limits, of course, as determined by how quickly the atoms can be pumped in relation to the relaxation processes. A saturation level exists beyond which the rate of stimulated emission cannot be increased. Spontaneously emitted photons can start the cascading process of stimulated emission.

The main components of a laser are a gain medium, an energy source (pump) for the gain medium and a resonant optical cavity with a partial transparency in one mirror. The resonant cavity is tuned to a particular frequency such that the photons of the desired laser frequency are coherent and, largely, isolated within the cavity. A partially transparent mirror on one end of the cavity allows some of the light to exit, thus producing the laser beam.

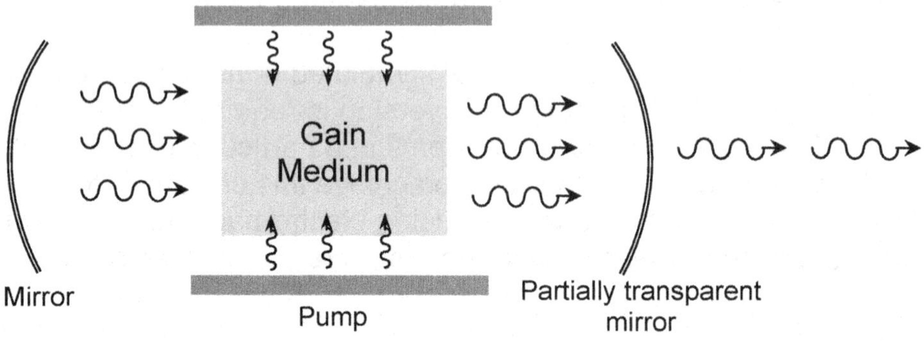

Masers, which operate at lower frequencies than lasers, generally rely on the same principles as lasers, although the types of gain media and resonant cavities may differ.

According to the theory of energy bands, as derived from quantum mechanics, electrons in the ground state reside in the valence band and are bound to their associated atoms or molecules. If the electrons gain sufficient energy, such as through heat, they can jump across the forbidden band (in which no electrons may exist) to the conduction band, thus becoming free electrons that can form a current in the presence of an applied field. For conductors, the valence band and conduction band meet or overlap, allowing electrons to easily jump to unoccupied conduction states. Thus, conductors have an abundance of free electrons. It is noteworthy that only two bands are shown in the diagram above, but that an infinite number of bands may exist at higher energies. A more general statement of the difference in band structures is that, for insulators and semiconductors, the valence electrons fill up all the states in a particular band, leaving a gap between the highest energy valence electrons and the next available band. The difference between these two types of materials is simply a matter of the "size" of the forbidden band. For conductors, the band that contains the highest energy electrons has additional available states.

A nearly ideal conducting material is a superconductor. As a material increases in temperature, increased vibrational motion of the atoms or molecules leads to decreased charge carrier mobility and decreased conductivity. In the case of semiconductors, the increase in free carrier population outweighs the loss in mobility of the charge carriers, meaning that the semiconductor increases in conductivity as temperature increases. Superconductors, on the other hand, reach their peak conductivity at extremely low temperatures (although there are currently numerous efforts to achieve superconductivity at higher and higher temperatures, with room temperature or higher being the ultimate goal). The critical temperature of the material is the temperature at which superconducting properties emerge. At this temperature, the material has a nearly infinite conductivity and maintains an almost perfect equipotential across its surface when in the presence of a static electric field. Inside a superconductor, the electric field is virtually zero at all times. As a result, the time derivative of the electric flux density is zero, and, by Maxwell's equations, the magnetic flux density must likewise be zero.

$$\nabla \times \mathbf{H} = \frac{\partial \mathbf{D}}{\partial t} + \mathbf{J}$$

Since the electric field is also zero, the current density **J** inside the superconductor must also be zero (or very nearly so). **This elimination of the magnetic flux density inside a superconductor is called the Meissner effect**.

DOMAIN IX. SCIENCE LEARNING, INSTRUCTION, AND ASSESSMENT

COMPETENCY 33.0 THE TEACHER UNDERSTANDS RESEARCH-BASED THEORETICAL AND PRACTICAL KNOWLEDGE ABOUT TEACHING SCIENCE, HOW STUDENTS LEARN SCIENCE, AND THE ROLE OF SCIENTIFIC INQUIRY IN SCIENCE INSTRUCTION.

Skill 33.1 Knows research-based theories about how students develop scientific understanding and how developmental characteristics, prior knowledge, experience, and attitudes of students influence science learning.

There is a lot of debate about how students develop scientific understanding and how various factors such as developmental characteristics influence their learning of science.

In simple terms, explanations lead to understanding. The more science is explained to the students in simple, plain terms, the better they understand it. We conclude that students develop understanding of science through proper explanation. Too often it happens that an activity is done without explaining the theoretical part of it - answering the "Why" and "How" questions. When an activity is done, it has to be explained - otherwise one is simply entertaining the students.

Various factors influence understanding of science. The most important thing is that if we want good scientists or science learners, the foundation has to be laid at an early stage. Whatever they are taught has to be correct and contextualized.

Students' prior knowledge is important to their learning of science, as is their experience. Experience can be explained in different ways. The students must have good experiences in science learning. That is, they must relate learning science to positive experience.

Good attitude is important to learning anything, including science. Many students believe that science is difficult and boring. It falls upon the teacher to be creative and make science interesting.

Skill 33.2 Understands the importance of respecting student diversity by planning activities that are inclusive and selecting and adapting science curricula, content, instructional materials, and activities to meet the interests, knowledge, understanding, abilities, and experiences of all students, including English Language Learners.

The term diversity is defined as the presence of a wide range of variation in the qualities or attributes under discussion. In the human context, particularly in a social context, the term diversity refers to the presence in one population of a variety of cultures, ethnic groups, languages, physical features, socio-economic backgrounds, religious faiths, sexuality, gender identity and neurology. At the international level, diversity refers to the existence of many peoples contributing their unique experiences to humanity's culture.

The teacher, as an adult, is responsible for recognizing the diversity of the students in a class, respecting their cultures and abilities and planning lessons keeping in mind that in some cases their first language is not English.

Cultural and ethnic diversity

The teacher must be both politically correct when handling such students and compassionate and empathetic, since it is a challenge to settle in a different country and call it home. It is important to incorporate different cultural identities into lesson plans and connect them with science. For example, studying the contributions made to science by Latino scientists, African-American scientists, Native American scientists, Asian scientists, etc. In February, for instance, we may study about famous scientists of African-American origin. When this is done, the students appreciate the time and effort taken to recognize their heritage.

Decorating the classroom using ethnic material is both intellectually interesting and also creates an atmosphere of acceptance. Such efforts require a bit of time and ingenuity, but they go a long way in establishing good relationships with students and their families.

Students with disabilities

Increasingly, educators have noticed that learning disabilities must be attended to. Such disabilities may include auditory processing disabilities, attention deficit hyperactivity disorder, visual processing disabilities (including varying degrees of blindness), autism, etc.

If a teacher notices abnormal behavioral or physical attributes of a student, it would be important to determine if this student has an IEP (Individualized Education Program) or other records that would determine whether or not this student has been evaluated. If not, the teacher might have a brief discussion with a specialist (such as the school's Special Education teacher) to determine if the noticed behavioral or physical attribute should be of concern.

Usually, at this point, if the Special Education teacher feels there is a problem, he or she will recommend the next course of action. Typically, interventions consist of instructional modifications (where the regular teacher keeps the student but modifies instruction to meet the student's needs), pull-out instruction (where a Resource Specialist, for example, may collaborate with the teacher and pull the student out to assist in certain academic areas where needed), full-day Special Education instruction (where a student is not in class with the other students, but rather in a special class designed to better assist that student), or removal from the current school (for example, students may be sent to schools that are better able to assist them with their particular needs, such as blind students).

Skill 33.3 Knows how to plan and implement strategies to encourage student self motivation and engagement in their own learning (e.g., linking inquiry-based investigations to students' prior knowledge, focusing inquiry-based instruction on issues relevant to students, developing instructional materials using situations from students' daily lives, fostering collaboration among students).

There are two important components of teaching science. The first is theory: explaining the lesson and answering why, how, when, what, and which; out of these, why and how are the most fundamental. The second component is practical exploration, which is *doing* – gaining knowledge through experimentation. In addition to other forms of diversity, students learn material differently. Some learn better by reading, others by listening, still others by doing. Theory and practical exploration must be balanced in order for a student to understand science fully.

Students become motivated when they can contribute to their own learning. Inquiry-based instruction, in which students are encouraged to ask questions and create opportunities to find answers for those questions, is becoming popular as a means to get students more involved in science. The best way to find answers for some of their questions is to let them investigate, experiment, and find the answers themselves.

Linking of ideas is very important because the students' prior knowledge is taken into consideration. Each step of instruction is planned based on the prior knowledge of the students. When students have not got the knowledge they are expected to have for their grade level, remedial work must be done; otherwise, the students will not be able to benefit from subsequent lessons. The teacher should be aware of what has been taught in previous science courses the students are likely to have taken, as well as what is currently being taught in their other science, math, and reading courses. The more these disciplines can be linked together, the more the knowledge will be reinforced.

Another way to motivate students is to ensure that the lesson has some practical relevance to their lives. Science has to be contextualized. For example, take the question of which fertilizer is best for their lawn. This might be made into an investigation that tells them which one is best and also teaches them a bit of economics – if a fertilizer is very good and very expensive, it may not be worth it for the average person. If the fertilizer is good *and* is reasonably priced, that would be the best choice. By contrast, studying about exotic plants in a remote area in Asia probably would not generate as much interest.

Skill 33.4 Knows how to use a variety of instructional strategies to ensure all students comprehend content-related texts, including how to locate, retrieve, and retain information from a range of texts and technologies.

The word strategy means a careful plan or method. Instructional strategies are plans and methods used in teaching students. Success in teaching lies in using a variety of strategies to keep the students interested.

Let's examine some of the instructional strategies employed by educators:

Lecture

This is an activity in which the teacher presents the information and knowledge orally through a series of organized and structured explanations. Student involvement is the lowest of all the strategies, and therefore, this approach should be used sparingly and in combination with other activities. Lectures can be either formal or informal. In formal lectures, the student interaction is non-existent. In informal lectures there is more interaction, and there is an increase of 20% in student retention of information.

There are ways to make the lecture method more interesting and beneficial to the students.

1. Feedback lecture – lecture lasts for only 10 minutes and then the students are divided into study groups to talk about what was discussed; students also have an opportunity to study the notes before the lecture.

2. Guided lecture – lecture lasts for 20 minutes, followed by discussion in small groups.

3. Responsive lecture – answering open-ended questions and student-generated questions.

4. Demonstration lecture – a lab/activity is demonstrated as part of the lecture.

5. Pause procedure lecture – after every 5-6 minutes of lecture, the students are given 2 minutes to compare notes with their peers and fill in any missing information.

6. Think / Write / Discuss lecture – starts with a critical thinking question; students are questioned about the topic during the lecture and at the end they are questioned again to find out how much they understood.

Mnemonic strategy

Mnemonic strategies are memory aids that provide a very systematic approach for organizing and remembering facts that have no apparent link or connection of their own. An example of a mnemonic strategy is a way of remembering the order of the planets from the sun, where the first letter of each word corresponds to the letter of a planet: My Very Excellent Mother Just Served Us Nine Pies (others can be used if you wish to leave off Pluto). These can be funny and engage the imagination of students to help them remember material that must be memorized.

Re Quest

Re Quest is a strategy that fosters active rather than passive reading of a text. This strategy provides a structure for the students to ask questions about the learning. Both teacher and student ask each other questions about what they are doing in terms of the lesson.

Reciprocal teaching

This is an instructional activity in the form of interactive dialogue between the teacher and the students regarding the segments of the text. The dialogue involves four strategies: 1) summarizing, 2) question generating, 3) clarifying, and 4) predicting. The teacher guides the students in all of the steps.

Technology can be an important component of learning. The Internet-linked computer helps us in locating information and retrieving it. It is important for the students to be familiar with various websites which give academic information. However it is equally important for the students to locate information in books and journals. Students should be introduced to various journals in their core subjects, which provide the latest research. A school or library subscription to Science, Nature, or National Geographic would be a good place to start for finding up-to-date and accessible articles on a variety of topic areas. They can be encouraged to bring interesting articles to class that they have found on the Internet or in journals, and write or speak about them.

Skill 33.5 Understands the science teacher's role in developing the total school program by planning and implementing science instruction that incorporates schoolwide objectives and the statewide curriculum as defined in the Texas Essential Knowledge and Skills (TEKS).

Science needs to be contextualized in order for the students to relate to it. It is much easier for the students to be interested in it when it is connected to their everyday lives. The role of the science teacher is to facilitate this by planning and participating in school-wide science programs. These programs cross various disciplines and are group learning activities, though they seem simply like fun activities. Science teachers may also participate in supporting other school-wide learning goals, such as interdisciplinary reading, writing, and math competencies.

The following are some ideas which could be used as school wide science programs:

Butterfly garden: Butterflies are attractive and most of the students would be interested in them. Their beauty and movement are worth observing and their biological transformations interesting. A committee consisting of student representatives from all grade levels may be formed. Initial planning includes raising money to carry out this project; organizing a group of students across the school; preparing a small area to grow plants that would attract butterflies to lay eggs; and monitoring progress daily, observing and recording. This project lasts a few months. Another way to do this project would be to separate the students into a number of groups that could pool their observations and come up with a more complete and statistically robust set of observations.

Weather station: This is another great idea for involving the whole school. A group of students can measure and record the rainfall in a year, wind speed and direction, temperature, humidity, atmospheric pressure and any other variables of interest. Correlations can be looked for, such as a relationship between atmospheric pressure and certain weather types.

Preserving natural resources: Students are made aware of the implications of using natural resources meaningfully and economically. For instance, preserving forests starts with using paper economically, though we can afford to buy plenty of it. Awards could be instituted to reward students who come up with good ideas for school-wide conservation of resources. Another example of a way to teach students about the natural environment might be to count species diversity in a number of vacant lots or backyards, and then do the same in the same number of natural habitats.

A science teacher should also take advantage of natural phenomena such as hurricanes, volcanic eruptions, and tornadoes and teach the students about their causes. Students can track hurricane pathways and monitor volcanic eruptions.

Skill 33.6 Knows how to design and manage the learning environment (e.g., individual, small-group, whole-class settings) to focus and support student inquiries and to provide the time, space, and resources for all students to participate in field, laboratory, experimental, and nonexperimental scientific investigation.

The learning environment is a very important factor in teaching any discipline. Ideally, the learning environment should be designed in such a way that the students are motivated to learn. When we talk about a learning environment, we are talking about the learning environment for the regular students. There is always a small number of students who are disruptive and interfere with the learning of other students, and these students should be handled in a different manner depending on their individual needs. It is the responsibility of the teacher to make the environment in the classroom suitable for various types of learning – individuals, pairs, small groups, or whole class setting.

The physical environment, such as the arrangement of furniture, should be modified to support the activity, e.g., if the teacher has planned collaborative pairs for learning, the tables can be moved around to achieve that. This should be done before the lesson or activity starts.

Individual teaching or independent study is a good approach for exceptional students and those who need more attention than the regular student. A few minutes of explaining the lesson or the task on hand will be very helpful. In the case of pair share or collaborative pairs, a small assignment could be given and a time frame set, at the end of which students will share as a class what they have learned. This is very good if an exceptional student and a bright student are paired. A small group is very productive as it teaches positive behaviors, including sharing information, waiting for one's turn, listening to others' ideas, views, and suggestions, and taking responsibility for doing a job in the group (writing/presenting/drawing etc.). Teaching in a whole-class setting involves traditional and modern methods such as lecture, lecture/demonstration, pause and lecture, etc. The same applies for experimental, field, and nonexperimental work, although some of these can be assigned as homework.

Today's learning, especially science, is largely inquiry-based. It is an important part of teaching to encourage the students to ask questions. Sufficient time must be given to students to ask these questions.

As a teacher, one must be a good manager, not only of the classroom but also of time, resources, and space. The teacher needs to plan how much time and energy should be given to exceptional students, bright students, regular students, and disruptive students. The exceptional and the disruptive students typically receive more of the teacher's time. Next will be the regular students, and last the bright students, since they typically need less supervision and help. If they finish work quickly, however, bright students need to be engaged, so some extra work must be available. In terms of space the same things apply. Resources must be shared equally as far as possible, since everybody has the right to have equal opportunity. However, there may be modification of resources suitable for the exceptional students, if required and available.

One thing is most important – a teacher must use logic and be able to think laterally since all the answers are not in books. The best teaching is part original thinking and part innovation and ingenuity.

Skill 33.7 Understands the rationale for using active learning and inquiry methods in science instruction and how to model scientific attitudes such as curiosity, openness to new ideas, and skepticism.

Learning can be broadly divided into two kinds - active and passive. Active learning involves, as the name indicates, a learning atmosphere full of action whereas in passive learning students are taught in a nonstimulating and inactive atmosphere. Active learning involves and draws students into it, thereby interesting them to the point of participating and purposely engaging in learning.

It is crucial that students are actively engaged, not entertained. They should be taught the answers for "How" and "Why" questions and encouraged to be inquisitive and interested.

Active learning is conceptualized as follows:

A Model of Active Learning

Experience of	Dialogue with
Doing	Self
Observing	Others

This model suggests that all learning activities involve some kind of experience or some kind of dialogue. The two main kinds of dialogue are "dialogue with self" and "dialogue with others." The two main kinds of experience are "observing" and "doing."

Dialogue with self: Students think reflectively about a topic. They ask themselves a number of questions about the topic and try to find the answers.

Dialogue with others: When the students are listening to a book being read by another student or when the teacher is teaching, a partial dialogue takes place because the dialogue is only one-sided. When there is an exchange of ideas back and forth, either with other students or with the teacher, it becomes a dialogue with others.

Observing: This is a most important skill in science. This occurs when a learner is carefully watching or observing someone else doing an activity or experiment. This is a good experience, although it is not quite like doing the experiment personally. Observing can be made into a richer activity by enlisting the observer to record something about the experiment and using the recorded information as part of the analysis of data after the experiment.

Doing: This refers to any activity where a learner actually does something, giving the learner a firsthand experience that is very valuable.

Inquiry is invaluable to teaching in general and to teaching science, especially. The steps involved in scientific inquiry are discussed in Skill 16.8.

The scientific attitude is to be curious, open to new ideas, and skeptical. In science, there are always new research results, new discoveries, and new theories proposed. Sometimes, old theories are disproved. To view these changes rationally, one must have such openness, curiosity, and skepticism. Skepticism is a Greek word, meaning a method of obtaining knowledge through systematic doubt and continual testing. A scientific skeptic is one who refuses to accept certain types of claims without subjecting them to a systematic investigation. This should be differentiated from irrational skepticism not based on evidence.

The students may not have these attitudes inherently, but it is the responsibility of the teacher to encourage, nurture, and practice these attitudes so that students will have a good role model and gain experience applying them.

Skill 33.8 Knows principles and procedures for designing and conducting an inquiry based scientific investigation (e.g., making observations; generating questions; researching and reviewing current knowledge in light of existing evidence; choosing tools to gather and analyze evidence; proposing answers, explanations, and predictions; and communicating and defending results).

Science investigations in the classroom environment are very important because they are something a student can do by himself/herself with assistance from the teacher. Through such "active learning," the student gains experience and knowledge.

Scientific investigations are carried out by the method generally known as the Scientific Method (see Skill 2.1). The Scientific Method is composed of a series of steps to solve a problem. We use this method in order to eliminate, to the extent possible, our preconceived ideas, prejudices, and bias. When students are explained the purpose of the method, they will be able to appreciate it better.

The Scientific Method is an inquiry-based method. It consists of the following steps:

1. Problem / Question: In order to investigate, we must have a problem or question to begin with. A problem or question may come from observing, one of the most important skills in science, or from theory. The problem needs to be communicated in clear terms and in simple language, so that anybody reading it will be able to understand it.

Research / Gathering information

There are number of resources available to students - websites, scientific journals, magazines, books, and people who are knowledgeable and experienced, which may help them research what is already known about the topic.

Formulating a hypothesis

This is also known as making an educated guess. An informed guess is made regarding a possible solution to the problem, which will be tested as part of the experiment.

Experimental design

Conducting an experiment is very exciting and interesting, but designing the experiment well is a challenging task. It is important that students understand this. They should not rush to do an experiment, but must have a clear understanding of the different elements of an experiment: identifying the control/standard, determining the constants, and deciding on the independent and dependent variables. Involving students in the experimental design rather than having it pre-determined will greatly increase their understanding and appreciation of the experiment.

An ideal experiment at the high school level should not last more than 12-14 days.

Collection of data

Through conducting an experiment, we acquire data. The data should be organized and visually presented, in tabular form and/or graphically (as shown in Skill 1.5).

Analysis of the data

The data should be analyzed to test the hypothesis and identify interesting patterns. Numbers are important, but they are typically not as useful or enlightening as patterns in the data.

Drawing conclusions

In the conclusion the investigator attempts to provide plausible answers to the initial question, determine whether the hypothesis was correct, summarize any trends or patterns observed, and make suggestions regarding subsequent research.

Communication

Oral and written communication skills are a necessity for anybody pursuing research. Effective oral communication is needed to present the research in front of the classroom or school. Written communication is needed to present a report on the research. It should be made clear to students that, in this age of communication, those who cannot communicate effectively will be left behind. Accordingly, the evaluation system should make provision for communication skills and activities.

Defending results

Defending results is as important as conducting an experiment. One can honestly defend one's own results only if the results are reliable, and experiments must be well-controlled and repeated at least twice to be considered reliable. It must be emphasized to the students that *honesty and integrity are the foundation for any type of investigation.*

Skill 33.9 Knows how to assist students with generating, refining, focusing, and testing scientific questions and hypotheses.

Scientific questions are the starting point for learning. Students should be encouraged, provoked and challenged to ask good questions.

First, students need to learn to **frame** questions. The following are just a few ways in which the students can be encouraged to do this:

1. Brainstorming the topic under study
2. Discussing a topic in the class and inviting students to ask questions
3. Having students discuss the topic in small groups and come up with their own questions (this is extremely useful to students who are introverts and shy by nature).

In the case of students who are not immediately curious and inquisitive, the teacher needs to show patience in encouraging these traits, which are almost always achievable over time.

The next step is teaching students to **refine** their questions. By now the students have learned to ask questions, but the questions may not be well-designed to generate subsequent discussion. It is the responsibility of the teacher to take these questions and to convert them to "how" and "why" type, open-ended questions. For example, students may begin by asking closed-ended questions, such as, "Who landed on the moon?" This sort of question does not really generate a great amount of knowledge or provoke further thought. The teacher can modify this question to, "What did the missions to the moon accomplish?" With this type of question, a lot of subsequent discussion will be generated.

The next step is **focusing** the questions on a specific topic under discussion or investigation. Focusing is important because students are easily carried away or sidetracked. For a scientific investigation, the question also needs to focused in such a way as to generate a testable hypothesis.

The last step is **testing** scientific questions and their potential answers (hypotheses). Not all questions can be tested, but many questions can be tested and answers found by research or experimentation – for example, the question "Which fertilizer is best for rose cuttings?" Once research or experimentation has been conducted, hypotheses are either supported or disproved. If the hypothesis is disproved it must be modified.

Skill 33.10 Knows strategies for assisting students in learning to identify, refine, and focus scientific ideas and questions guiding an inquiry-based scientific investigation; to develop, analyze, and evaluate different explanations for a given scientific result; and to identify potential sources of error in an inquiry-based scientific investigation.

Some strategies for assisting students in designing effective investigations and evaluating hypotheses are:

- **Brainstorming:** This is very effective when done in a relaxed atmosphere with no pressure on the students. Most students want to contribute and be recognized for their contributions.

- **Induction and deduction:** Induction is drawing conclusions based on facts or observations. Deduction is drawing conclusions based on generalizations.

- **Abstraction:** This very important strategy helps students to recognize and identify patterns and to connect them to their prior knowledge.

- **Questioning:** Allowing students to evaluate relevant questions helps them learn the technique of questioning. Focusing and refining questions are discussed in Skill 16.9. The end result of focused and refined questions is a good understanding of the problem on hand, which in turns produces an effective science investigation.

To identify errors in a science experiment, the student should be encouraged to do the following:

- Check whether all the steps of the scientific method were followed in the correct order.
- Check that the control, constants, independent variables, and dependent variable were correctly identified. Many times the variables and constants may be mixed up.

Review other sources of error discussed in Skill 3.2 and double-checks each step, such as calibration or measurement errors, transcription errors, calculation errors, and presentation errors.

Skill 33.11 Understands how to implement inquiry strategies designed to promote the use of higher-level thinking skills, logical reasoning, and scientific problem solving in order to move students from concrete to more abstract understanding.

Inquiry learning provides opportunities for students to experience and acquire thought processes through which they can gather information about the world. This requires a higher level of interaction among the learner, the teacher, the area of study, available resources, and the learning environment. Students become actively involved in the learning process as they:

1. Act upon their curiosity and interests
2. Develop questions that are relevant
3. Think their way through controversies or dilemmas
4. Analyze problems
5. Develop, clarify, and test hypotheses
6. Draw conclusions
7. Find possible solutions

The most important element in inquiry-based learning is questioning. Students must ask relevant questions and develop ways to search for answers and generate explanations. Higher-order thinking is encouraged.

Here are some **inquiry strategies**:

Deductive inquiry

The main goal of this strategy is moving the student from a generalized principle to specific instances. The process of testing general assumptions, applying them, and exploring the relationships between specific elements is stressed. The teacher coordinates the information and presents important principles, themes, or hypotheses. Students are actively engaged in testing generalizations, gathering information, and applying it to specific examples.

Inductive inquiry

The information-seeking process of the inductive inquiry method helps students to establish facts, determine relevant questions, and develop ways to pursue these questions and build explanations. Students are encouraged to develop and support their own hypotheses. Through inductive inquiry, students experience the thought processes which require them to move from specific facts and observations to more general inferences.

Interactive instruction

This strategy relies heavily on discussion and sharing among participants. Students develop social skills, learning from teacher and peers. They also learn organizational skills. Examples are debates, brainstorming, discussion, laboratory groups, etc.

Direct instruction

This is highly teacher-oriented and is among the most commonly used strategies. It is effective for providing information or developing step-by-step skills. Examples are lectures, demonstrations, explicit teaching, etc.

Indirect instruction

This is mainly student-centered. Direct and indirect instruction strategies can compliment each other. Indirect instruction seeks a high level of student involvement such as observing, investigating, drawing inferences from data, or forming hypotheses. In this strategy, the role of the teacher shifts from that of teacher/lecturer to that of facilitator, supporter, and resource person. Examples are problem solving, inquiry, concept formation, etc.

Independent study

Independent study refers to the range of instructional methods which are purposely provided to foster the development of individual student initiative, self reliance, and self improvement. Examples are independent research projects, homework, etc.

The previously mentioned strategies promote higher-level thinking skills such as problem solving, synthesizing (hypothesizing), designing (identifying the problem), analyzing data in an experiment, and connecting (logical thinking).

Skill 33.12 Knows how to guide students in making systematic observations and measurements.

The starting point for any science is systematic observation. Systematic observation is observing and recording the occurrence of certain specific (naturally occurring or experimental) behaviors.

There are four descriptive observation methods:

1. Naturalistic observation: observers record occurrence of naturally occurring behavior.

2. Systematic observation: observers record the occurrence of certain specific behaviors.

3. Case study: gather detailed information about one individual.

4. Archival research: use existing information to establish occurrence of behavior.

With each type of approach, there are potential problems and limitations. We discuss here systematic observation.

Systematic observation emphasizes gathering quantitative data on certain specific behaviors. The researcher is interested in a limited set of behaviors. This allows them to study and test specific hypotheses.

The first step is to develop a coding system. The coding system is a description of behaviors and how they will be recorded. The key idea is to delimit the range of behaviors that are observed. The operational definitions of each behavior that will be recorded are defined, and occurrence of each behavior and its duration is recorded.

For example, consider the recording of animal behavior and its comparison to laboratory experiments in chemistry. Potential problems that can occur during the observation include:

1. Remaining vigilant: Following an animal in its natural habitat is difficult, especially when human presence is required. Recording devices (audio, video) are used to deal with this problem. In a laboratory experiment such as titration, a similar difficulty can arise – the titration point can be missed if the observer is not vigilant or uses too large a volume of titrant for each change.

2. Reactivity: Humans and animals often change their behavior when they are being observed. The observer must take steps to remain unobtrusive or become a participant observer. This problem should not be an issue in high school chemistry labs, but is a fundamental feature of the behavior of quantum particles.

3. Reliability: Ensuring that the coding of behavior is accurate (two or more observers are used consistently and their results compared). This is known as inter-rater or inter-observer reliability. In the laboratory, students should be sure of what variables they are measuring, what instruments they should use to perform the measures, and what units will be used to record the data.

4. Sampling: Setting up a schedule of observation intervals, using multiple observations over a range of time. This measures the behavior of an animal over a period of time and is considered to be reliable. Similarly, in physical/chemical systems, random error can be reduced by conducting measurements multiple times and taking the average of the measurements.

It is the responsibility of the teacher to introduce the process of systematic observation to students in a meaningful way. It is also the responsibility of the teacher to instruct students on how to take correct scientific measurements.

Skill 33.13 Knows how to sequence learning activities in a way that uncovers common misconceptions, allows students to build upon their prior knowledge, and challenges them to expand their understanding of science.

There are many common misconceptions about science. The following are a few scientific misconceptions that are or have been common in the past:

- The Earth is the center of the solar system.
- The Earth is the largest object in the solar system
- Rain comes from the holes in the clouds
- Acquired characteristics of species can be inherited
- The eye receives upright images
- Heat is not energy

Some strategies to uncover and dispel misconceptions include:

1. Planning appropriate activities, so the students can see for themselves where there are misconceptions.

2. Web search is a very useful tool to dispel (or reinforce) misconceptions. Students should be guided in how to look for answers on the Internet and learn to differentiate between "junk" science and useful information. If necessary the teacher should explain scientific literature to help the students understand it.

3. Science journals are a great source of information. Recent research is highly beneficial for the senior science students.

4. Critical thinking and reasoning are two important skills that the students should be encouraged to use to discover facts – for example, that heat is a form of energy. Here, the students have to be challenged to use their critical thinking skills to reason that heat can cause change – for example, causing water to boil – and so it is not a thing but a form of energy, since only energy can cause change.

COMPETENCY 34.0 THE TEACHER KNOWS HOW TO MONITOR AND ASSESS SCIENCE LEARNING IN LABORATORY, FIELD, AND CLASSROOM SETTINGS.

Skill 34.1 Knows how to use formal and informal assessments of student performance and products (e.g., projects, laboratory and field journals, rubrics, portfolios, student profiles, checklists) to evaluate student participation in and understanding of inquiry-based scientific investigations.

Assessment is the act of observing an event and making a judgment about its status. The purpose of assessment in teaching is to help the teacher determine how well students are learning and progressing.

There are four main kinds of assessment.:

1. **Observation:** Watching someone or something and judging their actions.

2. **Informal continuous assessment** (less structured): Informal continuous assessment is informal because it is not formal like a test or exam. It is continuous because it occurs regularly, on a daily or weekly basis.

3. **Informal continuous assessment** (more structured): More structured informal continuous assessment consists of setting up *assessment situations* periodically. An assessment situation is an activity that is organized specifically so that the learners can be assessed. It could be a quiz, for example, or a monitored group activity.

4. **Formal assessment:** A structured, infrequent measure of learner achievement, the use of tests and exams to measure the learner's progress.

Informal assessment can be applied to homework assignments, field journals, or daily classwork, all of which are good indicators of student progress and comprehension.

There are four steps involved in informal continuous assessment:

1. **Observation:** Noticing whether there is a problem (as in homework).

2. **Diagnosis:** Identifying problems observed (e.g., the learner can correctly solve problems from two weeks ago but not from yesterday's homework, indicating a recent problem).

3. **Consolidation:** Helping the learner develop knowledge or skills needed to complete the task. With help from the teacher, the learner completes the task successfully.

4. **Follow up assessment:** Continuing to evaluate the learner's work.

Evaluating students' understanding of inquiry-based learning is an informal assessment. This involves a high degree of organizational skills from the teacher. A notebook recording observations, problems, and progress may be maintained, student participation and understanding continuously monitored, and help given as necessary.

Formal assessment, on the other hand, is highly structured. It must be done at regular intervals and if progress is not satisfactory, parental involvement is essential. Tests, exams, and graded science projects are good examples of formal assessment.

Skill 34.2 Understands the relationship between assessment and instruction in the science curriculum (e.g., designing assessments to match learning objectives, using assessment results to inform instructional practice).

Assessment and instruction are very closely connected because when educators are better informed of the learning progress of their students, they can make better decisions about what a student needs to learn next and how to teach that material in a manner that will maximize the students' learning outcomes.

Educators make three types of decisions using assessment skills:

1. **Instructional placement decisions:** What the student knows and where he or she should be in the instructional sequence, i.e., what to teach next.

2. **Formative evaluation decisions:** Information used to monitor a student's learning while an instructional program is underway, e.g., how quickly progress is being made, whether the instructional program is effective, whether a change in instructional program is needed to promote the learning of the student.

3. **Diagnostic decisions:** Finding out which specific problems are responsible for a student's inadequate progress and designing an instructional plan that is more effective in enhancing the learner's progress.

There are various methods of linking assessment to instruction, including behavioral assessment, mastery learning, and curriculum-based measurement.

Behavioral assessment

This approach is based on observing target behaviors, using repeated observations, e.g., behavior during science activities, following laboratory rules, cooperating, social skills such as sharing equipment, working as a team, etc. This assessment contributes in part to the overall assessment. It is an excellent assessment tool for students of science, which requires many such social skills.

Mastery learning

The curriculum is broken down into a set of subskills, which are prioritized according to their importance. Each student is assessed according to how well he or she has mastered each subskill. This method has limitations and restricts instructional methods, and is often impractical for obtaining a broader understanding of science.

Curriculum-based measurement (CBM)

This method allows students to know in advance how they will be assessed. Teachers use a variety of techniques and the assessment methods are manageable. In a science class, this approach could be implemented for daily lessons. It is especially useful for evaluating large assignments such as science projects, writing assignments, etc., where the students are given the criteria for assessment beforehand. This helps the students monitor and complete their assignments keeping in mind the assessment criteria. Though this method is traditionally intended for assessments over longer periods of time, it can be successfully applied to lengthier science projects.

Performance assessment

This is a relatively new but promising method. It incorporates three important elements: students construct responses, assessment allows teachers to observe student behavior on tasks, and scoring reveals patterns in students' learning and thinking.

- Students are encouraged to come up with their own responses, which gives them the freedom to think laterally and critically to analyze a problem and to present a solution.
- Teachers observe student behavior closely, which is very important in a science class, where experiments are sometimes done using chemicals, live animals, etc. In these situations, students need to behave properly since their own safety and that of others are involved. It is important that the students be observed very carefully and any unwanted and inappropriate behavior corrected immediately.
- Scoring reveals patterns in students' learning and thinking. This is similar to formal assessment, where students are tested for their comprehension of concepts and retention of knowledge.

The key element in using assessment methods is not to be restricted by a certain approach. If necessary, teachers should combine more than one method and try new methods as they become available. At the end of the day, what is important is not how students were taught but how much they understood and retained, and this can only be known through assessment.

Skill 34.3 Knows the importance of monitoring and assessing students' understanding of science concepts and skills on an ongoing basis by using a variety of appropriate assessment methods (e.g., performance assessment, self-assessment, peer assessment, formal/informal assessment).

Assessment is evaluating student progress. To be fair and effective, assessment has to be continuous and the teacher needs to use a variety of assessment methods to eliminate assessment bias.

Any assessment approach should meet the following seven criteria:

1. Measure important learning outcomes.
2. Address each of the following three purposes: 1) students construct rather than select responses, 2) assessment formats allow teachers to observe student behavior on tasks that are of applied nature, and 3) scoring reveals patterns of student learning and thinking.
3. Provide clear descriptions of student performance that can be linked to instructional actions.
4. Be compatible with a variety of instructional models.
5. Be easily administered and scored by teachers.
6. Communicate the goals of learning to teachers and students.
7. Generate accurate, meaningful (reliable and valid) information.

In addition to **performance assessment** and the various methods of formal and informal assessment discussed in Skill 34.1, teachers may use self assessment and peer assessment methods.

Self and peer assessment are often combined or considered together. Peer assessment is assessment of students by other students. It includes providing feedback and summative grading. Peer assessment is one form of innovative assessment which aims to improve the quality of learning and empower learners, where traditional assessment methods can bypass learners' needs. Peer assessment can be considered part of peer tutoring.

Advantages of self and peer assessment include:

1. Giving students a sense of ownership and thereby motivation and responsibility.
2. Treating assessment as part of learning so mistakes become opportunities rather than failures.
3. Learning evaluation skills.
4. Encouraging deep learning.
5. Using external evaluation as model for internal self assessment of a student's own learning.

Self and peer evaluation provide lifelong learning by helping students evaluate their own and peers' achievements realistically.

Teachers need to be aware of various assessment instruments and use them meaningfully. They should be adventurous and experiment with assessment methods.

Skill 34.4 Understands the purposes, characteristics, and uses of various types of assessment in science, including formative and summative assessments, and the importance of limiting the use of an assessment to its intended purpose.

"Teaching and learning are reciprocal processes that depend on and affect one another. Thus, the assessment component deals with how well the students are learning and how well the teacher is teaching" (Kellough and Kellough 1999, p. 417).

There are seven purposes of assessment:

1. To assist student learning
2. To identify students' strengths and weaknesses
3. To assess the effectiveness of a particular instructional strategy
4. To assess and improve the effectiveness of curriculum programs
5. To assess and improve teaching effectiveness
6. To provide data that assists in decision making
7. To communicate with and involve parents

Good assessment has the following attributes:

- Good educational values
- Clear and well-defined goals
- Continuous
- Contextualized
- Makes educators responsible to stakeholders
- Involves parents, educators, administration, and students

There are 3 main types of assessment:

1. Diagnostic assessment: The purpose of diagnostic assessment is to find out the strengths, weaknesses, knowledge, and skills of students prior to instruction. Diagnostic tests are administered in the first week of the beginning of the academic year.

2. Formative assessment: Formative assessment is an approach in which the assessment is an integral part of instruction that informs and guides teachers as they make instructional decisions. Assessment is done for students to guide and enhance their learning. Formative assessment should be done on a regular basis throughout the academic year. In theory, formative tests are not graded but are used solely as an ongoing diagnostic tool; hence the results are primarily used for modifying and adjusting instruction to benefit the students. In a science class, formative assessment could be accomplished through quizzes, daily classwork, or homework.

3. Summative assessment: Summative assessment occurs in the form of a test usually given at the end of a semester, term, chapter, or year. High stakes tests like the ACT and SAT are also examples of summative assessments. Though they are helpful to teachers in organizing their methods of teaching and course curricula, summative assessments such as standardized tests can adversely affect students by overly formalizing and limiting the scope of curriculum. However, in the current K-12 system, summative assessment cannot be avoided. A number of suggestions are provided to maximize the effectiveness of summative assessment:

- **Authenticity:** Assessment must be based on classroom objectives and that which has real-world application. For example, the concepts of speed, instantaneous speed, velocity, and acceleration have real-world applications, such as in driving; and the concept of cause and effect relates to the influence of alcohol and drugs on the central nervous system and decision-making capacity.
- **Variety:** A variety of assessment techniques must be used. Assessment should include all three domains of learning – cognitive, affective, and psychomotor. Assessment of the cognitive domain includes Higher Order Thinking Skills (HOTS) such as synthesis, evaluation, analyzing, etc. Using a variety of assessment techniques such as portfolios, cooperative research projects (science projects), papers (research), and performance tests gives a more complete picture of students' strengths and weaknesses and also eliminates assessment bias against at-risk and language minority students.
- **Volume:** The number of assessments should be limited, since students – and teachers – lose interest and lose focus on the real learning objectives. When teachers lose interest it may result in creation of test papers for ease of grading rather than for assessment purposes.
- **Reliability:** Proper rubrics must be established. This guarantees that anyone assessing the test will get the same result.

Skill 34.5 Understands strategies for assessing students' prior knowledge and misconceptions about science and how to use these assessments to develop effective ways to address these misconceptions.

Assessments are used to guide programs and to document student development. Science assessments focus on science concepts, processes, and applications. Before planning for instructional options, it is important that teachers plan pre-assessment strategies. It is essential to evaluate student interest, readiness, and learning needs. Formative assessment provides ongoing monitoring of student progress and provides information for targeting subsequent instruction.

Some strategies for pre-assessment within a science classroom include:

1. Design your own investigation to answer a question: Students independently plan a well-designed investigation to answer a question. Analysis of the plan will indicate the logical reasoning, critical and creative thinking, and planning ability of the student.

2. Graphic organizer: Students independently organize information on a graphic organizer. This strategy helps construct meaning and demonstrate their understanding of science concepts, processes, and applications.

3. Individual KWL: Students independently write what they Know, what they Want to know, and what they have Learned about a topic. Analysis of KWL reveals student interests, content knowledge, and misconceptions.

4. Interview: Students are asked to explain their understanding of a topic. Probing questions by the teacher help determine students' understanding of science concepts.

5. Writing assignment: Students are given a writing assignment that relates to a science topic. The independent response gives evidence of student development regarding processes, concepts, and applications, and their ability to communicate what they know.

6. Teacher observation: Student performance is assessed through teacher observation of student problem solving, interest, questioning, and content knowledge.

There are many more pre-assessment strategies, and each is useful in its own way for determining students' prior knowledge and misconceptions. Once this information is gathered, strategies for dispelling misconceptions include planning assignments or research addressing the relevant topics.

Skill 34.6 Understands characteristics of assessments, such as reliability, validity, and the absence of bias in order to evaluate assessment instruments and their results.

The characteristics of a good assessment are:

1. Reliability: The key to assessment reliability is the reproducibility of assessment results. A test can be a valid measure to assess the progress of students, but may not be reliable. However, it can be made reliable by designing a well-thought-out rubric and letting the students have copies of it in advance. In this case, even if somebody else grades the paper using the same rubric, the grading would be the same. This method applies mainly to projects and assignments where the students are notified in advance of the grading criteria.

2. Validity: The validity of the assessment refers to the extent to which the assessment measures performance on the aspects of the course which are important. That is, the test must clearly reflect the course's objectives. To explain this principle of assessment in the simplest terms, if a student is to be assessed in chemistry, he/she should be given a test in chemistry only.

3. Absence of bias: Bias is something that must be avoided by teachers at all cost. For instance, bias against a bilingual student due to his/her relative weakness in English could hamper a fair assessment of the student's content knowledge. Bias of any kind does not help the teacher or student. Assessment bias should be eliminated by using fair methods of assessment, an objective outlook, and an open mind.

4. Variety: Another method of ensuring quality assessment is to use a variety of assessment techniques. Traditionally, true/false and selected-response test items have been popular methods of assessing students. However, these are limited in scope and encourage rote memorization by students; in addition, there is a probability factor (i.e., guessing of the answers). Assessment should include all three domains of learning – cognitive, affective, and psychomotor. Assessment in the cognitive domain should include high-order processes such as synthesis and evaluation. Portfolios, cooperative research projects, papers, and performance tests add to the variety of potential assessment tools.

Variety is not only conducive to fair assessment, it is also more interesting to students. Students will enjoy doing cooperative research projects, investigations, etc. Variety is also effective in reducing assessment bias against at-risk groups, because it provides a more comprehensive picture of student learning.

5. Volume: Teachers often require more summative assessments than are necessary. Students tend to resent such over-assessment and that may affect their performance on tests. In addition, over-assessment takes its toll on teachers. Thus assessments should be limited to the minimum required.

6. Authenticity: Assessment which addresses the course objectives and has real-life applications is called authentic assessment. In science, a good example of an authentic assessment would be an investigation that requires students to produce rather than select the answer.

Skill 34.7 Understands the role of assessment as a learning experience for students and strategies for engaging students in meaningful self-assessment.

Assessment includes a number of things – collecting information, choosing the right assessment instruments, and implementing these in an effective and fair manner. Assessment can be used as a learning experience for the student as well as a tool for the teacher.

A number of strategies may be employed to assess students' progress and teach them simultaneously. These include:

- **Questioning:** Probing questions are very effective in teaching and understanding students' comprehension.
- **Interpreting a graphic:** Students are asked to interpret a graph, photograph, or diagram.
- **Multiple-choice, fill in the blanks, and matching terms with statements:** These are useful, but are very limited and hence should be sparingly used.
- **Essay questions:** These are very useful, as they are good indicators of students' comprehension.
- **Projects and written assignments:** These are very useful since they require higher-order thinking skills.

Peer assessment

One way to help students understand their quality of work is by having them evaluate the work of their peers. If they are to offer any helpful feedback to other students, they themselves need to understand evaluation first. It falls to the teacher to explain expectations in very clear terms. The teacher can give students a practice assessment session. Rubrics and well designed checklists are very helpful.

For peer assessment to work effectively, the learning environment must be conducive for that. Students must be able to trust each other and feel comfortable. The teacher may need to coach the students on effective ways to present assessments and constructive criticism.

Any assessment method has wider effects than simple measurement. It can support the achievement of the planned learning outcomes or undermine them. Peer assessment in this sense is authentic: conducting peer assessments involves using discipline, knowledge, and skills, while accepting peer assessments engages others' knowledge and skills.

The use of peer assessment encourages students to believe they are part of a community of scholarship. In peer assessment, we encourage students to take part in a key aspect of education, making critical judgments regarding the work of others.

Students are a classroom resource that is not always fully utilized. Peer assessment offers possible gains in cost-effectiveness, since teachers can be managing peer assessment processes rather than assessing large numbers of students. This helps teachers to assess quality rather than quantity, since the bulk of assessment is entrusted to students.

One possible problem is the validity and reliability of peer assessment. In the case of formative assessment (peer review), the validity of peer assessment can be questioned. Another possible problem is the accuracy of grades given by peers. These problems can be handled by using double anonymity, wherein students feel free to express their opinions fearlessly and honestly. In addition, the teacher should provide students with clear measurement indicators and grading instructions.

Self assessment

There are three key elements involved in successful self assessment: goal setting, guided practice with assessment tools, and portfolios.

1) Goal setting: Goal setting is essential because students can evaluate their progress more clearly when they have targets against which to measure their performance. In addition, a student's motivation to learn increases when he or she has self-defined and relevant learning goals.
2) Guided practice with assessment tools: Students should be taught strategies for self monitoring and self assessment. Well-designed checklists and rubrics are used for the purpose of assessment, and practice should be given prior to beginning use of the tools.
3) Portfolios: These are powerful, organized, systematic collections of student work that tell the story of a student's effort. Portfolio assessment emphasizes evaluation of students' progress, processes, and performance over time. There are two types of portfolios: a **process portfolio** serves the purpose of a classroom-level assessment; a **product portfolio** is more summative in nature. Often, an oral presentation accompanies a product portfolio.

Reflection is the key word in self assessment. Students must be able to step back and reflect upon their learning and come up with strategies to learn more effectively. Such self assessment can help students become independent, motivated learners for life.

Skill 34.8 Recognizes the importance of selecting assessment instruments and methods that provide all students with adequate opportunities to demonstrate their achievements.

The importance of assessment is to know how much the students have comprehended, retained, and applied the knowledge they acquired during their time in the class. In order to assess these aspects, the teacher must use appropriate and reliable assessment instruments and methods.

There are a number of **assessment instruments** available for teachers. Assessment instruments are of two main types:

1. Content-oriented: These are specifically oriented to the assessment of student mastery in a particular discipline such as chemistry. Concept tests, challenge problems, lab reports, and other projects such as writing assignments are found in this category.
2. Process-oriented: These are oriented towards assessing the process by which the students learned. These include classroom surveys, interview protocols, etc.

Assessment methods in use by the educator community include quizzes and tests, regular homework assignments, 3D models, short assignments, classwork, etc. These assessment instruments and methods provide a variety of opportunities for students to show their comprehension and mastery of a subject.

The advantage of using a variety of assessment instruments and methods is that assessment bias is minimized and all students have a fair chance of being assessed properly.

Skill 34.9 Recognizes the importance of clarifying teacher expectations by sharing evaluation criteria and assessment results with students.

Evaluation is a process that must be done openly and in fair manner. Teachers should communicate their methods of evaluation at the beginning of the school year, preferably on the first day at school. The teacher needs to recognize the value and importance of communicating the evaluation criteria to the students and other stakeholders, such as parents.

Teacher expectations are to be clarified and communicated. A good way of doing this is to give out an information packet outlining all the aspects of evaluation that will be used along with the evaluation criteria.

Evaluation criteria may include:

1. **Comprehension:** Students' understanding of the material.
2. **Processing information:** How students process information by finding patterns, etc.
3. **Critical thinking and problem solving:** Students' ability to think critically, evaluate a given situation, and find ways to answer the problem at hand.
4. **Communication skills:** The ability to communication through both written and oral communication.

Effective and prompt communication of evaluation results is essential. Results should first be communicated to the students, then to parents in the form of reports and report cards, and finally to educational stakeholders such as administrators.

Sample Test

1. A projectile with a mass of 1.0 kg has a muzzle velocity of 1500.0 m/s when it is fired from a cannon with a mass of 500.0 kg. If the cannon slides on a frictionless track, it will recoil with a velocity of ____ m/s.

A. 2.4

B. 3.0

C. 3.5

D. 1500

2. The weight of an object on the earth's surface is designated *x*. When it is two earth's radii from the surface of the earth, its weight will be

A. $x/4$

B. $x/9$

C. $4x$

D. $16x$

3. If a force of magnitude *F* gives a mass *M* an acceleration *A*, then a force 3*F* would give a mass 3*M* an acceleration

A. A

B. $12A$

C. $A/2$

D. $6A$

4. A car (mass m_1) is driving at velocity *v*, when it smashes into an unmoving car (mass m_2), locking bumpers. Both cars move together at the same velocity. The common velocity will be given by

A. m_1v/m_2

B. m_2v/m_1

C. $m_1v/(m_1 + m_2)$

D. $(m_1 + m_2)v/m_1$

5. When acceleration is plotted versus time, the area under the graph represents

A. Time

B. Distance

C. Velocity

D. Acceleration

6. An inclined plane is tilted by gradually increasing the angle of elevation θ, until the block will slide down at a constant velocity. The coefficient of friction, μ_k, is given by

A. cos θ

B. sin θ

C. cosecant θ

D. tangent θ

7. Use the information on heats below to solve this problem. An ice block at 0° Celsius is dropped into 100 g of liquid water at 18° Celsius. When thermal equilibrium is achieved, only liquid water at 0° Celsius is left. What was the mass, in grams, of the original block of ice?

Given: Heat of fusion of ice = 80 cal/g
Heat of vaporization of ice = 540 cal/g
Specific Heat of ice = 0.50 cal/g°C
Specific Heat of water = 1 cal/g°C

A. 2.0

B. 5.0

C. 10.0

D. 22.5

8. The combination of overtones produced by a musical instrument is known as its

A. Timbre

B. Chromaticity

C. Resonant Frequency

D. Flatness

9. A long copper bar has a temperature of 60°C at one end and 0°C at the other. The bar reaches thermal equilibrium (barring outside influences) by the process of heat

A. Fusion

B. Convection

C. Conduction

D. Microwaving

10. The First Law of Thermodynamics takes the form dU = dW when the conditions are

A. Isobaric

B. Isochloremic

C. Isothermal

D. Adiabatic

11. Given a vase full of water, with holes punched at various heights. The water squirts out of the holes, achieving different distances before hitting the ground. Which of the following accurately describes the situation?

A. Water from higher holes goes farther, due to Pascal's Principle.

B. Water from higher holes goes farther, due to Bernoulli's Principle.

C. Water from lower holes goes farther, due to Pascal's Principle.

D. Water from lower holes goes farther, due to Bernoulli's Principle.

12. A stationary sound source produces a wave of frequency *F*. An observer at position A is moving toward the horn, while an observer at position B is moving away from the horn. Which of the following is true?

A. $F_A < F < F_B$

B. $F_B < F < F_A$

C. $F < F_A < F_B$

D. $F_B < F_A < F$

13. The electric force in Newtons, on two small objects (each charged to – 10 microCoulombs and separated by 2 meters) is

A. 1.0

B. 9.81

C. 31.0

D. 0.225

14. A 10 ohm resistor and a 50 ohm resistor are connected in parallel. If the current in the 10 ohm resistor is 5 amperes, the current (in amperes) running through the 50 ohm resistor is

A. 1

B. 50

C. 25

D. 60

15. Fahrenheit and Celsius thermometers have the same temperature reading at

A. 100 degrees

B. - 40 degrees

C. Absolute Zero

D. 40 degrees

16. Which of the following apparatus can be used to measure the wavelength of a sound produced by a tuning fork?

A. A glass cylinder, some water, and iron filings

B. A glass cylinder, a meter stick, and some water

C. A metronome and some ice water

D. A comb and some tissue

17. When the current flowing through a fixed resistance is doubled, the amount of heat generated is

A. Quadrupled

B. Doubled

C. Multiplied by pi

D. Halved

18. The current induced in a coil is defined by which of the following laws?

A. Lenz's Law

B. Burke's Law

C. The Law of Spontaneous Combustion

D. Snell's Law

19. If an object is 20 cm from a convex lens whose focal length is 10 cm, the image is:

A. Virtual and upright

B. Real and inverted

C. Larger than the object

D. Smaller than the object

20. A cooking thermometer in an oven works because the metals it is composed of have different

A. Melting points

B. Heat convection

C. Magnetic fields

D. Coefficients of expansion

21. In an experiment where a brass cylinder is transferred from boiling water into a beaker of cold water with a thermometer in it, we are measuring

A. Fluid viscosity

B. Heat of fission

C. Specific heat

D. Nonspecific heat

22. A temperature change of 40 degrees Celsius is equal to a change in Fahrenheit degrees of

A. 40

B. 20

C. 72

D. 112

23. The number of calories required to raise the temperature of 40 grams of water at 30°C to steam at 100°C is

A. 7500

B. 23,000

C. 24,400

D. 30,500

24. The boiling point of water on the Kelvin scale is closest to

A. 112 K

B. 212 K

C. 373 K

D. 473 K

25. The kinetic energy of an object is __________ proportional to its ____________.

A. Inversely…inertia

B. Inversely…velocity

C. Directly…mass

D. Directly…time

26. A hollow conducting sphere of radius R is charged with a total charge Q. What is the magnitude of the electric field at a distance r (given r<R) from the center of the sphere? (k is the electrostatic constant)

A. 0

B. $k\,Q/R^2$

C. $k\,Q/(R^2 - r^2)$

D. $k\,Q/(R - r)^2$

27. A quantum of light energy is called a

A. Dalton

B. Photon

C. Curie

D. Heat Packet

28. The following statements about sound waves are true *except*

A. Sound travels faster in liquids than in gases.

B. Sound waves travel through a vacuum.

C. Sound travels faster through solids than liquids.

D. Ultrasound can be reflected by the human body.

29. The greatest number of 100 watt lamps that can be connected in parallel with a 120 volt system without blowing a 5 amp fuse is

A. 24

B. 12

C. 6

D. 1

30. A monochromatic ray of light passes from air to a thick slab of glass (n = 1.41) at an angle of 45° from the normal. At what angle does it leave the air/glass interface?

A. 45°

B. 30°

C. 15°

D. 55°

31. The magnitude of a force is

A. Directly proportional to mass and inversely to acceleration

B. Inversely proportional to mass and directly to acceleration

C. Directly proportional to both mass and acceleration

D. Inversely proportional to both mass and acceleration

32. A semi-conductor allows current to flow

A. Never

B. Always

C. As long as it stays below a maximum temperature

D. When a minimum voltage is applied

33. One reason to use salt for melting ice on roads in the winter is that

A. Salt lowers the freezing point of water.

B. Salt causes a foaming action, which increases traction.

C. Salt is more readily available than sugar.

D. Salt increases the conductivity of water.

34. Automobile mirrors that have a sign, "objects are closer than they appear" say so because

A. The real image of an obstacle, through a converging lens, appears farther away than the object.

B. The real or virtual image of an obstacle, through a converging mirror, appears farther away than the object.

C. The real image of an obstacle, through a diverging lens, appears farther away than the object.

D. The virtual image of an obstacle, through a diverging mirror, appears farther away than the object.

35. A gas maintained at a constant pressure has a specific heat which is greater than its specific heat when maintained at constant volume, because

A. The Coefficient of Expansion changes.

B. The gas enlarges as a whole.

C. Brownian motion causes random activity.

D. Work is done to expand the gas.

36. Consider the shear modulus of water, and that of mercury. Which of the following is true?

A. Mercury's shear modulus indicates that it is the only choice of fluid for a thermometer.

B. The shear modulus of each of these is zero.

C. The shear modulus of mercury is higher than that of water.

D. The shear modulus of water is higher than that of mercury.

37. A skateboarder accelerates down a ramp, with constant acceleration of two meters per second squared, from rest. The distance in meters, covered after four seconds, is

A. 10

B. 16

C. 23

D. 37

38. Which of the following units is not used to measure torque?

A. slug ft

B. lb ft

C. N m

D. dyne cm

39. In a nuclear pile, the control rods are composed of

A. Boron

B. Einsteinium

C. Isoptocarpine

D. Phlogiston

40. All of the following use semi-conductor technology, except a(n):

A. Transistor

B. Diode

C. Capacitor

D. Operational Amplifier

41. Ten grams of a sample of a radioactive material (half-life = 12 days) were stored for 48 days and re-weighed. The new mass of material was

A. 1.25 g

B. 2.5 g

C. 0.83 g

D. 0.625 g

42. When a radioactive material emits an alpha particle only, its atomic number will

A. Decrease

B. Increase

C. Remain unchanged

D. Change randomly

43. The sun's energy is produced primarily by

A. Fission

B. Explosion

C. Combustion

D. Fusion

44. A crew is on-board a spaceship, traveling at 60% of the speed of light with respect to the earth. The crew measures the length of their ship to be 240 meters. When a ground-based crew measures the apparent length of the ship, it equals

A. 400 m

B. 300 m

C. 240 m

D. 192 m

45. Which of the following pairs of elements are not found to fuse in the centers of stars?

A. Oxygen and Helium

B. Carbon and Hydrogen

C. Beryllium and Helium

D. Cobalt and Hydrogen

46. A calorie is the amount of heat energy that will

A. Raise the temperature of one gram of water from 14.5° C to 15.5° C.

B. Lower the temperature of one gram of water from 16.5° C to 15.5° C

C. Raise the temperature of one gram of water from 32° F to 33° F

D. Cause water to boil at two atmospheres of pressure.

47. Bohr's theory of the atom was the first to quantize

A. Work

B. Angular Momentum

C. Torque

D. Duality

48. A uniform pole weighing 100 grams, that is one meter in length, is supported by a pivot at 40 centimeters from the left end. In order to maintain static position, a 200 gram mass must be placed _____ centimeters from the left end.

A. 10

B. 45

C. 35

D. 50

49. A classroom demonstration shows a needle floating in a tray of water. This demonstrates the property of

A. Specific Heat

B. Surface Tension

C. Oil-Water Interference

D. Archimedes' Principle

50. Two neutral isotopes of a chemical element have the same numbers of

A. Electrons and Neutrons

B. Electrons and Protons

C. Protons and Neutrons

D. Electrons, Neutrons, and Protons

51. A mass is moving at constant speed in a circular path. Choose the true statement below:

A. Two forces in equilibrium are acting on the mass.

B. No forces are acting on the mass.

C. One centripetal force is acting on the mass.

D. One force tangent to the circle is acting on the mass.

52. A light bulb is connected in series with a rotating coil within a magnetic field. The brightness of the light may be increased by any of the following except:

A. Rotating the coil more rapidly.

B. Using more loops in the coil.

C. Using a different color wire for the coil.

D. Using a stronger magnetic field.

53. The use of two circuits next to each other, with a change in current in the primary circuit, demonstrates

A. Mutual current induction

B. Dielectric constancy

C. Harmonic resonance

D. Resistance variation

54. A brick and hammer fall from a ledge at the same time. They would be expected to

A. Reach the ground at the same time

B. Accelerate at different rates due to difference in weight

C. Accelerate at different rates due to difference in potential energy

D. Accelerate at different rates due to difference in kinetic energy

55. The potential difference across a five Ohm resistor is five Volts. The power used by the resistor, in Watts, is

A. 1

B. 5

C. 10

D. 20

56. An object two meters tall is speeding toward a plane mirror at 10 m/s. What happens to the image as it nears the surface of the mirror?

A. It becomes inverted.

B. The Doppler Effect must be considered.

C. It remains two meters tall.

D. It changes from a real image to a virtual image.

57. The highest energy is associated with

A. UV radiation

B. Yellow light

C. Infrared radiation

D. Gamma radiation

58. The constant of proportionality between the energy and the frequency of electromagnetic radiation is known as the

A. Rydberg constant

B. Energy constant

C. Planck constant

D. Einstein constant

59. A simple pendulum with a period of one second has its mass doubled. If the length of the string is quadrupled, the new period will be

A. 1 second

B. 2 seconds

C. 3 seconds

D. 5 seconds

60. A vibrating string's frequency is _______ proportional to the _______.

A. Directly; Square root of the tension

B. Inversely; Length of the string

C. Inversely; Squared length of the string

D. Inversely; Force of the plectrum

61. When an electron is "orbiting" the nucleus in an atom, it is said to possess an intrinsic spin (spin angular momentum). How many values can this spin have in any given electron?

A. 1

B. 2

C. 3

D. 8

62. Electrons are

A. More massive than neutrons

B. Positively charged

C. Neutrally charged

D. Negatively charged

63. Rainbows are created by

A. Reflection, dispersion, and recombination

B. Reflection, resistance, and expansion

C. Reflection, compression, and specific heat

D. Reflection, refraction, and dispersion

64. In order to switch between two different reference frames in special relativity, we use the ________ transformation.

A. Galilean

B. Lorentz

C. Euclidean

D. Laplace

65. A baseball is thrown with an initial velocity of 30 m/s at an angle of 45°. Neglecting air resistance, how far away will the ball land?

A. 92 m

B. 78 m

C. 65 m

D. 46 m

66. If one sound is ten decibels louder than another, the ratio of the intensity of the first to the second is

A. 20:1

B. 10:1

C. 1:1

D. 1:10

67. A wave has speed 60 m/s and wavelength 30,000 m. What is the frequency of the wave?

A. 2.0×10^{-3} Hz

B. 60 Hz

C. 5.0×10^{2} Hz

D. 1.8×10^{6} Hz

68. An electromagnetic wave propagates through a vacuum. Independent of its wavelength, it will move with constant

A. Acceleration

B. Velocity

C. Induction

D. Sound

69. A wave generator is used to create a succession of waves. The rate of wave generation is one every 0.33 seconds. The period of these waves is

A. 2.0 seconds

B. 1.0 seconds

C. 0.33 seconds

D. 3.0 seconds

70. In a fission reactor, heavy water

A. Cools off neutrons to control temperature

B. Moderates fission reactions

C. Initiates the reaction chain

D. Dissolves control rods

71. Heat transfer by electromagnetic waves is termed

A. Conduction

B. Convection

C. Radiation

D. Phase Change

72. Solids expand when heated because

A. Molecular motion causes expansion

B. PV = nRT

C. Magnetic forces stretch the chemical bonds

D. All material is effectively fluid

73. Gravitational force at the earth's surface causes

A. All objects to fall with equal acceleration, ignoring air resistance

B. Some objects to fall with constant velocity, ignoring air resistance

C. A kilogram of feathers to float at a given distance above the earth

D. Aerodynamic objects to accelerate at an increasing rate

74. An office building entry ramp uses the principle of which simple machine?

A. Lever

B. Pulley

C. Wedge

D. Inclined Plane

75. The velocity of sound is greatest in

A. Water

B. Steel

C. Alcohol

D. Air

76. All of the following phenomena are considered "refractive effects" except for

A. The red shift

B. Total internal reflection

C. Lens dependent image formation

D. Snell's Law

77. Static electricity generation occurs by

A. Telepathy

B. Friction

C. Removal of heat

D. Evaporation

78. The wave phenomenon of polarization applies only to

A. Longitudinal waves

B. Transverse waves

C. Sound

D. Light

79. A force is given by the vector 5 N x + 3 N y (where x and y are the unit vectors for the x- and y- axes, respectively). This force is applied to move a 10 kg object 5 m, in the x direction. How much work was done?

A. 250 J

B. 400 J

C. 40 J

D. 25 J

80. A satellite is in a circular orbit above the earth. Which statement is false?

A. An external force causes the satellite to maintain orbit.

B. The satellite's inertia causes it to maintain orbit.

C. The satellite is accelerating toward the earth.

D. The satellite's velocity and acceleration are not in the same direction.

Answer Key

1.	B	**17.**	A	**33.**	A	**49.**	B	**65.**	A
2.	B	**18.**	A	**34.**	D	**50.**	B	**66.**	B
3.	A	**19.**	B	**35.**	D	**51.**	C	**67.**	A
4.	C	**20.**	D	**36.**	B	**52.**	C	**68.**	B
5.	C	**21.**	C	**37.**	B	**53.**	A	**69.**	C
6.	D	**22.**	C	**38.**	A	**54.**	A	**70.**	B
7.	D	**23.**	C	**39.**	A	**55.**	B	**71.**	C
8.	A	**24.**	C	**40.**	C	**56.**	C	**72.**	A
9.	C	**25.**	C	**41.**	D	**57.**	D	**73.**	A
10.	D	**26.**	A	**42.**	A	**58.**	C	**74.**	D
11.	D	**27.**	B	**43.**	D	**59.**	B	**75.**	B
12.	B	**28.**	B	**44.**	D	**60.**	A	**76.**	A
13.	D	**29.**	C	**45.**	D	**61.**	B	**77.**	B
14.	A	**30.**	B	**46.**	A	**62.**	D	**78.**	B
15.	B	**31.**	C	**47.**	B	**63.**	D	**79.**	D
16.	B	**32.**	D	**48.**	D	**64.**	B	**80.**	B

Rationales with Sample Questions

1. A projectile with a mass of 1.0 kg has a muzzle velocity of 1500.0 m/s when it is fired from a cannon with a mass of 500.0 kg. If the cannon slides on a frictionless track, it will recoil with a velocity of ____ m/s.

A. 2.4

B. 3.0

C. 3.5

D. 1500

Answer:

B. 3.0

To solve this problem, apply Conservation of Momentum to the cannon-projectile system. The system is initially at rest, with total momentum of 0 kg m/s. Since the cannon slides on a frictionless track, we can assume that the net momentum stays the same for the system. Therefore, the momentum forward (of the projectile) must equal the momentum backward (of the cannon). Thus:

$p_{projectile} = p_{cannon}$

$m_{projectile}\ v_{projectile} = m_{cannon}\ v_{cannon}$

(1.0 kg)(1500.0 m/s) = (500.0 kg)(x)

x = 3.0 m/s

Only answer (B) matches these calculations.

2. The weight of an object on the earth's surface is designated *x*. When it is two earth's radii from the surface of the earth, its weight will be

A. $x/4$

B. $x/9$

C. $4x$

D. $16x$

Answer:

B. $x/9$

To solve this problem, apply the universal Law of Gravitation to the object and Earth:

$F_{gravity} = (GM_1M_2)/R^2$
Because the force of gravity varies with the square of the radius between the objects, the force (or weight) on the object will be decreased by the square of the multiplication factor on the radius. Note that the object on Earth's surface is *already* at one radius from Earth's center. Thus, when it is two radii from Earth's surface, it is three radii from Earth's center. R^2 is then nine, so the weight is $x/9$.
Only answer (B) matches these calculations.

3. If a force of magnitude *F* gives a mass *M* an acceleration *A*, then a force 3*F* would give a mass 3*M* an acceleration

A. A

B. $12A$

C. $A/2$

D. $6A$

Answer:

A. A

To solve this problem, apply Newton's Second Law, which is also implied by the first part of the problem:
Force = (*Mass*)(*Acceleration*)
$F = MA$
Then apply the same law to the second case, and isolate the unknown:
$3F = 3M x$
$x = (3F)/(3M)$
$x = F/M$
$x = A$ (by substituting from our first equation)
Only answer (A) matches these calculations.

4. A car (mass m_1) is driving at velocity v, when it smashes into an unmoving car (mass m_2), locking bumpers. Both cars move together at the same velocity. The common velocity will be given by

A. m_1v/m_2

B. m_2v/m_1

C. $m_1v/(m_1 + m_2)$

D. $(m_1 + m_2)v/m_1$

Answer:

C. $m_1v/(m_1 + m_2)$

In this problem, there is an inelastic collision, so the best method is to assume that momentum is conserved. (Recall that momentum is equal to the product of mass and velocity.)
Therefore, apply Conservation of Momentum to the two-car system:
Momentum at Start = Momentum at End
(Mom. of Car 1) + (Mom. of Car 2) = (Mom. of 2 Cars Coupled)
$m_1v + 0 = (m_1 + m_2)x$
$x = m_1v/(m_1 + m_2)$
Only answer (C) matches these calculations.

Watch out for the other answers, because errors in algebra could lead to a match with incorrect answer (D), and assumption of an elastic collision could lead to a match with incorrect answer (A).

5. When acceleration is plotted versus time, the area under the graph represents

A. Time

B. Distance

C. Velocity

D. Acceleration

Answer:

C. Velocity

The area under a graph will have units equal to the product of the units of the two axes. (To visualize this, picture a graphed rectangle with its area equal to length times width.)
Therefore, multiply units of acceleration by units of time:
(length/time^2)(time)
This equals length/time, i.e. units of velocity.

6. An inclined plane is tilted by gradually increasing the angle of elevation θ, until the block will slide down at a constant velocity. The coefficient of friction, μ_k, is given by

A. cos θ

B. sin θ

C. cosecant θ

D. tangent θ

Answer:

D. tangent θ

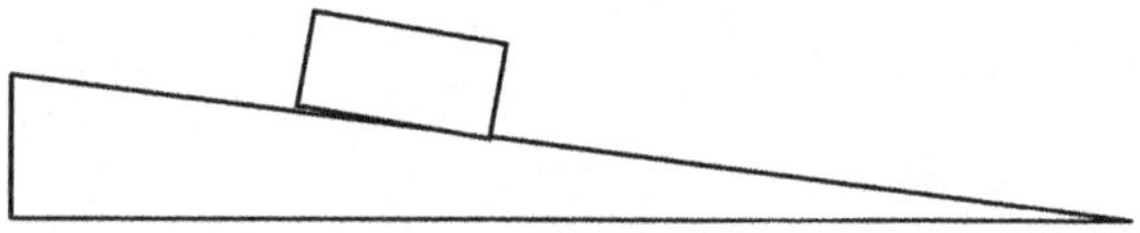

When the block moves, its force upstream (due to friction) must equal its force downstream (due to gravity).

The friction force is given by
$F_f = \mu_k N$
where μ_k is the friction coefficient and N is the normal force.

Using similar triangles, the gravity force is given by
$F_g = mg \sin \theta$
and the normal force is given by
$N = mg \cos \theta$

When the block moves at constant velocity, it must have zero net force, so set equal the force of gravity and the force due to friction:
$F_f = F_g$
$\mu_k mg \cos \theta = mg \sin \theta$
$\mu_k = \tan \theta$

Answer (D) is the only appropriate choice in this case.

7. Use the information on heats below to solve this problem. An ice block at 0° Celsius is dropped into 100 g of liquid water at 18° Celsius. When thermal equilibrium is achieved, only liquid water at 0° Celsius is left. What was the mass, in grams, of the original block of ice?

Given: **Heat of fusion of ice = 80 cal/g**
Heat of vaporization of ice = 540 cal/g
Specific Heat of ice = 0.50 cal/g°C
Specific Heat of water = 1 cal/g°C

A. 2.0

B. 5.0

C. 10.0

D. 22.5

Answer:

D. 22.5

To solve this problem, apply Conservation of Energy to the ice-water system. Any gain of heat to the melting ice must be balanced by loss of heat in the liquid water. Use the two equations relating temperature, mass, and energy:
$Q = m\,C\,\Delta T$ (for heat loss/gain from change in temperature)
$Q = m\,L$ (for heat loss/gain from phase change)
where Q is heat change; m is mass; C is specific heat; ΔT is change in temperature; L is heat of phase change (in this case, melting, also known as "fusion").

Then
$Q_{\text{ice to water}} = Q_{\text{water to ice}}$
(Note that the ice only melts; it stays at 0° Celsius—otherwise, we would have to include a term for warming the ice as well. Also the information on the heat of vaporization for water is irrelevant to this problem.)
$m\,L = m\,C\,\Delta T$
$x\,(80\text{ cal/g}) = 100\text{g}\ 1\text{cal/g°C}\ 18°\text{C}$
$x\,(80\text{ cal/g}) = 1800\text{ cal}$
$x = 22.5\text{ g}$

Only answer (D) matches this result.

8. The combination of overtones produced by a musical instrument is known as its

A. Timbre

B. Chromaticity

C. Resonant Frequency

D. Flatness

Answer:

A. Timbre

To answer this question, you must know some basic physics vocabulary. "Timbre" is the combination of tones that make a sound unique, beyond its pitch and volume. (For instance, consider the same note played at the same volume, but by different instruments.) Answer (A) is therefore the only appropriate choice. "Resonant Frequency" is relevant to music, because the resonant frequency of a wave will give the dominant sound tone. "Chromaticity" is an analogous word to "timbre," but it describes color tones. "Flatness" is unrelated, and incorrect.

9. A long copper bar has a temperature of 60ºC at one end and 0ºC at the other. The bar reaches thermal equilibrium (barring outside influences) by the process of heat

A. Fusion

B. Convection

C. Conduction

D. Microwaving

Answer:

C. Conduction

To answer this question, recall the different methods of heat transfer. (Note that since the bar is warm at one end and cold at the other, heat must transfer through the bar from warm to cold, until temperature is equalized.) "Convection" is the heat transfer via fluid currents. "Conduction" is the heat transfer via connected solid material. "Fusion" and "Microwaving" are not methods of heat transfer. Therefore the only appropriate answer is (C).

10. The First Law of Thermodynamics takes the form dU = dW when the conditions are

A. Isobaric

B. Isochloremic

C. Isothermal

D. Adiabatic

Answer:

D. Adiabatic

To answer this question, recall the First Law of Thermodynamics:
Change in Internal Energy = Work Done + Heat Added
$dU = dW + dQ$

Thus in the form we are given, dQ has been set to zero, i.e. there is no heat added.
"Adiabatic" refers to a case where there is no heat exchange with surroundings, so answer (D) is the appropriate choice. "Isobaric" means at a constant pressure, "Isothermal" means at a constant temperature, and "Isochloremic" is an imaginary word, as far as I can tell.

It might be tempting to choose "Isothermal," thinking that no heat added would require the same temperature. However, work and internal energy changes can change temperature within the system analyzed, even when no heat is exchanged with the surroundings.

11. Given a vase full of water, with holes punched at various heights. The water squirts out of the holes, achieving different distances before hitting the ground. Which of the following accurately describes the situation?

A. Water from higher holes goes farther, due to Pascal's Principle.

B. Water from higher holes goes farther, due to Bernoulli's Principle.

C. Water from lower holes goes farther, due to Pascal's Principle.

D. Water from lower holes goes farther, due to Bernoulli's Principle.

Answer:

D. Water from lower holes goes farther, due to Bernoulli's Principle.

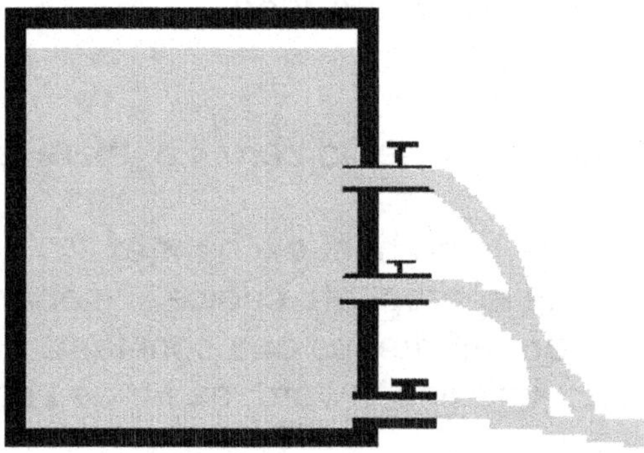

To answer this question, consider the pressure on the water in the vase. The deeper the water, the higher the pressure. Thus, when a hole is punched, the water stream will achieve higher velocity as it equalizes to atmospheric pressure. The lower streams will therefore travel farther before hitting the ground. This eliminates answers (A) and (B). Then recall that Pascal's Principle provides for immediate pressure changes throughout a fluid, while Bernoulli's Principle translates pressure, velocity, and height energy into each other. In this case, the pressure energy is being transformed into velocity energy, and Bernoulli's Principle applies. Therefore, the only appropriate answer is (D).

12. A stationary sound source produces a wave of frequency *F*. An observer at position A is moving toward the horn, while an observer at position B is moving away from the horn. Which of the following is true?

A. $F_A < F < F_B$

B. $F_B < F < F_A$

C. $F < F_A < F_B$

D. $F_B < F_A < F$

Answer:

B. $F_B < F < F_A$

To answer this question, recall the Doppler Effect. As a moving observer approaches a sound source, s/he intercepts wave fronts sooner than if s/he were standing still. Therefore, the wave fronts seem to be coming more frequently. Similarly, as an observer moves away from a sound source, the wave fronts take longer to reach him/her. Therefore, the wave fronts seem to be coming less frequently. Because of this effect, the frequency at B will seem lower than the original frequency, and the frequency at A will seem higher than the original frequency. The only answer consistent with this is (B). Note also, that even if you weren't sure of which frequency should be greater/smaller, you could still reason that A and B should have opposite effects, and be able to eliminate answer choices (C) and (D).

13. The electric force in Newtons, on two small objects (each charged to –10 microCoulombs and separated by 2 meters) is

A. 1.0

B. 9.81

C. 31.0

D. 0.225

Answer:

D. 0.225

To answer this question, use Coulomb's Law, which gives the electric force between two charged particles:
$F = k\ Q_1Q_2/r^2$
Then our unknown is F, and our knowns are:
$k = 9.0 \times 10^9\ Nm^2/C^2$
$Q_1 = Q_2 = -10 \times 10^{-6}\ C$
$r = 2\ m$

Therefore
$F = (9.0 \times 10^9)(-10 \times 10^{-6})(-10 \times 10^{-6})/(2^2)\ \ N$
$F = 0.225\ N$

This is compatible only with answer (D).

14. A 10 ohm resistor and a 50 ohm resistor are connected in parallel. If the current in the 10 ohm resistor is 5 amperes, the current (in amperes) running through the 50 ohm resistor is

A. 1

B. 50

C. 25

D. 60

Answer:

A. 1

To answer this question, use Ohm's Law, which relates voltage to current and resistance:
V = IR
where V is voltage; I is current; R is resistance.

We also use the fact that in a parallel circuit, the voltage is the same across the branches.

Because we are given that in one branch, the current is 5 amperes and the resistance is 10 ohms, we deduce that the voltage in this circuit is their product, 50 volts (from V = IR).

We then use V = IR again, this time to find I in the second branch. Because V is 50 volts, and R is 50 ohm, we calculate that I has to be 1 ampere.

This is consistent only with answer (A).

15. Fahrenheit and Celsius thermometers have the same temperature reading at

A. 100 degrees

B. - 40 degrees

C. Absolute Zero

D. 40 degrees

Answer:

B. - 40 degrees

To answer this question, use the relationship between Fahrenheit and Celsius temperature scales:
$F = 9/5\ C + 32$

Then, in a case where both ^{o}F and ^{o}C are equal, $F = C$, so
$C = 9/5\ C + 32$
$-4/5\ C = 32$
$C = -40$

Only answer (B) is consistent with this result.

16. Which of the following apparatus can be used to measure the wavelength of a sound produced by a tuning fork?

A. A glass cylinder, some water, and iron filings

B. A glass cylinder, a meter stick, and some water

C. A metronome and some ice water

D. A comb and some tissue

Answer:

B. A glass cylinder, a meter stick, and some water

To answer this question, recall that a sound will be amplified if it is reflected back to cause positive interference. This is the principle behind musical instruments that use vibrating columns of air to amplify sound (e.g. a pipe organ).

Therefore, presumably a person could put varying amounts of water in the cylinder, and hold the vibrating tuning fork above the cylinder in each case. If the tuning fork sound is amplified when put at the top of the column, then the length of the air space would be an integral multiple of the sound's wavelength. This experiment is consistent with answer (B). Although the experiment would be tedious, none of the other options for materials suggest a better alternative.

17. When the current flowing through a fixed resistance is doubled, the amount of heat generated is

A. Quadrupled

B. Doubled

C. Multiplied by pi

D. Halved

Answer:

A. Quadrupled

To answer this question, recall that heat generated will occur because of the power of the circuit (power is energy per time). For a circuit with a fixed resistance:

$P = I V$
where P is power; I is current; V is voltage. Then use Ohm's Law:
$V = I R$
where V is voltage; I is current; R is resistance, and substitute:
$P = I^2 R$
and so the doubling of the current I will lead to a quadrupling of the power, and therefore the a quadrupling of the heat.

This is consistent only with answer (A). If you weren't sure of the equations, you could still deduce that with more current, there would be more heat generated, and therefore eliminate answer choice (D) in any case.

18. The current induced in a coil is defined by which of the following laws?

A. Lenz's Law

B. Burke's Law

C. The Law of Spontaneous Combustion

D. Snell's Law

Answer:

A. Lenz's Law

Lenz's Law states that an induced electromagnetic force always gives rise to a current whose magnetic field opposes the original flux change. There is no relevant "Snell's Law," "Burke's Law," or "Law of Spontaneous Combustion" in electromagnetism. (In fact, only Snell's Law is a real law of these three, and it refers to refracted light.) Therefore, the only appropriate answer is (A).

19. If an object is 20 cm from a convex lens whose focal length is 10 cm, the image is:

A. Virtual and upright

B. Real and inverted

C. Larger than the object

D. Smaller than the object

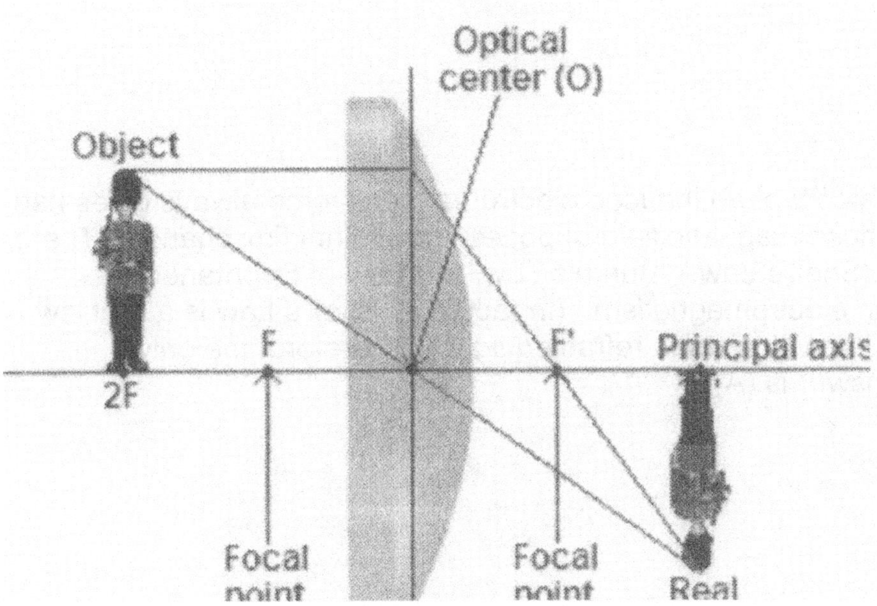

Answer:

B. Real and inverted

To solve this problem, draw a lens diagram with the lens, focal length, and image size.

The ray from the top of the object straight to the lens is focused through the far focus point; the ray from the top of the object through the near focus goes straight through the lens; the ray from the top of the object through the center of the lens continues. These three meet to form the "top" of the image, which is therefore real and inverted. This is consistent only with answer (B).

20. A cooking thermometer in an oven works because the metals it is composed of have different

A. Melting points

B. Heat convection

C. Magnetic fields

D. Coefficients of expansion

Answer:

D. Coefficients of expansion

A thermometer of the type that can withstand oven temperatures works by having more than one metal strip. These strips expand at different rates with temperature increases, causing the dial to register the new temperature. This is consistent only with answer (D). If you did not know how an oven thermometer works, you could still omit the incorrect answers: It is unlikely that the metals in a thermometer would melt in the oven to display the temperature; the magnetic fields would not be useful information in this context; heat convection applies in fluids, not solids.

21. In an experiment where a brass cylinder is transferred from boiling water into a beaker of cold water with a thermometer in it, we are measuring

A. Fluid viscosity

B. Heat of fission

C. Specific heat

D. Nonspecific heat

Answer:

C. Specific heat

In this question, we consider an experiment to measure temperature change of water (with the thermometer) as the cylinder cools and the water warms. This information can be used to calculate heat changes, and therefore specific heat. Therefore, (C) is the correct answer. Even if you were unable to deduce that specific heat is being measured, you could eliminate the other answer choices: viscosity cannot be measured with a thermometer; fission takes place at much higher temperatures than this experiment, and under quite different conditions; there is no such thing as "nonspecific heat".

22. A temperature change of 40 degrees Celsius is equal to a change in Fahrenheit degrees of

A. 40

B. 20

C. 72

D. 112

Answer:

C. 72

To answer this question, recall the equation for Celsius and Fahrenheit:
°F = 9/5 °C + 32

Therefore, whatever temperature difference occurs in °C, it is multiplied by a factor of 9/5 to get the new °F measurement:
new°F = 9/5(old°C + 40) + 32
(whereas old°F = 9/5(old°C) + 32)
Therefore the difference between the old and new temperatures in Fahrenheit is 9/5 of 40, or 72 degrees. This is consistent only with answer (C).

23. The number of calories required to raise the temperature of 40 grams of water at 30°C to steam at 100°C is

A. 7500

B. 23,000

C. 24,400

D. 30,500

Answer:

C. 24,400

To answer this question, apply the equations for heat transfer due to temperature and phase changes:
$Q = mC\Delta T + mL$
where Q is heat; m is mass; C is specific heat; ΔT is temperature change; L is heat of phase change.

In this problem, we are trying to find Q, and we are given:
m = 40 g
C = 1 cal/g°C for water (this should be memorized)
ΔT = 70 °C
L = 540 cal/g for liquid to gas change in water (this should be memorized)

thus Q = (40 g)(1 cal/g°C)(70 °C) + (40 g)(540 cal/g)
Q = 24,400 cal
This is consistent only with answer (C).

24. The boiling point of water on the Kelvin scale is closest to

A. 112 K

B. 212 K

C. 373 K

D. 473 K

Answer:

C. 373 K

To answer this question, recall that Kelvin temperatures are equal to Celsius temperatures plus 273.15. Since water boils at 100ºC under standard conditions, it will boil at 373.15 K. This is consistent only with answer (C).

25. The kinetic energy of an object is __________ proportional to its ___________.

A. Inversely…inertia

B. Inversely…velocity

C. Directly…mass

D. Directly…time

Answer:

C. Directly…mass

To answer this question, recall that kinetic energy is equal to one-half of the product of an object's mass and the square of its velocity:
$KE = \frac{1}{2} m v^2$

Therefore, kinetic energy is directly proportional to mass, and the answer is (C). Note that although kinetic energy is associated with both velocity and momentum (a measure of inertia), it is not *inversely* proportional to either one.

26. A hollow conducting sphere of radius R is charged with a total charge Q. What is the magnitude of the electric field at a distance r (given $r<R$) from the center of the sphere? (k is the electrostatic constant)

A. 0

B. $k\,Q/R^2$

C. $k\,Q/(R^2 - r^2)$

D. $k\,Q/(R - r)^2$

Answer:

A. 0

You may be tempted to use the equation for the electric field:
$E = F/Q$ (E = electric field; F = electric force; Q = charge)

and the Coulomb's Law expression for electric force:
$F = k\,Q_1Q_2/r^2$ (k = constant; Q_1 and Q_2 = charges; r = distance apart),

which usually would give
$E = k\,Q/(R-r)^2$ in a similar context.

However, this question addresses a special case, i.e. a hollow conductor. Inside a hollow conductor, no electric field exists, because if there were an electric field inside, the conductor's free electrons would be forced to move (by the electric force) until the electric field became zero. Therefore, the only correct answer is (A).

27. A quantum of light energy is called a

A. Dalton

B. Photon

C. Curie

D. Heat Packet

Answer:

B. Photon

The smallest "packet" (quantum) of light energy is a photon. "Heat Packet" does not have any relevant meaning, and while "Dalton" and "Curie" have other meanings, they are not connected to light. Therefore, only (B) is a correct answer.

28. The following statements about sound waves are true *except*

A. Sound travels faster in liquids than in gases.

B. Sound waves travel through a vacuum.

C. Sound travels faster through solids than liquids.

D. Ultrasound can be reflected by the human body.

Answer:

B. Sound waves travel through a vacuum.

Sound waves require a medium to travel. The sound wave agitates the material, and this occurs fastest in solids, then liquids, then gases. Ultrasound waves are reflected by parts of the body, and this is useful in medical imaging. Therefore, the only correct answer is (B).

29. The greatest number of 100 watt lamps that can be connected in parallel with a 120 volt system without blowing a 5 amp fuse is

A. 24

B. 12

C. 6

D. 1

Answer:

C. 6

To solve fuse problems, you must add together all the drawn current in the parallel branches, and make sure that it is less than the fuse's amp measure. Because we know that electrical power is equal to the product of current and voltage, we can deduce that:
I = P/V (I = current (amperes); P = power (watts); V = voltage (volts))

Therefore, for each lamp, the current is 100/120 amperes, or 5/6 ampere. The highest possible number of lamps is thus six, because six lamps at 5/6 ampere each adds to 5 amperes; more will blow the fuse.

This is consistent only with answer (C).

30. A monochromatic ray of light passes from air to a thick slab of glass (n = 1.41) at an angle of 45° from the normal. At what angle does it leave the air/glass interface?

A. 45°

B. 30°

C. 15°

D. 55°

Answer:

B. 30°

To solve this problem use Snell's Law:
$n_1 \sin\theta_1 = n_2 \sin\theta_2$ (where n_1 and n_2 are the indexes of refraction and θ_1 and θ_2 are the angles of incidence and refraction).

Then, since the index of refraction for air is 1.0, we deduce:
$1 \sin 45° = 1.41 \sin x$
$x = \sin^{-1} ((1/1.41) \sin 45°)$
$x = 30°$

This is consistent only with answer (B). Also, note that you could eliminate answers (A) and (D) in any case, because the refracted light will have to bend at a smaller angle when entering glass.

31. The magnitude of a force is

A. Directly proportional to mass and inversely to acceleration

B. Inversely proportional to mass and directly to acceleration

C. Directly proportional to both mass and acceleration

D. Inversely proportional to both mass and acceleration

Answer:

C. Directly proportional to both mass and acceleration

To solve this problem, recall Newton's 2nd Law, i.e. net force is equal to mass times acceleration. Therefore, the only possible answer is (C).

32. A semi-conductor allows current to flow

A. Never

B. Always

C. As long as it stays below a maximum temperature

D. When a minimum voltage is applied

Answer:

D. When a minimum voltage is applied

To answer this question, recall that semiconductors do not conduct as well as conductors (eliminating answer (B)), but they conduct better than insulators (eliminating answer (A)). Semiconductors can conduct better when the temperature is higher (eliminating answer (C)), and their electrons move most readily under a potential difference. Thus the answer can only be (D).

33. One reason to use salt for melting ice on roads in the winter is that

A. Salt lowers the freezing point of water.

B. Salt causes a foaming action, which increases traction.

C. Salt is more readily available than sugar.

D. Salt increases the conductivity of water.

Answer:

A. Salt lowers the freezing point of water.

In answering this question, you may recall that salt is used for road traction because it has large particles to increase traction (analogous to sand). This is true, but salt also has the potential to lower the freezing point of water, thus melting the ice with which it has contact. This is consistent with answer (A). Answer (B) is untrue, and the other two choices, while usually true, are irrelevant in this case.

34. Automobile mirrors that have a sign, "objects are closer than they appear" say so because

A. The real image of an obstacle, through a converging lens, appears farther away than the object.

B. The real or virtual image of an obstacle, through a converging mirror, appears farther away than the object.

C. The real image of an obstacle, through a diverging lens, appears farther away than the object.

D. The virtual image of an obstacle, through a diverging mirror, appears farther away than the object.

Answer:

D. The virtual image of an obstacle, through a diverging mirror, appears farther away than the object.

To answer this question, first eliminate answer choices (A) and (C), because we have a mirror, not a lens. Then draw ray diagrams for diverging (convex) and converging (concave) mirrors, and note that because the focal point of a diverging mirror is behind the surface, the image is smaller than the object. This creates the illusion that the object is farther away, and therefore (D) is the correct answer.

35. A gas maintained at a constant pressure has a specific heat which is greater than its specific heat when maintained at constant volume, because

A. The Coefficient of Expansion changes.

B. The gas enlarges as a whole.

C. Brownian motion causes random activity.

D. Work is done to expand the gas.

Answer:

D. Work is done to expand the gas.

To answer this question, recall that the specific heat is a measure of how much energy it takes to raise the temperature of a given mass of gas. Thus, you can reason that when a gas is maintained at constant pressure, some energy is used to expand the volume of the gas, and less is left for temperature changes. In fact, this is the case, and (D) is the correct answer. If you were not able to figure that out, you could still eliminate the other answers, because they are not strictly relevant to a change in specific heat.

36. Consider the shear modulus of water, and that of mercury. Which of the following is true?

A. Mercury's shear modulus indicates that it is the only choice of fluid for a thermometer.

B. The shear modulus of each of these is zero.

C. The shear modulus of mercury is higher than that of water.

D. The shear modulus of water is higher than that of mercury.

Answer:

B. The shear modulus of each of these is zero.

To answer this question, recall that shear modulus is meaningful only for solids, and that liquids, instead, have a bulk modulus. The only reasonable answer is therefore (B).

37. A skateboarder accelerates down a ramp, with constant acceleration of two meters per second squared, from rest. The distance in meters, covered after four seconds, is

A. 10

B. 16

C. 23

D. 37

Answer:

B. 16

To answer this question, recall the equation relating constant acceleration to distance and time:
$x = \frac{1}{2} a t^2 + v_0 t + x_0$ where x is position; a is acceleration; t is time; v_0 and x_0 are initial velocity and position (both zero in this case)

thus, to solve for x:
$x = \frac{1}{2} (2\ m/s^2)(4^2 s^2) + 0 + 0$
$x = 16\ m$

This is consistent only with answer (B).

38. Which of the following units is not used to measure torque?

A. slug ft

B. lb ft

C. N m

D. dyne cm

Answer:

A. slug ft

To answer this question, recall that torque is always calculated by multiplying units of force by units of distance. Therefore, answer (A), which is the product of units of mass and units of distance, must be the choice of incorrect units. Indeed, the other three answers all could measure torque, since they are of the correct form. It is a good idea to review “English Units” before the teacher test, because they are occasionally used in problems.

39. In a nuclear pile, the control rods are composed of

A. Boron

B. Einsteinium

C. Isoptocarpine

D. Phlogiston

Answer:

A. Boron

Nuclear plants use control rods made of boron or cadmium, to absorb neutrons and maintain “critical” conditions in the reactor. However, if you did not know that, you could still eliminate choice (D), because “phlogiston” is the word for the imaginary element in an ancient structure of earth-air-water-fire.

40. All of the following use semi-conductor technology, except a(n):

A. Transistor

B. Diode

C. Capacitor

D. Operational Amplifier

Answer:

C. Capacitor

Semi-conductor technology is used in transistors and operational amplifiers, and diodes are the basic unit of semi-conductors. Therefore the only possible answer is (C), and indeed a capacitor does not require semi-conductor technology.

41. Ten grams of a sample of a radioactive material (half-life = 12 days) were stored for 48 days and re-weighed. The new mass of material was

A. 1.25 g

B. 2.5 g

C. 0.83 g

D. 0.625 g

Answer:

D. 0.625 g

To answer this question, note that 48 days is four half-lives for the material. Thus, the sample will degrade by half four times. At first, there are ten grams, then (after the first half-life) 5 g, then 2.5 g, then 1.25 g, and after the fourth half-life, there remains 0.625 g. You could also do the problem mathematically, by multiplying ten times $(\frac{1}{2})^4$, i.e. ½ for each half-life elapsed.

42. When a radioactive material emits an alpha particle only, its atomic number will

A. Decrease

B. Increase

C. Remain unchanged

D. Change randomly

Answer:

A. Decrease

To answer this question, recall that in alpha decay, a nucleus emits the equivalent of a Helium atom. This includes two protons, so the original material changes its atomic number by a decrease of two.

43. The sun's energy is produced primarily by

A. Fission

B. Explosion

C. Combustion

D. Fusion

Answer:

D. Fusion

To answer this question, recall that in stars (such as the sun), fusion is the main energy-producing occurrence. Fission, explosion, and combustion all release energy in other contexts, but they are not the right answers here.

44. A crew is on-board a spaceship, traveling at 60% of the speed of light with respect to the earth. The crew measures the length of their ship to be 240 meters. When a ground-based crew measures the apparent length of the ship, it equals

A. 400 m

B. 300 m

C. 240 m

D. 192 m

Answer:

D. 192 m

To answer this question, recall that a moving object's size seems contracted to the stationary observer, according to the equation:
Length Observed = (Actual Length) $(1 - v^2/c^2)^{½}$
where v is speed of motion, and c is speed of light.

Therefore, in this case,
Length Observed = (240 m) $(1 - 0.6^2)^{½}$
Length Observed = (240 m) (0.8) = 192 m

This is consistent only with answer (D). If you were unsure of the equation, you could still reason that because of length contraction (the flip side of time dilation), you must choose an answer with a smaller length, and only (D) fits that description. Note that only the dimension in the direction of travel is contracted. (The length in this case.)

45. Which of the following pairs of elements are not found to fuse in the centers of stars?

A. Oxygen and Helium

B. Carbon and Hydrogen

C. Beryllium and Helium

D. Cobalt and Hydrogen

Answer:

D. Cobalt and Hydrogen

To answer this question, recall that fusion is possible only when the final product has more binding energy than the reactants. Because binding energy peaks near a mass number of around 56, corresponding to Iron, any heavier elements would be unlikely to fuse in a typical star. (In very massive stars, there may be enough energy to fuse heavier elements.) Of all the listed elements, only Cobalt is heavier than iron, so answer (D) is correct.

46. A calorie is the amount of heat energy that will

A. Raise the temperature of one gram of water from 14.5° C to 15.5° C.

B. Lower the temperature of one gram of water from 16.5° C to 15.5° C

C. Raise the temperature of one gram of water from 32° F to 33° F

D. Cause water to boil at two atmospheres of pressure.

Answer:

A. Raise the temperature of one gram of water from 14.5° C to 15.5° C.

The definition of a calorie is, "the amount of energy to raise one gram of water by one degree Celsius," and so answer (A) is correct. Do not get confused by the fact that 14.5° C seems like a random number. Also, note that answer (C) tries to confuse you with degrees Fahrenheit, which are irrelevant to this problem.

47. Bohr's theory of the atom was the first to quantize

A. Work

B. Angular Momentum

C. Torque

D. Duality

Answer:

B. Angular Momentum

Bohr was the first to quantize the angular momentum of electrons, as he combined Rutherford's planet-style model with his knowledge of emerging quantum theory. Recall that he derived a "quantum condition" for the single electron, requiring electrons to exist at specific energy levels.

48. A uniform pole weighing 100 grams, that is one meter in length, is supported by a pivot at 40 centimeters from the left end. In order to maintain static position, a 200 gram mass must be placed _____ \centimeters from the left end.

A. 10

B. 45

C. 35

D. 50

Answer:

D. 50

In answering this question, do not be tricked into calculating the position of the mass to create balance on the pole's pivot. (This calculation, with equal torques, would lead to incorrect answer (B).) A careful read of the question reveals that we want to "maintain static position" i.e. keep the pole from moving. Because it is already tilted toward the right side, that side (or anywhere to the right of the 45 cm pivot balance answer) is the correct place to put the additional weight without causing the pole to move. Thus, only answer (D) can be correct.

49. A classroom demonstration shows a needle floating in a tray of water. This demonstrates the property of

A. Specific Heat

B. Surface Tension

C. Oil-Water Interference

D. Archimedes' Principle

Answer:

B. Surface Tension

To answer this question, note that the only information given is that the needle (a small object) floats on the water. This occurs because although the needle is denser than the water, the surface tension of the water causes sufficient resistance to support the small needle. Thus the answer can only be (B). Answer (A) is unrelated to objects floating, and while answers (C) and (D) could be related to water experiments, they are not correct in this case. There is no oil in the experiment, and Archimedes' Principle allows the equivalence of displaced volumes, which is not relevant here.

50. Two neutral isotopes of a chemical element have the same numbers of

A. Electrons and Neutrons

B. Electrons and Protons

C. Protons and Neutrons

D. Electrons, Neutrons, and Protons

Answer:

B. Electrons and Protons

To answer this question, recall that isotopes vary in their number of neutrons. (This fact alone eliminates answers (A), (C), and (D).) If you did not recall that fact, note that we are given that the two samples are of the same element, constraining the number of protons to be the same in each case. Then, use the fact that the samples are neutral, so the number of electrons must exactly balance the number of protons in each case. The only correct answer is thus (B).

51. A mass is moving at constant speed in a circular path. Choose the true statement below:

A. Two forces in equilibrium are acting on the mass.

B. No forces are acting on the mass.

C. One centripetal force is acting on the mass.

D. One force tangent to the circle is acting on the mass.

Answer:

C. One centripetal force is acting on the mass.

To answer this question, recall that by Newton's 2nd Law, F = ma. In other words, force is mass times acceleration. Furthermore, acceleration is any change in the velocity vector—whether in size or direction. In circular motion, the direction of velocity is constantly changing. Therefore, there must be an unbalanced force on the mass to cause that acceleration. This eliminates answers (A) and (B) as possibilities. Recall then that the mass would ordinarily continue traveling tangent to the circle (by Newton's 1st Law). Therefore, the force must be to cause the turn, i.e. a centripetal force. Thus, the answer can only be (C).

52. A light bulb is connected in series with a rotating coil within a magnetic field. The brightness of the light may be increased by any of the following except:

A. Rotating the coil more rapidly.

B. Using more loops in the coil.

C. Using a different color wire for the coil.

D. Using a stronger magnetic field.

Answer:

C. Using a different color wire for the coil.

To answer this question, recall that the rotating coil in a magnetic field generates electric current, by Faraday's Law. Faraday's Law states that the amount of emf generated is proportional to the rate of change of magnetic flux through the loop. This increases if the coil is rotated more rapidly (A), if there are more loops (B), or if the magnetic field is stronger (D). Thus, the only answer to this question is (C).

53. The use of two circuits next to each other, with a change in current in the primary circuit, demonstrates

A. Mutual current induction

B. Dielectric constancy

C. Harmonic resonance

D. Resistance variation

Answer:

A. Mutual current induction

To answer this question, recall that changing current induces a change in magnetic flux, which in turn causes a change in current to oppose that change (Lenz's and Faraday's Laws). Thus, (A) is correct. If you did not remember that, note that harmonic resonance is irrelevant here (eliminating (C)), and there is no change in resistance in the circuits (eliminating (D)).

54. A brick and hammer fall from a ledge at the same time. They would be expected to

A. Reach the ground at the same time

B. Accelerate at different rates due to difference in weight

C. Accelerate at different rates due to difference in potential energy

D. Accelerate at different rates due to difference in kinetic energy

Answer:

A. Reach the ground at the same time

This is a classic question about falling in a gravitational field. All objects are acted upon equally by gravity, so they should reach the ground at the same time. (In real life, air resistance can make a difference, but not at small heights for similarly shaped objects.) In any case, weight, potential energy, and kinetic energy do not affect gravitational acceleration. Thus, the only possible answer is (A).

55. The potential difference across a five Ohm resistor is five Volts. The power used by the resistor, in Watts, is

A. 1

B. 5

C. 10

D. 20

Answer:

B. 5

To answer this question, recall the two relevant equations for potential difference and electric power:
$V = IR$ (where V is voltage; I is current; R is resistance)
$P = IV = I^2R$ (where P is power; I is current; R is resistance)

Thus, first calculate the current from the first equation:
$I = V/R = 1$ Ampere

And then use the second equation:
$P = I^2R = 5$ Watts

This is consistent only with answer (B).

56. An object two meters tall is speeding toward a plane mirror at 10 m/s. What happens to the image as it nears the surface of the mirror?

A. It becomes inverted.

B. The Doppler Effect must be considered.

C. It remains two meters tall.

D. It changes from a real image to a virtual image.

Answer:

C. It remains two meters tall.

Note that the mirror is a plane mirror, so the image is always a virtual image of the same size as the object. If the mirror were concave, then the image would be inverted until the object came within the focal distance of the mirror. The Doppler Effect is not relevant here. Thus, the only possible answer is (C).

57. The highest energy is associated with

A. UV radiation

B. Yellow light

C. Infrared radiation

D. Gamma radiation

Answer:

D. Gamma radiation

To answer this question, recall the electromagnetic spectrum. The highest energy (and therefore frequency) rays are those with the lowest wavelength, i.e. gamma rays. (In order of frequency from lowest to highest are: radio, microwave, infrared, red through violet visible light, ultraviolet, X-rays, gamma rays.) Thus, the only possible answer is (D). Note that even if you did not remember the spectrum, you could deduce that gamma radiation is considered dangerous and thus might have the highest energy.

58. The constant of proportionality between the energy and the frequency of electromagnetic radiation is known as the

A. Rydberg constant

B. Energy constant

C. Planck constant

D. Einstein constant

Answer:

C. Planck constant

Planck estimated his constant to determine the ratio between energy and frequency of radiation. The Rydberg constant is used to find the wavelengths of the visible lines on the hydrogen spectrum.
The other options are not relevant options, and may not actually have physical meaning. Therefore, the only possible answer is (C).

59. A simple pendulum with a period of one second has its mass doubled. If the length of the string is quadrupled, the new period will be

A. 1 second

B. 2 seconds

C. 3 seconds

D. 5 seconds

Answer:

B. 2 seconds

To answer this question, recall that the period of a pendulum is given by:

$T = 2\pi (L/g)^{1/2}$ where T is period; L is length; g is gravitational acceleration (This is derived from balancing the forces and making small-angle approximation for small angles.)

Note that this equation is independent of mass, so that change is irrelevant. Since the length is quadrupled, and all other quantities on the right side of the equation are constant, the new period will be increased by a factor of two (the square root of four).
This is consistent only with answer (B).

60. A vibrating string's frequency is _______ proportional to the _______.

A. Directly; Square root of the tension

B. Inversely; Length of the string

C. Inversely; Squared length of the string

D. Inversely; Force of the plectrum

Answer:

A. Directly; Square root of the tension

To answer this question, recall that
$f = (n\ v) / (2\ L)$ where f is frequency; v is velocity; L is length

and

$v = (F_{tension} / (m / L))^{1/2}$ where $F_{tension}$ is tension; m is mass; others as above

so

$f = (n / 2\ L)\ ((F_{tension} / (m / L))^{1/2})$

indicating that frequency is directly proportional to the square root of the tension force. This is consistent only with answer (A). Note that in the final frequency equation, there is an inverse relationship with the square root of the length (after canceling like terms). This is not one of the options, however.

61. When an electron is "orbiting" the nucleus in an atom, it is said to possess an intrinsic spin (spin angular momentum). How many values can this spin have in any given electron?

A. 1

B. 2

C. 3

D. 8

Answer:

B. 2

To answer this question, recall that electrons fill orbitals in pairs, and the two electrons in any pair have opposite spin from one another. Thus, (B) is correct. Note that answer (D) is trying to mislead you into thinking of the number of valence electrons in an atom.

62. Electrons are

A. More massive than neutrons

B. Positively charged

C. Neutrally charged

D. Negatively charged

Answer:

D. Negatively charged

Electrons are negatively charged particles that have a tiny mass compared to protons and neutrons. Thus, answer (D) is the only correct alternative.

63. Rainbows are created by

A. Reflection, dispersion, and recombination

B. Reflection, resistance, and expansion

C. Reflection, compression, and specific heat

D. Reflection, refraction, and dispersion

Answer:

D. Reflection, refraction, and dispersion

To answer this question, recall that rainbows are formed by light that goes through water droplets and is dispersed into its colors. This is consistent with both answers (A) and (D). Then note that refraction is important in bending the differently colored light waves, while recombination is not a relevant concept here. Therefore, the answer is (D).

64. In order to switch between two different reference frames in special relativity, we use the ________ transformation.

A. Galilean

B. Lorentz

C. Euclidean

D. Laplace

Answer:

B. Lorentz

The Lorentz transformation is the set of equations to scale length and time between inertial reference frames in special relativity, when velocities are close to the speed of light.

The Galilean transformation is a parallel set of equations, used for 'classical' situations when velocities are much slower than the speed of light. Euclidean geometry is useful in physics, but not relevant here. Laplace transforms are a method of solving differential equations by using exponential functions. The correct answer is therefore (B).

65. A baseball is thrown with an initial velocity of 30 m/s at an angle of 45°. Neglecting air resistance, how far away will the ball land?

A. 92 m

B. 78 m

C. 65 m

D. 46 m

Answer:

A. 92 m

To answer this question, recall the equations for projectile motion:
$y = \frac{1}{2} a t^2 + v_{0y} t + y_0$
$x = v_{0x} t + x_0$

where x and y are horizontal and vertical position, respectively; t is time; a is acceleration due to gravity; v_{0x} and v_{0y} are initial horizontal and vertical velocity, respectively; x_0 and y_0 are initial horizontal and vertical position, respectively.

For our case:

x_0 and y_0 can be set to zero
both v_{0x} and v_{0y} are (using trigonometry) = $(\sqrt{2} / 2)$ 30 m/s
$a = -9.81 \text{ m/s}^2$

We then use the vertical motion equation to find the time aloft (setting y equal to zero to find the solution for t):
$0 = \frac{1}{2} (-9.81 \text{ m/s}^2) t^2 + (\sqrt{2} / 2) 30 \text{ m/s } t$
Then solving, we find:
t = 0 s (initial set-up) or t = 4.324 s (time to go up and down)

Using t = 4.324 s in the horizontal motion equation, we find:
$x = ((\sqrt{2} / 2) 30 \text{ m/s}) (4.324 \text{ s})$
$x = 91.71$ m

This is consistent only with answer (A).

66. If one sound is ten decibels louder than another, the ratio of the intensity of the first to the second is

A. 20:1

B. 10:1

C. 1:1

D. 1:10

Answer:

B. 10:1

To answer this question, recall that a decibel is defined as ten times the log of the ratio of sound intensities:
(decibel measure) = 10 log (I / I_0) where I_0 is a reference intensity.

Therefore, in our case,
(decibels of first sound) = (decibels of second sound) + 10
$10 \log (I_1 / I_0) = 10 \log (I_2 / I_0) + 10$
$10 \log I_1 - 10 \log I_0 = 10 \log I_2 - 10 \log I_0 + 10$
$10 \log I_1 - 10 \log I_2 = 10$
$\log (I_1 / I_2) = 1$
$I_1 / I_2 = 10$

This is consistent only with answer (B).
(Be careful not to get the two intensities confused with each other.)

67. A wave has speed 60 m/s and wavelength 30,000 m. What is the frequency of the wave?

A. 2.0×10^{-3} Hz

B. 60 Hz

C. 5.0×10^{2} Hz

D. 1.8×10^{6} Hz

Answer:

A. 2.0×10^{-3} Hz

To answer this question, recall that wave speed is equal to the product of wavelength and frequency. Thus:
60 m/s = (30,000 m) (frequency)
frequency = 2.0×10^{-3} Hz

This is consistent only with answer (A).

68. An electromagnetic wave propagates through a vacuum. Independent of its wavelength, it will move with constant

A. Acceleration

B. Velocity

C. Induction

D. Sound

Answer:

B. Velocity

Electromagnetic waves are considered always to travel at the speed of light, so answer (B) is correct. Answers (C) and (D) can be eliminated in any case, because induction is not relevant here, and sound does not travel in a vacuum.

69. A wave generator is used to create a succession of waves. The rate of wave generation is one every 0.33 seconds. The period of these waves is

A. 2.0 seconds

B. 1.0 seconds

C. 0.33 seconds

D. 3.0 seconds

Answer:

C. 0.33 seconds

The definition of a period is the length of time between wave crests. Therefore, when waves are generated one per 0.33 seconds, that same time (0.33 seconds) is the period. This is consistent only with answer (C). Do not be trapped into calculating the number of waves per second, which might lead you to choose answer (D).

70. In a fission reactor, heavy water

A. Cools off neutrons to control temperature

B. Moderates fission reactions

C. Initiates the reaction chain

D. Dissolves control rods

Answer:

B. Moderates fission reactions

In a nuclear reactor, heavy water is made up of oxygen atoms with hydrogen atoms called 'deuterium,' which contain two neutrons each. This allows the water to slow down (moderate) the neutrons, without absorbing many of them. This is consistent only with answer (B).

71. Heat transfer by electromagnetic waves is termed

A. Conduction

B. Convection

C. Radiation

D. Phase Change

Answer:

C. Radiation

To answer this question, recall the different ways that heat is transferred. Conduction is the transfer of heat through direct physical contact and molecules moving and hitting each other. Convection is the transfer of heat via density differences and flow of fluids. Radiation is the transfer of heat via electromagnetic waves (and can occur in a vacuum). Phase Change causes transfer of heat (though not of temperature) in order for the molecules to take their new phase. This is consistent, therefore, only with answer (C).

72. Solids expand when heated because

A. Molecular motion causes expansion

B. $PV = nRT$

C. Magnetic forces stretch the chemical bonds

D. All material is effectively fluid

Answer:

A. Molecular motion causes expansion

When any material is heated, the heat energy becomes energy of motion for the material's molecules. This increased motion causes the material to expand (or sometimes to change phase). Therefore, the answer is (A). Answer (B) is the ideal gas law, which gives a relationship between temperature, pressure, and volume for gases. Answer (C) is a red herring (misleading answer that is untrue). Answer (D) may or may not be true, but it is not the best answer to this question.

73. Gravitational force at the earth's surface causes

A. All objects to fall with equal acceleration, ignoring air resistance

B. Some objects to fall with constant velocity, ignoring air resistance

C. A kilogram of feathers to float at a given distance above the earth

D. Aerodynamic objects to accelerate at an increasing rate

Answer:

A. All objects to fall with equal acceleration, ignoring air resistance

Gravity acts to cause equal acceleration on all objects, though our atmosphere causes air resistance that slows some objects more than others. This is consistent only with answer (A). Answer (B) is incorrect, because ignoring air resistance leads to the result of constant acceleration, not zero acceleration. Answer (C) is incorrect because all objects (except tiny ones in which random Brownian motion is more significant than gravity) eventually fall due to gravity. Answer (D) is incorrect because it is not related to the constant acceleration due to gravity.

74. An office building entry ramp uses the principle of which simple machine?

A. Lever

B. Pulley

C. Wedge

D. Inclined Plane

Answer:

D. Inclined Plane

To answer this question, recall the definitions of the various simple machines. A ramp, which trades a longer traversed distance for a shallower slope, is an example of an Inclined Plane, consistent with answer (D). Levers and Pulleys act to change size and/or direction of an input force, which is not relevant here. Wedges apply the same force over a smaller area, increasing pressure—again, not relevant in this case.

75. The velocity of sound is greatest in

A. Water

B. Steel

C. Alcohol

D. Air

Answer:

B. Steel

Sound is a longitudinal wave, which means that it shakes its medium in a way that propagates as sound traveling. The speed of sound depends on both elastic modulus and density, but for a comparison of the above choices, the answer is always that sound travels faster through a solid like steel, than through liquids or gases. Thus, the answer is (B).

76. All of the following phenomena are considered "refractive effects" except for

A. The red shift

B. Total internal reflection

C. Lens dependent image formation

D. Snell's Law

Answer:

A. The red shift

Refractive effects are phenomena that are related to or caused by refraction. The red shift refers to the Doppler Effect as applied to light when galaxies travel away from observers. Total internal reflection is when light is totally reflected in a substance, with no refracted ray into the substance beyond (e.g. in fiber optic cables). It occurs because of the relative indices of refraction in the materials. Lens dependent image formation refers to making images depending on the properties (including index of refraction) of the lens. Snell's Law provides a mathematical relationship for angles of incidence and refraction. Therefore, the only possible answer is (A).

77. Static electricity generation occurs by

A. Telepathy

B. Friction

C. Removal of heat

D. Evaporation

Answer:

B. Friction

Static electricity occurs because of friction and electric charge build-up. There is no such thing as telepathy, and neither removal of heat nor evaporation are causes of static electricity. Therefore, the only possible answer is (B).

78. The wave phenomenon of polarization applies only to

E. Longitudinal waves

F. Transverse waves

G. Sound

H. Light

Answer:

B. Transverse waves

To answer this question, recall that polarization is when waves are screened so that they come out aligned in a certain direction. (To illustrate this, take two pairs of polarizing sunglasses, and note the light differences when rotating one lens over another. When the lenses are polarizing perpendicularly, no light gets through.) This applies only to transverse waves, which have wave parts to align. Light can be polarized, but it is not the only wave that can be. Thus, the correct answer is (B).

79. A force is given by the vector 5 N x + 3 N y (where x and y are the unit vectors for the x- and y- axes, respectively). This force is applied to move a 10 kg object 5 m, in the x direction. How much work was done?

A. 250 J

B. 400 J

C. 40 J

D. 25 J

Answer:

D. 25 J

To find out how much work was done, note that work counts only the force in the direction of motion. Therefore, the only part of the vector that we use is the 5 N in the x-direction. Note, too, that the mass of the object is not relevant in this problem. We use the work equation:
Work = (Force in direction of motion) (Distance moved)
Work = (5 N) (5 m)
Work = 25 J
This is consistent only with answer (D).

80. A satellite is in a circular orbit above the earth. Which statement is false?

A. An external force causes the satellite to maintain orbit.

B. The satellite's inertia causes it to maintain orbit.

C. The satellite is accelerating toward the earth.

D. The satellite's velocity and acceleration are not in the same direction.

Answer:

B. The satellite's inertia causes it to maintain orbit.

To answer this question, recall that in circular motion, an object's inertia tends to keep it moving straight (tangent to the orbit), so a centripetal force (leading to centripetal acceleration) must be applied. In this case, the centripetal force is gravity due to the earth, which keeps the object in motion. Thus, (A), (C), and (D) are true, and (B) is the only false statement.

XAMonline, INC. 21 Orient Ave. Melrose, MA 02176

Toll Free number 800-509-4128

TO ORDER Fax 781-662-9268 OR www.XAMonline.com

TEXAS EXAMINATION OF EDUCATOR STANDARDS- EXAMINATION FOR THE CERTIFICATION OF EDUCATORS - TEXES/EXCET - 2007

P0# Store/School:

Address 1:

Address 2 (Ship to other):

City, State Zip

Credit card number_______-________-_________-________ **expiration**_______

EMAIL ________________________________

PHONE **FAX**

13# ISBN 2007	TITLE	Qty	Retail	Total
978-1-58197-925-1	ExCET ART SAMPLE TEST (ALL-LEVEL-SECONDARY) 005 006			
978-1-58197-949-7	TExES CHEMISTRY 8-12 140			
978-1-58197-938-1	TExES COMPUTER SCIENCE 141			
978-1-58197-933-6	TExES ENGLISH LANG-ARTS AND READING 4-8 117			
978-1-58197-935-0	TExES ENGLISH LANG-ARTS AND READING 8-12 131			
978-1-58197-926-8	ExCET FRENCH SAMPLE TEST (SECONDARY) 048			
978-1-58197-605-2	TExES GENERALIST 4-8 111			
978-1-58197-945-9	TExES GENERALIST EC-4 101			
978-1-58197-946-6	TExES SCIENCE 4-8 116			
978-1-58197-931-2	TExES SCIENCE 8-12 136			
978-1-58197-604-5	TExES LIFE SCIENCE 8-12 138			
978-1-58197-932-9	TExES MATHEMATICS 4-8 114-115			
978-1-58197-937-4	TExES MATHEMATICS 8-12 135			
978-1-58197-939-8	TExES MATHEMATICS-PHYSICS 8-12 143			
978-1-58197-948-0	TExES MATHEMATICS-SCIENCE 4-8 114			
978-1-58197-929-9	TExES PEDAGOGY AND PROFESSIONAL RESPONSIBILITIES 4-8 110			
978-1-58197-899-5	TExES PEDAGOGY AND PROFESSIONAL RESPONSIBILITIES 8-12 130			
978-1-58197-943-5	TExES PHYSICAL EDUCATON EC-12 158			
978-1-58197-928-2	TExES PRINCIPAL 068			
978-1-58197-941-1	TExES READING SPECIALIST 151			
978-1-58197-942-8	TExES SCHOOL COUNSELOR 152			
978-1-58197-940-4	TExES SCHOOL LIBRARIAN 150			
978-1-58197-934-3	TExES SOCIAL STUDIES 4-8 118			
978-1-58197-936-7	TExES SOCIAL STUDIES 8-12 132			
978-1-58197-927-5	ExCET SPANISH (SECONDARY) 047			
978-1-58197-944-2	TExES SPECIAL EDUCATION EC-12 161			
			SUBTOTAL	
	FOR PRODUCT PRICES GO TO WWW.XAMONLINE.COM		**Ship**	$8.25
			TOTAL	